Reactivity and Structure Concepts in Organic Chemistry

Volume 12

J. Fabian H. Hartmann

Light Absorption of Organic Colorants

Theoretical Treatment and Empirical Rules

With 76 Figures and 48 Tables

Springer-Verlag
Berlin Heidelberg New York 1980

Jürgen Fabian
Horst Hartmann

Sektion Chemie der Technischen Universität Dresden
Mommsenstr. 13
DDR-8027 Dresden

ISBN-13: 978-3-642-67589-8 e-ISBN-13: 978-3-642-67587-4
DOI: 10.1007/978-3-642-67587-4

Library of Congress Cataloging in Publication Data. Fabian Jürgen, 1936–. Light absorption of organic colorants. (Reactivity and structure ; v. 12). Bibliography: p. Includes index. 1. Dyes and dyeing – Chemistry. 2. Color. I. Hartmann, Horst, 1937– joint author. II. Title. III. Series. QD441. F3. 547.8′6. 80-12452

Softcover reprint of the hardcover 1st edition 1980

2152/3140-5 4 3 2 1 0

Preface

Although studies on synthetic dyes have been performed for more than loo years, their detailed elucidation requires further extensive research. The discovery of novel high polymers, the necessity of supplying a whole range of shades and increasing requirements for dyestuffs of high fastness properties give rise to a permanent search for new dyes. Extensive investigations on dyes were also occasioned by various applications in the field of spectral sensitization and of staining of biological specimens. Another more recent development concerns the lasing properties of some organic dyes. Most of the progress, however, was only achieved by time-consuming, purely empirical approaches and theoretical understanding of the dye properties is only at its very beginnings.

The color is the sine qua non of every dye. For this reason organic chemists and color chemists have looked for relations between the "color and constitution" of dye molecules for a long time. This knowlege as a whole is known as "theory of color". The classic theory of color was established about 100 years ago by Witt and was significantly extended 50 years later by W. König.

With the production of highly efficient UV-VIS spectrophotometers and the development of appropriate quantum chemical methods, the theory of color received new impulses. Some classic concepts were proved to be based on wrong assumptions, but many of the rules and relationships derived could be explained. Moreover, they were extended, and predictions even became possible about the color of structures for which any empirical experience is lacking. The better understanding of the electronic structure of the dye molecules, in turn, permits one to tackle theoretically further unsolved problems of color chemistry.

This book reviews the absorption features of organic dyes and recent progress in their explanation in terms of molecular structure. Originally we intended a more detailed description both of the basic chromophoric system and of the theoretical methods which render possible the derivation of color-structure relationships. Since the book became too voluminous we decided after fruitful discussions with the editor to restrict the introductory chapter to a brief outline.

In the first five chapters only basic problems and fundamental theoretical approaches are mentioned. Here a conception is advanced as to how the relate the different fields of experience and how to assess possibilities and limits of some qualitative and quantitative theories. For the sake of brevity, the application of the different approaches is not described in detail, but references are made to examples discussed later on.

The chapters VI to XV deal with the spectral characteristics of various organic colorants encompassing naturally occurring as well as synthetic dyes. As far as possible the spectral shifts are accounted for by theoretical models.

Although the compilation of spectral data had to be selective, we have tried to illustrate the full range of accessible absorption wavelengths by modifying the respective chromophoric system.

The book is mainly intended for organic chemists and color chemists which are interested in the spectral characteristics of dyes and their conceptual understanding. It is hoped that it will also be useful as source of data to spectroscopists and theoreticians, whose support is needed for the further development of the theory on color-structure relationships.

The writing of this book was suggested by Prof. Dr. R. Zahradník (Czechoslovak. Academy of Science) to whom we are indebted for encouragement and many helpful discussions.

For helpful dicussions and critical reading of the manuscript we would also like to express our gratitude to Dr. S. Dähne (Academy of Science, GDR), Dr. M. Nepraš (Institute of Organic Synthesis, Pardubice, ČSSR), Prof. Dr. R. Mayer and Dr. A. Mehlhorn (Technical University of Dresden). We are also grateful to Mrs. Brigitte Friebe for her skilled assistance in preparing the manuscript.

Dresden, Spring 1980

J. Fabian
J. Hartmann

Table of Contents

I. Phenomenological Conceptions on Color and Constitution

Dyestuffs have been used for thousands of years. Until the beginning of this century, indigo, alizarin, and further dyestuffs were produced from plant and animal materials. The origin of the color of organic substances could only be successfully explained when the molecular constitution of the first dyes had been established and the first syntheses carried out..

We shall not analyze in detail the various concepts which have arisen during the last 100 years in order to explain color, and the exciting pros and cons. Most textbooks of color chemistry describe the historical development and several review papers cover this subject [1–4]. We should mention here a few concepts which in our opinion have significantly contributed to the theory of color-structure relationships.

The first relations between color and molecular constitution can be traced back to Graebe and Liebermann [5], who related the color of organic substances to a system of unsaturated bonding. About one decade later, after the elucidation of the constitution of several dye molecules, Witt [6] founded the "classical theory of color". He attributed the visual color to the presence of particular groups, the "chromophores". Structures containing chromophores were termed as "chromogens". When the chromophores is colorless itself, so called "auxochromic groups" may give it color. Chromophores in the original sense were, e.g., N=O, N=N, C=C, C=O, and auxochromes NH_2, OH etc.

Witt's theory of color was extended and refined in the course of this century by Kaufmann [7], Dilthey, and Wizinger [8–11]. During this time the meaning of the terms chromophore and auxochrome underwent considerable change [12]. The term "anti-auxochromes" was newly introduced for groups which induce a brightening of the color [8]. Antiauxochromes are e.g. N=O or C=O, i.e. groups which formerly belonged to the chromophores. Wizinger attributed the color of organic colorants to the presence of intramolecular ionic states, which were modelled by coupling of auxochromes and antiauxochromes with conjugated systems.

These approaches obviously consist in a partitioning of the whole conjugated systems of the dye molecule into several molecular building blocks. Although the explanations given in the Witt-Dilthey-Wizinger's color theory later proved to be less satisfactory or even completely wrong, the idea of molecular partitioning retained much of its appeal. Thus Griffiths most recently has regarded important synthetic dyes as being composed of donor and acceptor fragments ("donor-acceptor chromogens") [13].

On the other hand, already soon after the foundation of Witt's theory of color also concepts were developed which clearly contrast with those just mentioned. In these concepts the color was attributed to larger conjugated bond systems as a whole. Thus Nietzky [14] and Armstrong [15] related the color to quinoid structures. According to them, triphenylmethane dyes owe their color to quinoid substructures. Von Baeyer [16] assumed, moreover, an oscillation between quinoid and benzenoid ring systems. Along the same lines, Willstätter and Piccard [17] defined the "meriquinoid state", attributing to each benzene ring of triphenylmethane some quinoid character. According to Iz'mailskii [18] dyes molecules exist in a "mesotropic equilibrium state" which gives rise to the color ("chromo-state"). This conception resembles to the later introduced and widely used mesomerism approach [19] and its equivalent presentation in terms of quantum mechanical resonance [20]. The color-determining conjugated system has now been generally designated the "chromophore" [12].

Color-determining conjugated systems, however, are not exclusively cyclic structures. Already early studies pointed out that some colored compounds contain chain molecules [21]. Later König [22] recognized the basic polymethine chromophore. He defined polymethines as chain molecules containing an odd number of methine or heteroanalogous groups ("mesochrome") and terminal groups ("perichromes"). These structures give rise to the "chromo-state". According to König the chromo-state is characterized by a high degree of bond equalization and charge alternation along the conjugated chain. The terminal perichromes are monovalent groups. They determine whether the dye molecule is neutral or ionic. The cationic streptopolymethine-cyanine and the anionic streptopolymethine-oxonols, for example, encompass equal perichromes, the neutral streptopolymethine-merocyanine two different ones.

More recently Dähne and coworkers [23–28, 28a] revived and refined König's theory of color. They gave a generalized definition of the polymethine state and directed the chemist's attention to the occurrence of polymethine building blocks in various organic dyes. According to them polymethinic subunits are the essential part of every organic dyestuff. Hünig and Quast independently demonstrated the importance of polymethine chains in an excellent review paper on newer colored systems [29].

All the above mentioned concepts of the classic theory of color are phenomenological in essence, based on the chemist's experience and intuition ("chemical theories"). Although each concept possesses its own features, there is one principle inherent in all conceptions: the deeply colored compounds contain conjugated bonds. The whole conjugated system is, in principle, responsible for the color. But the color can also often be rationalized alternatively by assembling the conjugated system from chromophoric subunits. The latter approach has an appreciable heuristic value: a great variety of chromophores can be conceived by combination of only a few basic chain and cyclic subchromophores.

II. UV/VIS Spectroscopy and Quantum Chemistry of Organic Colorants

It is well known that the visual colour of organic compounds results from absorption of light in the 350 to 750 nm region. Although the spectral response of the human eye is involved in color perception, the measurement of the spectral absorption feature in the ultraviolet (UV) and visible (VIS) region is of fundamental importance for discussing colour-structure relationships.

Absorption spectra are mostly obtained from dissolved colorants or colorants embedded in solid materials. They mostly consist of broad bands. Absorption bands are characterized by the absorption wavelength λ in nm (or in absorption wave numbers ν in cm^{-1}) and the molar absorptivities ϵ in $1\ mol^{-1}\ cm^{-1}$ at maximum absorption, and the direction of the polarization vectors. Additionally, proposals has been made to characterize the shape of the colour band [30].

Excellent textbooks about color chemistry [13, 31] and UV-VIS spectroscopy [32–35] cover the relationship between visual color and absorption spectrum and might be consulted for more details.

The first compilation of spectral absorption maxima of commercial dyes was already published at the beginning of this century [36]. Later spectral curves of dyes were collected [37, 38]. Spectral curves of some representative organic colorants are contained in the DMS Atlas of UV Spectra [39], and λ_{max} and lgϵ values of dyes can be found, among other data, in a comprehensive collection covering the period 1946–1971 [40]. However, caution is required, for many synthetic dyes are difficult to purify. Dyes also often undergo structural change on going into solution such as complex formation, isomerization or protonation [41].

In general, the wavelength scale has been favoured over the wavenumber scale. Since wavelengths are reciprocal to wavenumbers the spectrum in wavelength appears compressed at short wavelength and expanded at long wavelengths. The expanded spectrum in the visible region better corresponds to the high response of the eye in this spectral region. On the other hand, the $\tilde{\nu}$-representation is more suited for comparing experimental with theoretical results because wavenumbers are linearly related to excitation energies. This fact was taken into consideration in the DMS representation of spectra, accepted here for the sake of comparability.

Quantum chemistry has established that light absorption of organic colorants in the UV-VIS region arises from electronic excitation. The electronic excitation energy and transition probability determine the position and intensity, respectively, of the absorption band. Molecular vibration in the excited state account for the shape of the absorption band.

The first attempts in interpreting light absorption of organic colorants made use of the knowledge that long wavelength absorptions are observed for conjugated systems. This suggests that only electrons involved in conjugated bonds need to be considered. As theoretical counterpart of the empirical mesomerism conception, the valence bond (VB) theory of conjugated systems ("resonance theory"), seemed to be, at first, the most attractive approach in deriving colour-structure relationships [20, 42–44].

Förster described dye molecules in both the ground and lowest energy excited state as mixture of two interacting resonance structures [45]. Along these lines, the simplest streptocyanine dyes were studied by Herzfeld and Sklar [46] and Moffitt [47]. The wave functions of the two resonance structures, in which the positive charge is positioned either at the one or at the other terminal nitrogen, are used to define the state wave functions of the dye molecules. The symmetrical combination of the starting functions represents the ground state Ψ_0, the antisymmetrical combination the excited state Ψ_1. The energy difference of both states represents the transition energy ΔE.

However, according to Pauling [42] the two resonance structures do not overlap appreciably. The calculated resonance splitting is, therefore, too low and further structures must be included [46, 47]. The large number of resonance structures makes VB calculations very time-consuming and hardly feasible.

In spite of the fact that several relationships between color and chemical structure have been successfully interpreted (such as substitution and solvent effects [45, 48–50] and the so-called "Brooker deviation" [51]), and that more refined treatments became accessible [52], the valence bond theory lost its paramount importance in elucidating color-structure relationships. Even the validity of simple resonance rules for application to electronic excited states has been questioned [53–55].

Later the theory of molecular orbitals (MO theory) attracts more and more attention in accounting for absorption features of organic colorants. Within the framework of the MO theory the electronic states are described by means of molecular one-electron functions ψ_i (molecular orbitals, MOs).

At first sight this method seems to be less suited as no information is immediately obtained about the wave functions of the ground and lowest excited states Ψ_0 and Ψ_1, resp. However, some basic conclusions can be drawn without knowing explicitly these state functions.

Much attention in color chemistry was first paid to the free electron model (MO-FE theory) which was developed simultaneously by Kuhn [56], Bayliss [57], Platt [58], and Simpson [59]. Within this approach one assumes that the potential energy of the π-electrons, moving along a chain (chain structures) or along a perimeter (cyclic structures), is independent of their positions. In the simplest cases one examines the energies of an electron in an one-dimensional box. The MO's are then given in closed form shown, for example, for streptopolymethines. In the ground state of the molecule two electrons of antiparallel spin occupy the lowest-energy orbitals.

According to this model light absorption results from exciting an electron from one of the occupied to one of the unoccupied orbitals. Excitation from the highest occupied MO (HOMO) to the lowest unoccupied MO (LUMO), i.e. excitation between the frontier

MO's (FMO's), gives rise to the lowest energy $\pi \rightarrow \pi^*$ transition. The calculated transition energies can be compared directly with the experimental excitation energies.

This model has been successfully applied by Kuhn to rationalize the color of various natural and artifical dyes [60, 61], but its application is limited in scope. The assumption of an uniform potential breaks down for many structures of interest. Although it was shown how to take into account the irregularities of the potential, the branching of the conjugated system and the electron interaction and electron correlation [61–63], the MO-FE method could not compete with MO-LCAO methods. The MO-LCAO method is not only of similar conceptual simplicity, but also proved more easily to be handled in refined versions from the computational point of view.

In the MO-LCAO approximation each molecular orbital ψ_i is represented by a linear combination of atomic orbitals ϕ_r, where the expansion coefficients c_{ir} are algebraic numbers. If an electron occupies the MO ψ_i, c_{ir}^2 means the probability to find the electron in the region of the atom r. π-Type MO's are generated by combining p_π orbitals. This most simple approximation, put forward by Hückel (HMO method), starts from few atomic and bond parameter (Coulomb- and resonance integrals, resp.), which are defined empirically. The method has been described in several textbooks (e.g. Ref. [64, 65]).

LCAO-MO's show much resemblance to FE-MO's. The color determining lowest transition energy can be similarly calculated from the frontier orbital energies. The most simple cases transition energies are described by closed formulae [66–69, 69a].

In contrast to the MO-FE theory the resulting transition energies are not obtained in wavenumbers (wavelengths) but in β-units. β represents the carbon-carbon resonance integral. In order to calculate wavelengths the energy equivalent of β must be known. Appropriate definition of β first enabled Dewar 1950 to calculate successfully organic colorants in the framework of the π-approximation of the simple one-electron theory (Hückel MO theory) [70]. Later studies showed, however, that the correlation between HMO transition energies and experimental data is more complicated as formerly assumed. If experimental data are plotted against theoretical ones, linear correlations were obtained which do not pass through the origin, but intersect the ordinate axis at non-zero values. As extensively reported by Zahradník [71], the correlations split, moreover, into several straight lines, each involving data of compounds of related structure. Although the splitting can be avoided in some series by taking into account the bond length alternation of the carbon-carbon bonds [72, 73], this pattern mainly results from methodical deficiencies inherent to the HMO-approximation [71]. Additional problems arise if heteroatomic parameter must be introduced. Optimal parameters in calculating excitation energies within the simple Hückel theory [74] or within a refined Hückel theory [75] have been derived. HMO calculations are useful to make qualitative prediction for the longest-wavelength absorptions of organic colorants, and, if done in an extensively studied series of related compounds, to get numerical data by means of correlative equations.

The most serious drawback of the HMO method represents the explicit neglect of electron-electron interaction. Thus semi-empirical treatments have been developed which take into account electron-electron interaction explicitly, e.g. Pople's SCF-method [76]. In this framework transition energies are no longer simply equal to energy differences between occupied and unoccupied orbitals, but contain additional electron-electron interaction terms. This method has been further improved by adding the configuration interaction (CI) procedure, as described by Pariser and Parr [76]. The semiempirical SCF-CI treatment is known as PPP method, named after Pariser, Parr and Pople.

With introduction of the CI method the simple picture of pure one-electron excitation is abandoned. Since different excited states are allowed to mix, each electronic transition is now interpreted in terms of several electronic excitations between occupied and unoccupied orbitals. In order to avoid a high computational expenditure only a limited number of single excited configurations is generally considered. They are formed from a certain number of highest occupied and lowest unoccupied orbitals [71], or are chosen by energy criteria defined by perturbation theory [77]. In some cases doubly or even higher excited states are necessary to describe correctly the electronic excited states [78, 79]. This problem is now extensively under discussion for long chain polyenes [79]. A doubly excited configuration is assumed to be responsible for the lowest energy state or one of the lowest energy states in that case, giving rise to some spectral anomalies. Fortunately, higher excited configurations seem to be, in general, of low importance for the color determining lowest-energy transitions. Moreover, PPP calculations of various colorants showed that the singly excited configuration, which results from electron excitation from the HOMO to the LUMO, contribute dominantly to the resulting excited state. This finding accounts for the fact that correct qualitative predictions about lowest-energy excitations are still feasible in the framework of the HMO-approximation.

In contrast to HMO theory, PPP calculations also permit a discussion about the shorter-wavelength absorptions. Transition energies are now immediately obtained in wavenumber (or wavelength) and absorption intensities in oscillator strength. Provided the band widths are comparable, experimental molar absorptivities are approximately linearly related to oscillator strengths.

Integrals needed for PPP calculations are mostly determined empirically, for example by fitting atomic spectroscopic data, whereas the two-centre electron repulsion integrals usually follow from the Mataga-Nishimoto equation [71]. Resonance integrals are adjusted, at most, with molecular spectra. Only few data are required to fix these variable parameters and, based on it, a vast number of compounds can be calculated without added assumptions. The presumption of an idealized geometry is mostly sufficient, but some uncertainty may remain if several isomers are possible, or if some departure from the coplanarity occurs. Compounds with considerable bond length alternation are satisfactorily calculated by methods which iteratively correct resonance integrals in dependence on the calculated bond orders [80, 81]. Such so-called PPP-SC_β methods are especially useful in calculating polyene dyes. Similar procedures have been proposed for taking into account charge effects via calculated π-charge orders [81–84].

Considerable uncertainty still remained in choosing of parameters for charged heteroatomic compounds, as encountered in cyanine and oxonol dyes or in zwitterionic compounds. If the charge is considered to be localized at a single heteroatom, it should be added to the core charge. Then, an azonium nitrogen $=N-^+$ attains the effective nuclear charge +2. Consequently the azonium nitrogen corresponds to a pyrrole-type nitrogen. Similarly an oxygen bearing a full negative charge should have the effective nuclear charge +1 as that of the carbonyl oxygen. From this point of view the parametrization seems to be straightforward, and additional π-charge effects could be taken into account by an iterative procedure as mentioned above. In reality, however, the problem is more complicated, because the charges are partly delocalized over the σ-electronic system. Therefore, some proposals have been put forward to calculate σ-charge effects. Thus σ-charge distributions estimated via the electronegativities have been used for parameter correction [83, 84]. A more sophisticated procedure, put forward by Pancíř and Zahradník, considers σ-charge as well as bond alternation effects [84]. Unfortunately, up to now no dye molecules have been calculated along this line. There are only some calculations of polymethine dyes by more rough procedures [85, 86]. Whether these results are really better than those devoid of any charge correction has not yet been demonstrated, however. PPP calculations of simple streptopolymethines based on quantum chemically calculated all-valence-electron distributions provided different spectral effects in dependence on the charge correction procedure used [81].

In spite of the last mentioned difficulties PPP calculations proved to be highly useful in elucidating the chromophoric systems in all classes of organic colorants. As shown in Chapt. VI to XIV the main absorption features of the various chromophores are satisfactorily reproduced. The calculated absorption wavenumber differs, no more than 3,000 cm^{-1} from the experimentally observed absorption maximum. This is about 20 nm in the 250 nm region, but more than 100 nm in the 600 nm region. Therefore, the prediction of colour from calculated wavelengths in the visible region is rendered difficult by the sensitivity of the eye to small wavelength changes. Compounds absorbing at long wavelengths such as substituted azo, amino-anthraquinone, indigo, quinacridone and polymethine dyes are mostly calculated at shorter wavelengths than experimentally found. Since spectral data are not always available in non-polar solvents, it is not clear to what extent solvent effects contribute to the observed deviations. Although the only fair numerical agreement between calculated and experimental absorption wavelengths is a serious drawback for theoretical predictions about the position of the color bands, the PPP method nevertheless represents a powerful tool for colour chemists. Since often only change in color associated with structural modification is questioned, more reliable conclusions can be reached from calculated wavelength differences. In series of closely related compounds numerical accuracy can also be improved by a fully empirical parametrization [87].

Some problems mentioned above may suggest abandonment of the π-approximation and to calculate colour-determining $\pi \rightarrow \pi^*$ transitions by virtue of all-valence-electron methods. SCF-CI method in the so-called CNDO or INDO approximation has been modified for application in electronic spectroscopy (CNDO/S, INDO/S [88–91]). The extension of the procedure has two important consequences:

Firstly, $\pi - \pi^*$ transitions are not only influenced by σ-electron distribution but they also mix with certain $\sigma \rightarrow \sigma^*$ transitions. Thus, the simple classification schema of electronic transitions breaks down. Fortunately, however, $\sigma \rightarrow \sigma^*$ excited states play a very minor role and can be neglected in describing the lowest energy excited states of organic colorants.

Secondly, electronic transitions between the σ- and π-electronic systems are also now calculated. Transitions of the type $n - \pi^*$ belong to them. In the case of organic colorants $n \rightarrow \pi^*$ absorption bands are mostly hidden underneath the intense $\pi \rightarrow \pi^*$ absorptions. Although not important for the observed color in that case, they can heavily influence the photophysical and photochemical properties.

Although CNDO/S or INDO/S calculations may be useful in studying small π-electronic systems, only few important new information has been obtained for organic colorants so far. Moreover, the agreement between calculated an experimental wavelength is mostly not improved compared to that in π-approximation, inspite of the much larger computational expenditure. The same holds for ab initio calculations, which are available for only few $\pi \rightarrow \pi^*$ chromophoric systems (cf. porphin, p. 203).

All calculations mentioned above provide discrete excitation energies, whereas broad absorption band are observed experimentally. Since the occurrence of the absorption band is due to molecular vibrations, quantum chemical methods are needed which calculate both electronic and vibrational states. Some methods have been advanced [92–95], but the expenditure is still to high to permit calculations of spectral curves of large organic molecules.

On the other hand, all calculations mentioned above refer to the isolated molecule. But the observed color or spectral data are more or less influenced by the surrounding medium, giving rise, in many cases, to considerable bathochromic or hypsochromic shifts (positive or negative solvatochromism). Quantum chemical studies of the solvent effect on π-systems have been attempted in two directions:

Firstly, considering the molecule along with the surrounding substrate molecules as a complex supermolecule which is treated by all-valence-electron methods [96] or ab initio methods [97, 98] ("discrete molecule approach"). Due to the high expenditure spectral solvent effects of only few structures have yet been calculated [99, 100].

Secondly, considering the effect of the surrounding molecules as a perturbation on the molecular π-electron structure [101]. The solute molecules are treated in a solvent continuum. The reaction of this field on the molecule is then included in the quantum chemical calculation in π-approximation ("microstructural model of solvatochromism"). Contrary to the classic continuum models of solvatochromism the latter approach enables all change of molecular parameters to be covered which are brought about by solvent interaction. This way it has been shown why unsymmetrical merocyanines behave like symmetrical cyanines (oxonols) in polar solvents [101–103]. A more sophisticated approach is based on an all-valence-electron approximation and makes use of Klopman's solvaton conception [104].

III. Relation between Phenomenological and Quantum Chemical Theories

At first glance one may be tempted to believe that quantum chemical theories make phenomenological "theories" which involve doubtful or sometimes even completely wrong ideas about the origin of the phenomena superfluous. But looking at this problem more thoroughly this conclusion has proved to be rather superficial.

Quantum chemistry aims at the profound understanding of the experimental findings in terms of physical models. The more sophisticated the theoretical treatment, the better the physical understanding of the phenomenon. This goal of research necessarily leads to a separate examination of the molecular systems, and the view over a large number of electronic systems becomes more and more obscured. The multiplicity of computed numbers may even make it difficult to compare between related structures. Whereas the reliability of numerical predictions is improved with the level of sophistication, the intrinsic power of the quantum chemical methods generally becomes small. These theoretical methods are not longer adapted to fit the intuitive conceptual world of chemists. In this respect, chemist's are often disappointed with the outcome of quantum chemical studies and then favour a "chemical explanation" within phenomenological theories over a physical explanation.

In contrast to quantum chemical theories, phenomenological theories aim at a broad and simple understanding of the phenomena. This is achieved by supressing details and by introducing terms essential for an appropriate description. Comparison between individual experimental findings and structures then becomes of paramount importance. This kind of approach determines the nature of the chemical models. The phenomenological examination of color-structure relationships is based on the comparison of conjugated systems, either compared as a whole or considered in molecular parts. The physical model, however, could completely differ from this chemical model and no answer is then obtained in terms of chemical thinking.

On the other hand, serious drawbacks are also inherent in phenomenological theories. Since the deeper origin of light absorption is not recognized or intentionally omitted, rules are only valid in the close limits of the available experimental material. Moreover, classification and interpretation in terms of reference systems may appear rather contradictory. Many controversies have arisen from different phenomenological interpretations. Interestingly, controversial interpretations still remained though quantum chemical calculations were available [105]. Calculations yielded arguments in favour of both of the presented interpretations. This illustrates the difficulty of assessing the validity of chemical models from results of sophisticated quantum chemical calculations.

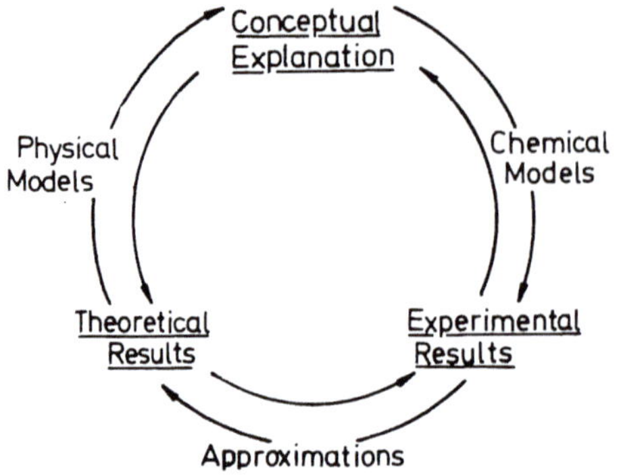

Fig. III.1. Schematic illustration of the interrelationship of experiment, theory and modelling. This schema is a modified version of one given by Csizmadia [106]

In order to make the above mentioned problems clearer and to find a proper solution, we should distinguish between distinct areas of scientific activity. This can be best done by modifying a schematic illustration recently presented by Csizmadia [106]. According to him three fields of activity should be recognized: theoretical observation (theoretical results), experimental observation (experimental results) and conceptual explanation (Fig. III.1).

With respect to the problems considered here the curved arrows between the three fields may be discussed as follows: Within certain approximations quantum chemical calculations are used to derive spectral data either for interpretation of the known absorption behaviour or for prediction. On the other hand, there is a feedback from experiment to theory. Thus, empirical data are needed in approximate semiempirical calculations for parameter adjustment (lower part of Fig. III.1).

Another field of interrelationships covers experimental results and conceptual explanation (right-hand part of Fig. III.1). The experimental data are analyzed in terms of similarity and analogy. The purely empirical examination of experimental absorption data with the aid of chemical models gives rise to color-structure relationships or "color rules", which, in turn, may stimulate experimental research. This is the field of phenomenological conceptions. Finally, there is a third field, which connects theoretical results and conceptual explanation (lefthand part of Fig. III.1). Since it is important for later discussions, it requires closer examination:

In principle, it should be possible to achieve a simple conceptual understanding in terms of purely physical models, which might completely differ from chemical ones. However, chemical models created during the last 100 years on the treatment of color-structure relationships provide considerable basis for a physical reinterpretation. In other words, physical models can be successfully adapted to chemical models.

The introduction of the quantum chemical π-approximation is one example illustrating the creation of a physical model based on chemical experience. There are numerous further examples for using quantum chemical methods for conceptual understanding. Thus Platt [107] directed much attention to a theoretical interpretation of absorption spectra in terms of similarity and analogy. On the other hand, calculations of a π-system as the whole and of substructures were performed to reveal the parent chromophore of dyes. This kind of examination has been impressively demonstrated by Klessinger and Lüttke, who defined, in this way, the basic parent chromophore of indigo by comparing calculated excitation energies (cf. Chapt. IX) [108]. The parent chromophore was found easily and without large computational effort, whereas the later synthesis of chromophoric parent structures was rather cumbersome.

However, a definition of parent chromophores along these lines can be similarly doubtful as if gained by purely empirical comparison of absorption data. The near coincidence of the long-wavelength absorptions between a considered dye structure and a particular substructure needs not necessarily imply that the omitted part of the π-system are not involved in the electronic excitation of the dye molecule. If the excitation is not mainly localized at a particular subunit ("parent chromophore"), substitution in the "extrachromophoric parts" can heavily influence the absorption wavelengths. In that case the former definition of the parent chromophore would be of limited value. Obviously, the definition of parent chromophores necessitates additional theoretical arguments.

Also the description of chromophores in terms of composed subchromophores has been recognized early. The molecular exciton theory considers the interaction between subchromophores completely neglecting the overlap between them ("energy delocalization") [109]. The splitting of the color band of dyes upon aggregation has been mainly discussed along this line, but there are also applications in interpreting long-wavelength absorptions of bichromophoric dyes [110]. Stronger coupling between molecular parts requires that orbital overlap be taken into account ("electron delocalization"). This was accomplished in the LHM method, named after Longuet-Higgins and Murrell [111]. This method differs, in principle, from the PPP method only by the fact that localized orbitals are used instead of delocalized ones. This method has often been considered as an alternative computational method in calculating π-electronic systems [112], while its heuristic value was underestimated. A remarkable exception mentioned above is a paper of Heilbronner and Grinter [54], who used the method to rationalize color rules.

If each of the lowest excited states is mainly described by a single configuration, localized at a particular part of the molecule, the influence of additional configurations can be considered as perturbation. On the other hand, if an electron is easily transferred upon excitation from one part of the molecule to another one, this charge transfer configuration determines the lowest excited state. This approach has been widely used in describing the absorption features of donator-acceptor complexes (Mulliken CT theory) [113].

However, more recent papers demonstrated the limited applicability of these simple "molecules-in-molecule" approaches. Thus, Heilbronner and coworkers [114] showed experimentally as well as theoretically that the classic LHM method completely fails in reproducing the term levels of the hydrocarbon fluoranthene when considered as naphthalene-ethylene composite molecule. Carefully examining the wave functions they found that the breakdown is originated from the number of the configurations, defined in terms of the subunits. If the number of these configuration is too low, the spectral absorption behaviour cannot be correctly interpreted.

In the view of these difficulties a theoretical method becomes desirable which enables results of PPP calculations to be analyzed in terms of their molecular subunits. A suited method, named as "configuration analysis", has been proposed by Baba, Suzuki and Takemura recently [115]. This method does not only provide alternative interpretations of theoretical results, but also permits a judgement of the appropriateness of the set of molecules-in-molecule type configuration functions chosen. As expected, the breakdown of the LHM calculation in the case of fluoranthene is well reflected in the results of a PPP configuration analysis, clearly showing the inadequacy of the method in the case of a strongly limited set of reference configurations [116].

The comparison of the electronic transitions between related compounds can also be done on the level of the simple one-electron theory. The delocalized MO's are composed of the localized MO's of fragments, now indicating a distinct degree of localization over the fragments (MO-LCMO method) [65]. Conversely, the delocalized MO's may analyzed in terms of localized MO's (Polansky's pars orbital technique) [117]. If the consideration is restricted to the lowest energy transition, only position and nature of the HOMO and LUMO are important. If the molecules consist of weakly composed molecular fragments, these molecular orbitals can be easily derived from fragment orbitals by perturbational MO theory (PMO theory) [65, 118]. Dewar was the first who demonstrated the high heuristic value of the PMO method in deriving color-structure relationships [119]. Later discussions were based on his study [13, 120–124]. But the question arises whether the application of the simple one-electron approximation is still justified.

In summary, there are obviously several theoretical approaches which are capable of connecting theoretical results and conceptual understanding in terms of chemical thinking. All we need is a straightforward strategy which appropriately considers the scope and limitations of the various methods.

The following approach proved to be useful in explaining the nature of the electronic excitation: The theoretical reproduction of the absorption spectrum is first attempted in the framework of the PPP method. Next the state wave functions are analyzed in terms of topologically equivalent systems as a whole or, alternatively, as composition of molecular fragments by applying the configuration analysis technique. Analyses with respect to different comparative structures of interest will reveal the most adequate reference system, or, in other words, the most attractive chemical model. If the model is very suitable, there is a good chance of deriving the term levels correctly in terms of the reference states considered. In that case the results of PPP and LHM calculations correspond closely and the excited states can be understood from their constituent parts. Tracing back the genesis of excited states obtained by LHM calculations makes it possible to recognize the most important interactions between the fragment orbitals. This examination will show whether color-structure relationships can be rationalized within the simple MO picture or only on the level of the configuration interaction. The curved arrow directed upward in the left-hand part of Fig. III.1 indicates this deductive approach.

For experimentalists, the inductive approach which is depicted by the downward directed arrow in Fig. III.1 becomes important. Based on the experience gathered in deductive treatments estimations can be made starting from some fundamental knowledge about delocalized reference MO's or MO's of molecular fragments. Here perturbation theoretical considerations are very useful. These considerations may show whether the long-wavelength absorption is bathochromically or hypsochromically shifted. If the qualitative considerations are continued on the level of the configuration interaction expansion, even a rough idea about the relative position of the lowest energy term level may be obtained. This way simple and transparent theoretical models can be creatively used by experimentalists without access to computer facilities. This approach yields insight into the essential feature of the electronic excitation and helps to explore structures with interesting color properties, or structural modifications which could give rise to a color change in a definite direction. If only selected structures are calculated later on for proving the correctness of the prediction, the computational expenditure will be low and expensive computer time is saved.

Since we intend to describe more details about the "analyzing" and "synthesizing" concepts elsewhere, only a short outline is given in the following section.

IV. Theoretical Methods for Deriving Color-Structure Relationships

1. PPP-CA Method

The Pariser-Parr-Pople (PPP) method enables one to define the ground state by the wave function Ψ_0 and a manifold of excited states by the wave functions Ψ_I. Each excited state wave function represents a mixture between excited electronic configurations, which result from one electron excitations from occupied to unoccupied MO's. The MO's, in turn, represent delocalized MO's, that means they involve, in general, all p_π-orbitals of the conjugated system. Due to the two series of expansion coefficients, arising from the MO-LCAO and CI expansion, excited state functions are not easily to interpret by inspection only.

Configuration analysis (CA) is a straightforward procedure for comparing the wave functions with other wave functions of the same multiplicity, provided they are defined by the same number of p_π-orbitals. Two types of reference systems are of interest:

Firstly, the considered π-system differs from the reference system by internal heteroatomic substitution or (and) by additional intramolecular bonds, whereas the topological equivalence and the number of π-electrons are retained. In that case the reference system is also described by delocalized orbitals which give rise to wave functions denoted hereafter as Ψ_0^0 or $\Psi_{i\to j}^0$ (entire molecule approaches).

Secondly, the considered π-system differs from the reference system by the presence of one or several bonds which connect molecular fragments. Each MO of this reference system is then fully localized at one of the fragments. Consequently, the ground state is described by the doubly occupied MO's of the fragments (no-bond ground state configuration Γ_0) and the excited states either by electron excitation within one fragment (locally excited configuration $\Lambda_{i\to j}$) or by electron excitation between two fragments (charge transfer configuration $T_{k\to 1}$) (composite molecule approach).

By configuration analysis the wave functions Ψ_0 and $\Psi_{i\to j}$ of the considered chromophoric π-electronic system are expressed by the wave functions of the reference system, as schematically shown in Fig. IV.1. Each electronic state is now correlated with the respective reference states. If few reference states predominate in the calculated configuration expansion, a clear-cut interpretation is obtained.

A second important outcome of a configuration analysis is a measure of the completeness of the configuration expansion. Since the number of the reference state functions is limited to the zero-order ground state and some singly excited states, the wave functions of the considered system are only approximately described by the reference

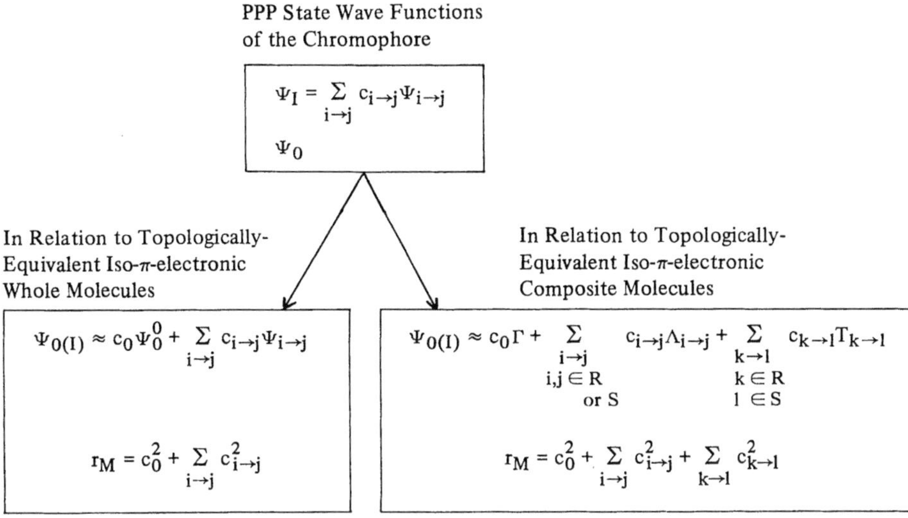

Fig. IV.1. State wave functions obtained by PPP calculations and their representation after configuration analyses

states. This becomes apparent in the expansion coefficients which, squared and summed up, yield a value r_M lower than 1 (or 100%) (cf. Fig. IV.1). Therefore, the sum r_M represents a measure of the correspondence of the wave functions under comparison. If the r_M value of the electronic state considered is nearly unity, the wave function is well described by those of the reference states, if not, important additional reference state wave functions were omitted. In the latter case the number of configurations is not sufficient to permit a profound interpretation.

This theoretical examination can be applied for interpreting chromophoric π-systems in terms of different chemical structural models, i.e. comparing chromophores as entireties or considering them as consisting of some molecular building blocks. Then r_M values allow one to assess scope and limitation of the different models.

In practice, the calculated state wave functions of the considered system are subsequently analyzed with respect to each reference system wanted. The results show, in general, a clear graduation of the appropriateness of the comparison. It has been found that delocalized reference systems are better than composite reference systems. Whereas r_M values in the first case are near to 1, they spread over large range in the latter case. On the other hand, the ground state is generally better described than excited states. Composite molecule models are appropriate models if the r_M value exceed 0.95 for the ground state and 0.80 for the excited states. As expected, these models are excellently applicable if molecular parts are weakly linked to one another by only a single bond.

Although PPP-CA indicates the adequacy of different approaches contributing to a conceptual understanding and facilitates the derivation of color-structure relationships, it cannot afford any direct help in numerical predictions. The resulting configuration expansion does not give any answer about the energetics. This question would only be answered if the energies of the wave functions, which result from the analysis, are known.

The less the wave functions correspond to each other, the more the energy values of the CA wave functions differ from those of the PPP calculation. The same holds for wave functions obtained by LHM calculations, although now the starting point is a quite different one.

Before pursuing these problems in the framework of the LHM method, an analysis technique should be mentioned which resembles that of the PPP-CA method under certain conditions. Without explicitly referring to wave functions of molecular fragments, electronic transitions can be characterized in terms of LE and CT excitation by procedures proposed by Ohta, Kuroda and Kunii [125] (hereafter termed OKK analysis) and Lusanov, Sukhorukov and Umanskii [126] (hereafter termed LSU analysis). Whereas in the first mentioned method the coefficient matrices of the wave function are examined, the last one defines the degree of LE and CT upon excitations from the squared transition density matrix. Only in the limiting case of very weak interfragmental coupling does the PPP-CA characterization of excited electronic state correspond to that for the respective electronic transition.

2. LHM Method

Instead of analyzing the state wave function in terms of the wave function of a reference system we can also start from wave functions of the reference system and combine them separately to linear combinations. Now the MO's of the reference systems form the basis set for the electronic configurations, which are then mixed by additionally taking into account internal heteroatomic substitution or a change in bonding. If the calculation starts from completely localized orbitals the method represents the Longuet-Higgins-Murrell (LHM) method [111], denoted also as "molecule-in-molecule" CI method [127]. Even if the configurations chosen are the same as those of the PPP-CA calculation, the LHM-CI expansion will differ from that obtained by the PPP-CA calculation. Summation of the squared expansion coefficients yields 1 (normalization condition), whereas a lower value is obtained by the PPP-CA calculation. The results of the PPP-CA calculations suggest that (in terms of the reference state functions) higher excited singly or multiply excited states are contained in the PPP-state wave functions, which appear as additional contributions of low-energy singly excited states in the LHM calculation. Consequently the results of both calculations differ. It is quite reasonable that the difference between the PPP-CA and LHM expansion coefficients and between the term levels of the PPP and LHM calculation decreases, if the r_M value of the CA calculation approaches unity.

The genesis of the term levels can easily be revealed within the LHM formalism. Since the topic has been recently reviewed in detail [112], only some qualitative aspects are outlined in the following.

If the lowest calculated excited state is predominantly localized at one part of the molecule, transition energy and intensity are only slightly changed compared to those of the isolated molecule corresponding to the molecular part (case *A* in Fig. IV.2, upper part). Then the excitation is approximately localized in one of the molecular parts (LE transition).

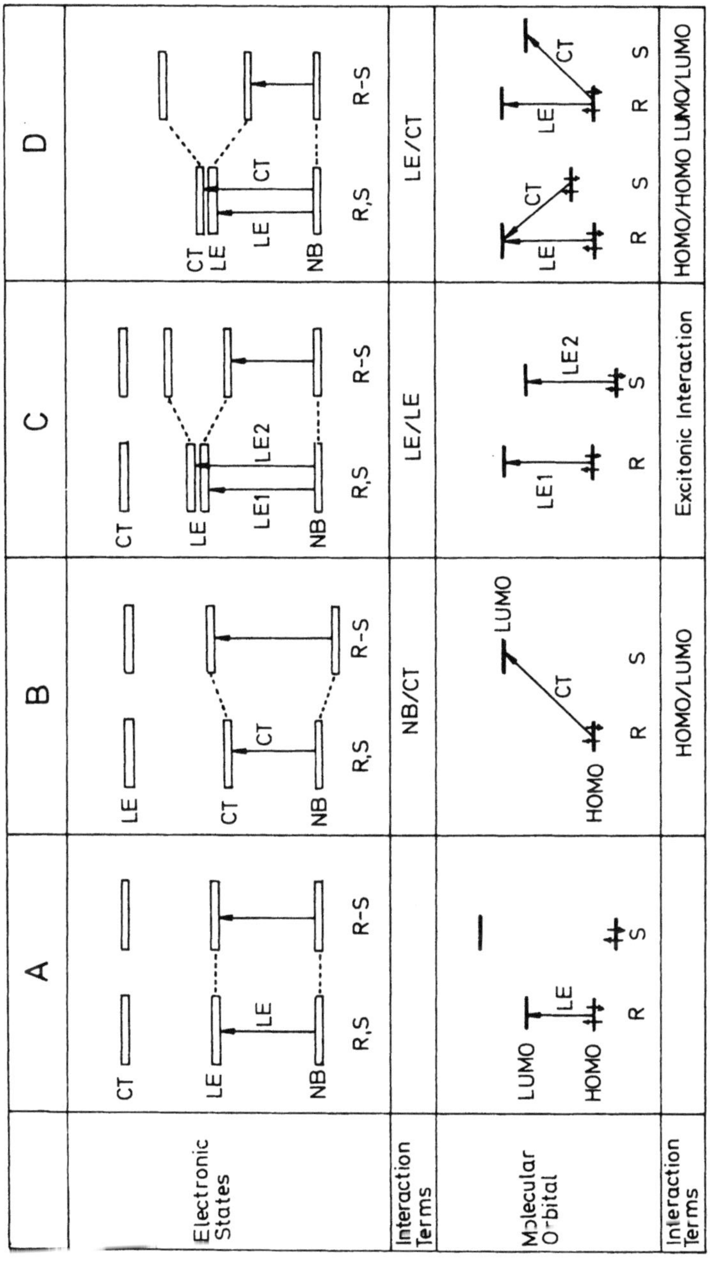

Fig. IV.2. Nature of excited states and electronic transitions of a composite molecule R–S in dependence on dominant configurations entering into the configuration interaction expansion (upper part), and the corresponding representation in the one-electron picture (lower part)

Apart from effects of additional configurations, the electron excitation can be affected by the polarization from other subunits. This polarization effect also occurs if electron delocalization is completely excluded.

On the other hand, a charge transfer (CT) configuration may predominate in the lowest energy excited state. In that case an intramolecular CT excitation occurs (case *B* in Fig. IV.2, upper part), whose transition energy is governed by the donor and acceptor capacity of the fragments and the strength of interaction between the no-bond (NB) configuration and the CT configuration. This interaction leads to a stabilization of the ground state and a destabilization of the excited state and, therefore, to an increase in transition energy. On the other hand, a non-zero transition probability requires some interfragmental interaction. In that case, intramolecular CT transitions can also become sufficiently intense [128].

More than one electronic configuration contributes to the lowest excited state. Two examples should be considered. The locally excited states of two fragments are assumed to have the same or nearly the same energy, whereas CT configurations are unimportant for energy reasons. Due to the electronic interaction the two LE states interact. Then they split into two clearly separated excited states, giving rise to distinct transition energies. This energy delocalization has been described in a point-dipole approximation by molecular excition theory [109]. Simple expressions describe the energy splitting and the intensity ratio of the two electronic transitions in dependence on the mutual geometrical arrangement of local transition moments (case *C* in Fig. IV.2, upper part).

Another case concerns the mixing between one LE and one CT configuration of comparable energy. Since the CT configuration interacts, in general, more strongly with the LE than with the NB configuration, the transition energy can be lowered considerably.

The estimation of the lowest electronic excitation energies requires, therefore, some knowledge about the energy of local excitations and CT excitations, but also some knowledge of the interaction terms. Since the interaction terms are dependent on the overlap between the fragment MO's, some fundamental conclusions can be reached on the simple one-electron level (cf. following section).

To summarize, LHM calculations permit us to establish the nature of the lowest excited states. If the excited state wave functions are described by few configurations only, we are able to interpret the phenomena plausibly, and to give predictions based on some elementary knowledge about the subsystems.

3. FMO Considerations

Some problems discussed in the last section can be easily elucidated in the HMO picture by considering the frontier molecular orbital (FMO) interaction (Fig. IV.2, lower part). The local excitation appears as one-electron transition within, say, the fragment *R* (case *A*), whereas CT excitation occurs from the donor part *R* to the acceptor part *S* (case *B*). The CT transition energy is determined by the position of the HOMO of the donor part *R* and the LUMO of the acceptor part *S*. If both orbitals interact, the energy of HOMO is lowered, whereas the LUMO is raised. HOMO-LUMO interaction raises, therefore, the transitions energy.

Whereas the phenomenon of exciton splitting cannot be understood in the framework of the HMO model (case *C* in Fig. IV.2), the interaction between the LE and CT configuration (case *D*) can now be specified. If we presume a local excitation at the fragment R, two types of CT excitation can be conceived of: Either the CT excitation occurs from the fragment S to the fragment R or, conversely, from R to S. Then the interaction between the LE and CT configurations is determined by the HOMO-HOMO interaction in the first case, and the LUMO-LUMO interaction in the second case. If these interactions are large, whereas the formerly mentioned HOMO-LUMO or LUMO-HOMO interaction are weak, low transition energies occur.

The magnitude of the orbital interaction may be assessed by second order perturbation calculation [65, 118]. Qualitatively, the interaction between the orbitals are the stronger, the larger the expansion coefficients of the localized MO's at the joined position and the lower the energy difference between the orbitals considered. If one of the fragment MO's has a node or a negligibly small expansion coefficient at the position of linking, the MO's cannot interact.

In the case of HOMO-HOMO or LUMO-LUMO interaction stabilization of the excited state occurs. If, however, the HOMO (LUMO) interacts with the LUMO (HOMO), the ground state is stabilized. Predominant ground state stabilization increases the excitation energy, giving rise to a hypsochromic shift. Blue shift caused by donor substituents, known as "inversion of auxochromes" [11], could not be satisfactorily explained by phenomenological theories.

Fragment orbitals can only interact if they display the same symmetry behaviour with respect to the symmetry elements of the composite molecule. Consequently, the transition energy is mainly determined by the fragment orbital symmetries in that case. Based on molecular symmetry color rules has been derived for symmetrically-bivalent joining of molecular fragments [122, 129, 130].

A frontier orbital rationale can also be achieved for effects of heteroatomic substitution. Upon replacement of atoms or atomic groups by atoms of more electronegative elements, e.g. –CH= by –N=, both the HOMO and LUMO are lowered in energy proportional to the squared expansion coefficients in the corresponding position (first order perturbation calculation [65, 118]). If the values are not equal, either an increase or a decrease of the excitation energy occurs. Conversely, replacement by atoms of less electronegative elements raises the MO energy proportional to their squared expansion coefficients. Thus, the lowering of the transition energy occurs if electronegative (electropositive) elements are introduced in positions where the electron density increases (decreases) upon excitation.

Similar considerations can be made for a change in the π-bond strength, which may arise from internal heteroatomic substitution, by torsion around bonds, or by introducing cross bonds. Shorter or longer excitation wavelengths are observed, according as the HOMO or the LUMO is more stabilized or destabilized. Lowering of the transition energy occurs if bonds are weakened (strengthened) in positions where the bond orders decrease (increase) upon excitation. The application of the HMO perturbation theory has been described and exemplified in detail in textbooks [65, 118].

Perturbational considerations presuppose a knowledge about the frontier orbitals of simple fragments or delocalized reference system. As to conjugated hydrocarbons, there are several sources of such data [65, 66, 131]. The MO's of the simple streptopolymethines have been presented more recently [68, 69, 132].

V. Classification of Organic Colorants

Chemistry is characterized by similarity and contiguity whereas formal classification schemes presuppose the existence of discontinuities. On the other hand, classification schemes help to recognize regularities and to derive relationships. Thus, a compromise should be reached by accepting some classifications, on the one hand, and by taking the liberty of looking at the phenomena from distinct points of view, on the other hand.

Based on theoretical arguments, we may distinguish between five groups:

1) Organic colorants which consist of a strongly delocalized π-electronic system, favouring its description as a whole

Such delocalized π-systems represent particular chromophores. They may be parent chromophores of series of related compounds. Fundamental chromophoric systems are straight and branched chain π-systems, as encountered in polyenes or polymethines or cyclic π-systems. They are considered as a whole and their spectral properties can be described in dependence on the chain length and ring size, resp. Regularities have been derived in the framework of the MO-FE, HMO or PPP method, as described in many textbooks. Also topologically equivalent iso-π-electronic systems, differing by hetero-atomic substitution, belong to this first group. The PPP-CA method permits their term levels to be correlated, and spectral effects caused by internal heteroatomic substitution may be explained by first order HMO perturbation calculations.

Although the entire π-electronic system is, in general, involved in the excitation, a partial localization may occur, for example, in the case of branched conjugated systems. It gives rise to so called "partial bands" or "partials" [2]. The occurrence of partials does not prove the individuality of molecular parts. They rather reflect the nodal properties of orbitals which are strongly involved in the electron excitation [133]. Partials can be readily recognized by analyzing electronic transition in the framework of the OKK [125] or LSU analysis [126].

2) Organic colorants which consist of a strongly delocalized π-electronic system, slightly perturbed by substituents

In that case a composite molecules approach can be most advantageous. The LE states of the parent chromophores determine the absorption properties. The substituents can be treated as exerting a perturbation effect on the chromophoric system. PPP-CA calculations reveal the appropriateness of the model and the nature of the excited states. Similar results will be obtained by LHM calculations. Qualitative relations can be derived by second order HMO-perturbation calculations or, correspondingly, by qualitative considerations.

3) Organic colorants which consist of a strongly delocalized π-electronic system, considerably affected by donor or acceptor substituents

In that case the composite molecule model is, in general, less suited than for case 2). The color determining lowest excited state represents a mixture of LE and CT states, but also the ground state may aquire some CT character. PPP-CA calculations reveal whether a limited number of configurations can still describe the composite molecule. Although in this case a changeover into a particular delocalized chromophoric system may be indicated, reasonable qualitative predictions can be generally made. Many commercial dyes belong to this group. Because of the LE/CT mixing the effectiveness of HOMO-HOMO and LUMO-LUMO interactions determines the spectral substituent effect.

4) Organic colorants which consist of two (or several) delocalized π-electronic systems, weakly coupled to each other

In this structures, which are also termed bi- and polychromophoric systems, the individuality of the subsystems ("subchromophores") is partly retained. We observe a rough additivity of the spectral properties of the constituent parts (mixed chromophore dyes) or a splitting of the color band in the case of identical subchromophores. PPP-CA and LHM calculations enable one to examine the relevance of the LE states of the subchromophores. Some qualitative relations can be obtained in the framework of the simple exciton theory.

5) Organic colorants which consist of a delocalized π-electronic donor and acceptor system, weakly coupled to each other

The long-wavelength absorption band is now determined by low-energy CT configurations and can be considered as arising from an intramolecular CT excitation.

The description of the electronic states in terms of a composite molecule can be given by PPP-CA and LHM calculations. The wavelengths may be estimated by taking into account donor and acceptor strength as well as HOMO-LUMO interaction. Conjugated betaines belong to this group of dyes.

One may be tempted to classify all colorants to one or another group. However, we should bear in mind that there is a continuous changeover. Any strict discrimination between slightly or stronger linked substituent or between substituent groups and additional chromophores is difficult to be made and any classification becomes then rather arbitrary. Therefore, a more flexible conception should be applied, permitting a different kind of comparison. Although quantitative treatments presuppose appropriate models, some qualitative insight can nevertheless be attained with less suited ones. Also the experimental material available dictates the kind of comparison.

Some examples may illustrate the problem. Theoretical and experimental results furnish striking evidence for considering 2.5-diamino-*para*-benzoquinones as consisting of two loosely coupled trimethine-merocyanines [134] (cf. p. 96). From this point of view not only their absorption features can be understood in terms of the chain structure but also their striking spectral resemblance with 4.5-diamino-*ortho*-benzoquinone (colorants of the group 4). This kind of consideration becomes, however, less suited when the whole series of substituted *para*-benzoquinones are compared. In that case a comparison starting from the benzoquinone substructure seems to be more advantageous. Although this reference system becomes extremely perturbed by amino substituents, making the derivation of quantitative data by composite molecule approaches impossible, the substituent effect can nevertheless be satisfactorily accounted for (colorants of the group 2 or 1).

A second example exemplifies the application of the concept of localization as against the concept of delocalization. Many azo dyes consist of donor-substituted benzene rings and the azo group. If these compounds are considered as donor-acceptor systems every donor-substituted azobenzene can be included in the comparative series (colorants of the group 3). If, however, one residue attached at the azo group is a donor and the second one an acceptor unit, another type of comparison may become more attractive (cf. p. 42 and 54). The absorption features of this compound can now be compared with that of an iso-π-electronic structure which belong either to the stilbene or hemicyanine dyes (colorants of the group 1).

A final example illuminates problems accompanied with the term "parent chromophore". Anthraquinone can be well understood as two benzene units which are bridged by carbonyl groups (cf. p. 90). But this reference may be less useful in rationalizing spectral data of a series of anthraquinones. Since all compounds contain the anthraquinone unit, the comparison may appropriately refer to this common unit. Then anthraquinone is considered as parent chromophore. Similarly, if 1-aminoanthraquinone represents the common constitutive part of a series of related compounds, regularities can be best understood by referring to this subunit. Now 1-aminoanthraquinone becomes the parent chromophore. Corresponding approaches are feasible for other organic colorants. Thus, in other series of colorants a reference to the parent compounds azobenzene, indigo or triphenylmethane may be useful. These structures are then considered as particular chromophores.

Approaching in this flexible way, different relationships can be revealed depending on the comparative series considered. Furthermore, color properties of unknown structures may be consistently predicted from the relationships in different comparative series.

Because of the limitation of a single classification schemata, the practically important colorants were arranged in this book in a rather conventional way. This arrangement may also permit a rapid access to the relevant data. We start with dyes containing polyene chains. The first discussed carotenoid dyes are important natural pigments, but derivatives of stilbene belong to the artificial dyes. The knowledge of the spectral feature of stilbene offers relations to the absorptions of the azobenzene, which represents the parent compound of the class of dyes at present most extensively applied for textile dyeing. Numerous azo dyes exhibit color properties which suggest a strong intramolecular charge transfer interaction between donor and acceptor groups. This feature also determines the electronic structure of the most important commercial nitro and quinoid dyes which are discussed subsequently. The indigoid structure directs attention to branched chromophores. Diphenyl- and triphenyl-methane dyes are strongly related to the polymethine dyes. Although xanthenes, thioxanthenes, acridine, oxazine, thiazine and diazine dyes are frequently considered separately, they are regarded here as derivatives of di- and triphenylmethane dyes. The subsequent section comprises the various polymethine dyes, which are of great practical and theoretical interest. After a brief outline of naturally occurring porphins and of synthetic phthalocyanines containing cyclic-conjugated systems some colorants are discussed which represent a zwitterionic structure. In the final chapter some dyes are exemplified which encompass distinct chromophoric systems.

VI. Polyene Dyes

1. Chromophoric System

The chromophoric system of polyene dyes is determined by the chain polyene chromophore. Polyenes (*1*) exhibit a strongly allowed long-wavelength absorption, which mainly arises from electronic excitation from the HOMO to the LUMO. The transition is polarized along the longitudinal molecular axis. The next two electronic transitions at higher energies are degenerate in the one-electron theory, but due to electron-electron interaction the degeneracy is removed and two distinct absorption bands may appear. In Platt's notation [135] the transitions are designed as A → B and A → C transitions (Fig. VI.1). In the case of all-*trans* polyenes (*1*) the A → C transitions are electronically forbidden.

According to more recent experimental and theoretical studies a symmetry-forbidden electronic transition is masked by the allowed long-wavelength absorption. For particular pentaene and hexaene compounds this electronic transition has even been found at lower energies than the respective intense transitions [136]. Although the forbidden transition does not appreciable affect the intensity of the long-wavelength absorptions and thus the visual color, its presence may be of great importance for the photophysical behaviour. Probably, this transition can only be described by taking into account doubly excited configurations [79].

The long-wavelength absorption of polyenes displays generally some vibrational fine structure. According to experimental and theoretical studies only few vibrational modes

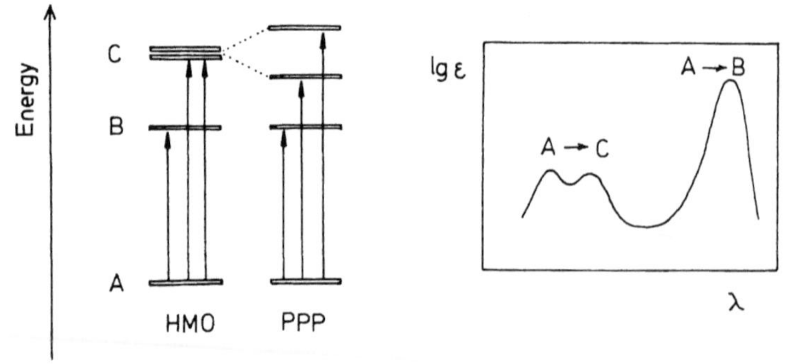

Fig. VI.1. Lowest-energy term levels and general absorption features of straight chain chromophores

Table VI.1. Longest-wavelength absorptions of polyenes and heteropolyenes in nm (lg ε)

n	*1*	*2*	*3*
0	217[a]		220 (4.17)[f]
1	257	228 (4.16)[b]	270 (4.43)
2	290	278 (4.94)[c]	312 (4.60)
3	318	328 (4.75)[d]	343 (4.60)
4	344	351 (4.80)[e]	370 (4.76)
5	368	380.5 (4.40)[e]	393 (4.81)
6	386	406 (4.72)[e]	415 (4.80)

[a] in hydrocarbons [139]
[b] in ethanol [235]
[c] in ethanol [236]
[d] in methanol [237]
[e] in ethanol [238]
[f] in dioxane [239]

are involved. Three fundamentals have been calculated recently for hexatriene from potential curves obtained by a refined PPP procedure [92].

The intense longest-wavelength absorption is bathochromically shifted with increasing chain length (Table VI.1).

Due to the bond alternation the vinylene shift decreases in passing to higher chain lengths. The wavelengths of the absorption maxima approximately obey the equation

$$\lambda = A\sqrt{n} + B \tag{VI.1}$$

where n denotes the number of double bonds and A and B are constants.

The decrease of the vinylene shift suggests that the energy of the A → B transition approaches at infinite chain length to definite value, which has been estimated to about 18,000 cm^{-1} for *1*. Remarkably, the intercept of the linear correlation obtained by plotting experimental wavelengths versus HMO-transition energies is in the same order of magnitude (about 16,500 cm^{-1}). More recent theoretical studies of the electronic structure of linear polyenes revealed, however, that although bond alternation plays a role in the spectral convergence of linear polyenes, the major contribution to the gap comes from electron-electron interaction [137].

The absorption wavelengths of polyenes have been satisfactorily reproduced by theoretical methods [138]. They mostly treated the π-electron system as a whole. PPP calculations require bond alternation to be considered [80, 81]. As reviewed in Ref. [112], there are also some papers in which polyenes are calculated based on ethylenic subunits, e.g. by LHM calculations [139].

Polyene diketones (*2*) and polyene aldehydes (*3*) display a similar increase in the absorption wavelengths with increasing chain length as polyenes (*1*) (cf. Table VI.1). However, polyene aldehydes *3* are distinguished from polyenes by the occurrence of a weak longest-wavelength absorption band, which arises from excitation of an oxygen lone pair electron to the lowest unoccupied π-MO (n → π* transition). The n → π* absorption band is clearly resolved only for the first two vinylogues of *3* and is submerged under the strong absorption in longer chain polyene aldehydes. More recent calculations on *3* are in complete agreement with this finding [140, 141]. As depicted in Fig. VI.2,

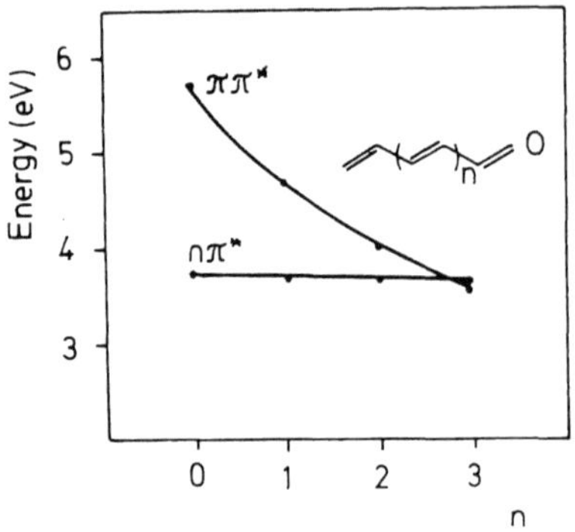

Fig. VI.2. Calculated lowest n → π* and π → π* state energies of *trans* polyene aldehyde (PPP approximation) [141]

n → π* transition energies decrease much more slowly with increasing chain length than π → π* transition energies. The calculation has been performed by a refined PPP method which additionally takes into account the n-orbital [141].

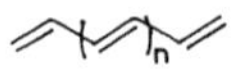

1

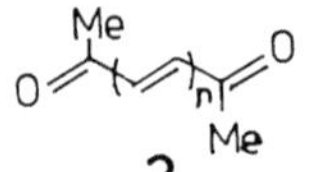

2

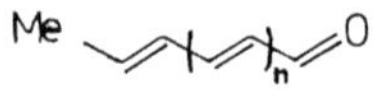

3

The sequence of the term levels is of some significance with respect to the photophysical processes of the excited states.

The absorption maxima of all-*trans* polyenes undergo a spectral shift upon isomerization. The transition energy decreases as the number of s-*cis* configurations increases [138]. The spectral shifts have been recently discussed in the framework of a refined PPP method [142]. Since various of the isomers have a non-planar structure due to relieving steric strain by twisting about single bonds, the calculations were based on the conformational Boltzmann distributions.

Since polyenes belong to alternant hydrocarbons, no change of charge order occurs upon excitation. Consequently, substitution of methine groups by heteroatomic groups of comparable bond properties should hardly affect the absorption wavelengths. A spectral shift is expected when internal substitution is accompanied by a marked change in bond strength. However, since changes of bond alternation and solvent effects additionally contribute to the spectral shifts, the above-mentioned perturbation theoretical arguments may fail.

More reliable qualitative conclusions can be reached by perturbational considerations when external substituents are introduced. Because of non-zero expansion coefficients in both HOMO and LUMO donor as well as acceptor substituents undergo orbital overlap. Since frontier orbital interaction either causes raising of the HOMO (donor substitution) or lowering of the LUMO (acceptor substitution), a decrease of transition energy occurs in both cases. External substitution at the polyene chromophore therefore brings about red shift.

Bulky substituents may cause a violation of coplanarity. In general, steric strain in polyenes may be relieved by twisting around an essentially single bond. As the bond orders of essentially single bonds increase upon excitation, twisting about these bonds should give rise to a blue shift.

2. Carotenes

Long chain polyenes are well known as carotenoids. The bright yellow, orange, red and violet carotenoid dyes form one of the most important group of natural pigments which are to be found in various vegetable and animal matters [143]. Although carotenoid dyes are not applied in textile dyeing, they also found practical application [144]. Thus, they were introduced for the coloring of foods and food-stuffs.

The molecular structure of several carotenoid dyes was carefully studied and the absorption feature in the ultraviolet and visible region was extensively reviewed [145–147]. Thus, only a brief outline of their spectral feature is given.

Since carotenoid dyes represent alkyl-substituted polyenes the absorption curves of them are closely related to those of the unsubstituted polyenes. The longest-wavelength absorption arises from the A → B transition and usually shows a well-defined fine struc-

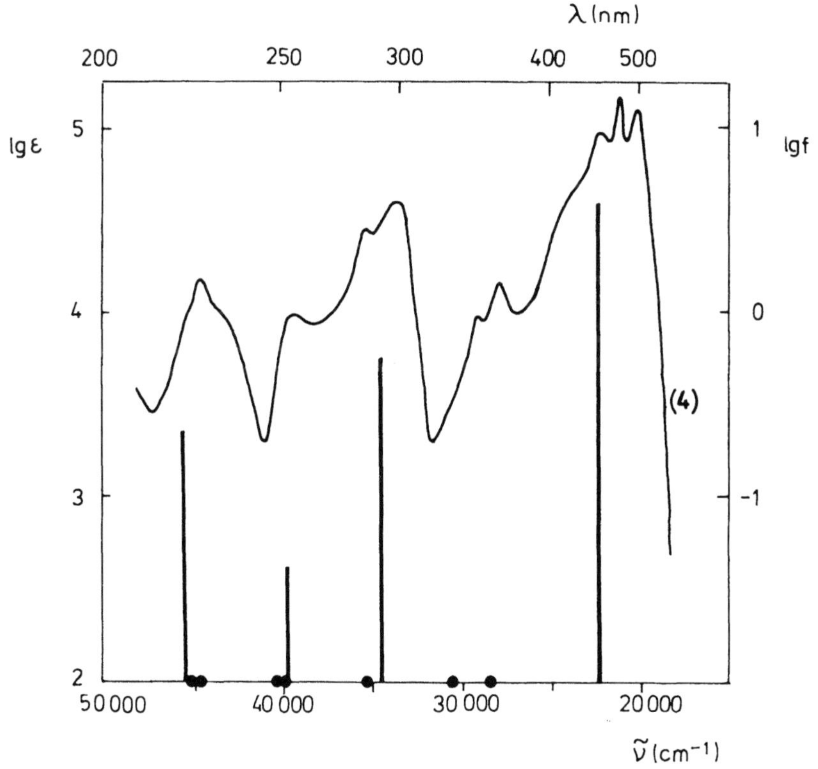

Fig. VI.3. Spectral absorption curve of lycopene *(4)* in cyclohexane [139] and results of the PPP calculation [81]. The vertical lines indicate the calculated transition energies, the full circles electronic transitions with lg f ⩽ −2

ture. Three distinct maxima of high intensity are mostly observed, as illustrated with lycopene *4* in Fig. VI.3.

The higher energy A → C transitions of all-*trans* isomers are forbidden, but may cause a weak absorption region at shorter wavelengths (cf. Fig. VI.3). Refined $\pi \rightarrow \pi^*$ methods, which take into account the bond fixation, provide a satisfactory interpretation of the spectra of carotenoids [81, 139, 148, 149]. For illustration, results of the PPP-calculation of *4* are depicted in Fig. VI.3 along with the spectral absorption curve.

A progressive hypsochromic and hypochromic shift of the absorption maxima occurs when one or both of the terminal groups are involved in cyclic structures, e.g. in passing from *4* to γ-carotene *5* or β-carotene *6*.

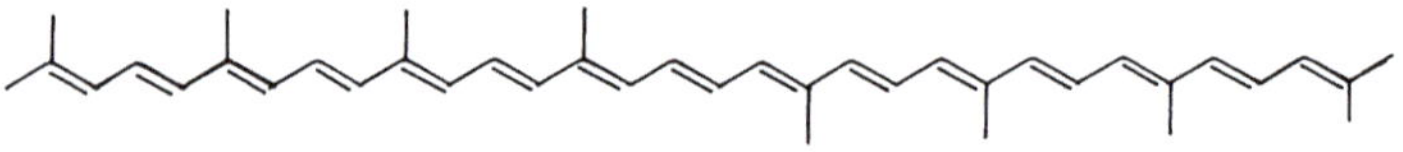

4

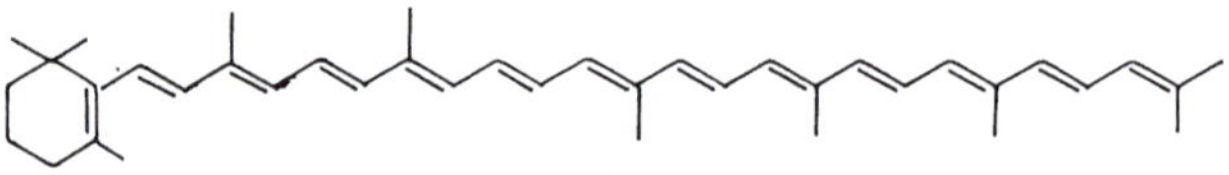

5

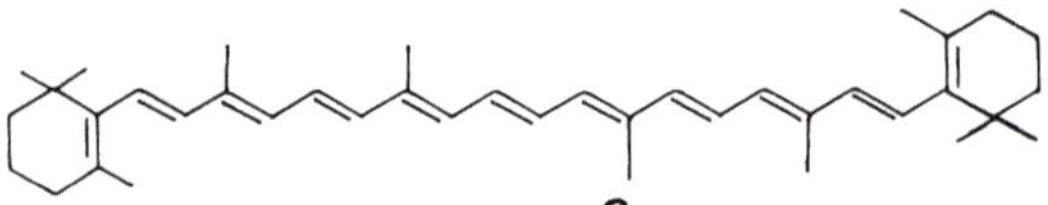

6

The spectral shift suggests that the double bonds of the β-ionone ring are inclined with respect to the plane of the polyene chain. The X-ray studies of various carotenoids show in fact a twisted s-*cis* arrangement to be preferred in the crystal structure [146, 150].

The experimentally found pronounced bond length alternation along the chain ($R_{C=C}$: 1.355 Å, R_{C-C}: 1.460 Å) provides a rationale for the decreasing red shift occurring with increasing chain length. Whereas the wavelength of most intense maximum of the A → B band increases by about 50 nm in going from β-carotene *6* (11 double bonds) to decapreno-β-carotene *7* (15 double bonds), the shift amounts only to about 35 nm in going from *7* to dodecapreno-β-carotene *8* (19 double bonds) (cf. Fig. VI.4).

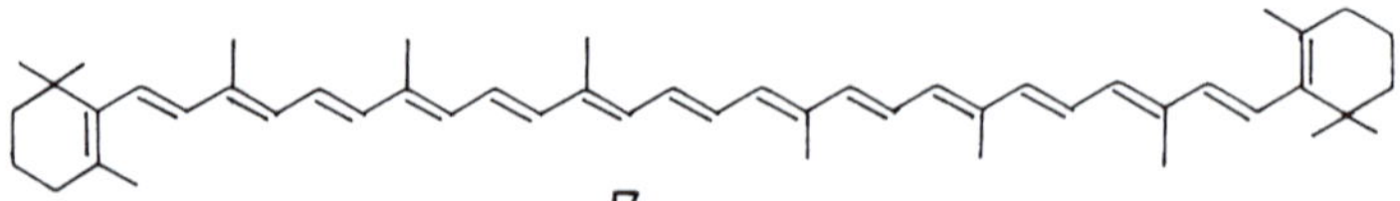

7

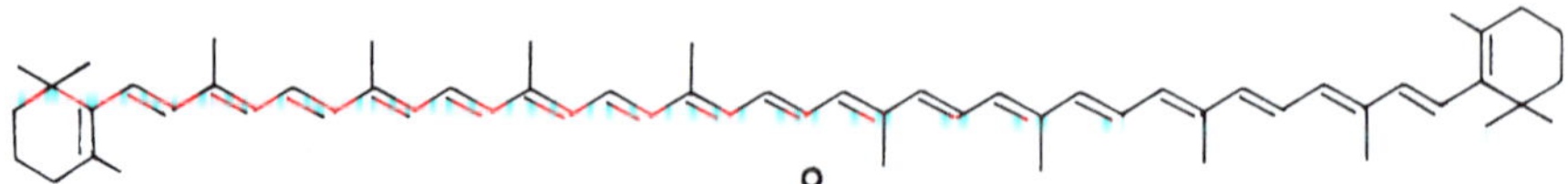

8

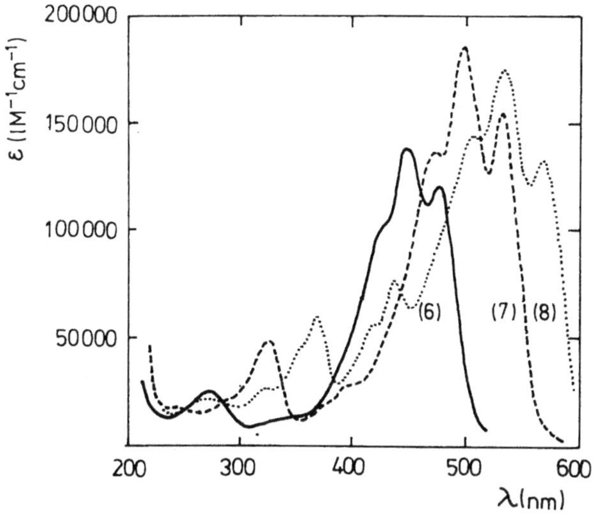

Fig. VI.4. Spectral absorption curves of β-carotene (*6*), decapreno-β-carotene (*7*) and dodecapreno-β-carotene (*8*) in petroleum ether [147]

In accord with PPP-type calculations the color band of β-carotene *6* is polarized along the long molecular axis [149, 151, 152].

Hiryama [153] attempted to systematize experimental data of carotenoids by the general equation:

$$\lambda^2 = A - B \cdot 0.920^N \qquad \text{(VI.2)}$$

A and B are constants which depend on the class of compounds (polyenes as well as derivatives) and on the solvent. The exponents N are calculated from increments which include the number of double bonds and the number and type of substituents. Substractions are made for double bonds which are not fully effective, as, for instance, for bonds in the terminal β-ionone rings. Equation (VI.2) predicts the absorption maximum for infinite chain length at 608 nm or 16,500 cm^{-1} (in ethanol). These values agree surprisingly well with theoretically estimated ones (cf. Sect. VI.1).

UV-VIS spectroscopy has turned out to be a powerful tool for proving the presence of *cis*-isomers [145]. *Cis*-carotenoids exhibit light absorption of lower intensity than the corresponding *trans*-isomers, and the maxima of the A → B bands are blue-shifted (about 2–5 nm for one and about 10 nm for two *cis* double bonds). The most noticeable feature of the spectra of mono-*cis*-carotenoids is, however, the appearance of a subsidiary absorption band in the near ultraviolet region, generally denoted as "*cis* band". This band has highest intensity when the *cis* double bond is centrally located within the molecule. If the linear shape of the all-*trans* form, however, becomes restored by introduction of a second linkage, the *cis* band disappears.

The *cis* band originates from the A → C transition which is allowed within the mono-*cis* configuration. This assignment has been corroborated by polarization spectra [152]. In accord with theoretical predictions linear dichroic spectra of *3* in stretched polyethylene films showed the *cis* band to be perpendicularly polarized relative to the longer wavelength band.

PPP-type calculations predict the A → C transition in the correct wavelength range. However, in disagreement with the experimental outcome the color band of the mono-*cis*-carotenoids is calculated at longer wavelength than that of the corresponding all-*trans*-isomer. This discrepancy may arise from a slight departure from molecular coplanarity due to steric hindrance of the methine groups adjacent to the *cis* C=C bond [153a].

3. Retinal and Derivatives

The absorption behaviour of the carotenoid retinal (vitamin A aldehyde) and some of its derivatives has attracted particular experimental and theoretical interest during the past few years [145, 154–156]. This interest has been stimulated by their importance in the process of vision [157]. 11-*cis*-retinal *9a* is the chromophore of the visual pigment rhodopsin.

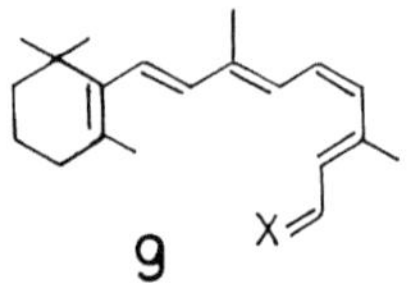

9

a: X = O
b: X = NR
c: X = NHR $^{\oplus}$

In the dark, 11-*cis*-retinal is covalently linked to a protein in the retina known as opsin. It has been shown that the retinal-opsin linkage is a Schiff base corresponding to formula *9b*. There is also some evidence for the formation of a protonated Schiff base linkage *9c*.

The primary step in the visual process is the photochemical isomerization of the 11-*cis*-retinal derivative to that of the corresponding all-*trans*-retinal [157] or, alternatively, a proton transfer in the photoexcited state [158]. The molecular and electronic structure and the spectral behaviour of retinal become, therefore, of fundamental importance in understanding the primary process of vision.

The molecular structure of 11-*cis* retinal has been extensively studied theoretically [159–163]. Karplus and coworkers [164–166] correctly predicted the most stable 12s-*cis* conformation, which was established later by X-ray study [167].

The absorption spectra of the *9a* and of its isomers have been sufficiently described by PPP calculations by appropriately considering bond alternation [163, 168–170], and most recently by CNDO/S calculations [171]. The long-wavelength absorption at around 370 nm in solution was assigned to the A → B transition, but like in polyenes a second band may be hidden beneath the intense absorption band [172].

In order to rationalize the origin of the diffuseness of the long-wavelength absorption of *9a* and the source of its marked temperature dependence, relative vibronic intensities (Frank-Condon factors) have also been calculated [166].

The Schiff base of retinal and its peptid complexes absorbs at slightly lower wavelengths than the parent compound (about 365 nm), but a considerable bathochromic effect occurs with protonation at the nitrogen (to about 450 nm) [173, 174, 174a]. Protonation obviously gives rise to a change of the polyene-type structure towards a polymethine type structure [132]. NMR studies and CNDO/2 calculations seem to support this interpretation [156].

The spectroscopic studies of retinal have been extended to the vinylogues [141].

Whereas the spectral properties of the retinal isomers and derivatives are well understood, the color of the visual pigments is less satisfactorily interpreted. Although the visual pigments contain the same chromophore and differ only in the nature of the protein, the main absorption of the visual pigments ranges from 345 nm up to 640 nm, whereas absorption maxima are displaced only to about 460 nm under normal solvent conditions. Longer wavelength absorptions have only been found in significantly polarizable solvents (e.g. trichloroethanol), which give rise to a shift up to about 500 nm [174–176]. From this result it was concluded that the polarizability of the microenvironment may exert a dominant effect.

Another explanation is based on the assumption that the anion controls the hybridization of the terminal nitrogen. If this nitrogen is close to a sp^3 hybrid, the longest-wavelength absorption is expected to be that of the retinylic anion (ca. 660 nm) [176, 177].

Theoretical estimates about protonation effect [163, 178–181] are difficult because charge and solvent effects must be appropriately taken into account. According to Suzuki et al. [178] a 200 nm shift is to be expected by assuming both a protonated Schiff base binding of retinal with opsin and an electrostatic interaction between both compounds. The origin of the electrostatic interaction was ascribed to a counterion such as COO^- of the protein in the vicinity of the protonated Schiff base binding. The red shift arising from anionic charges or negative point charges has also been discussed by means of PPP and CNDO calculations in Ref. [163] and [180]. Either negative or positive charges can produce shifts to longer or shorter wavelength depending on their position relative to the chain, for example if a carbon atom is less negative in the ground state than in the excited state, a negative charge near that atom would destabilize the former less than the latter thus giving a shift to longer wavelengths [154]. CNDO/S model type calculations suggest that a CT-type transition from environmental donor molecules to the protonated retinylidene Schiff base may play an important rôle [182]. A recent paper, however, casts doubt on protonation prior to the absorption [158]. According to this paper the red shift is solely be explained by invoking a hydrogen bond on the nitrogen atom and other intermolecular interactions, especially a polar interaction at the β-ionone end group. Also interaction between retinal chromophores may contribute to the observed red shift. The interaction between three chromophores has been recently discussed by exciton theory [183]. To find a reliable answer on the processes which actually occur, additional experimental information about details of the chromophore environment is required.

4. Diarylpolyenes

Phenylpolyenes form an another class of polyene dyes the spectral properties of which were extensively explored [145]. Stilbene represents the first member in the series of the diphenylpolyenes *10*, already carefully studied in the pioneering papers of Hauser et al. [184]. Derivatives of the stilbene find broad application as whiteners and laser dyes and are widely used for dyeing of cellulosic fibres, papers and leather [31]. Commercial stilbene dyes are of importance because of their bright shades, fairly good fastness properties and modest costs. Since photoisomerization is assumed to be the initial fading reaction of disperse stilbene dyes [185], the experimental and theoretical study of the isomerization process of stilbene becomes of fundamental importance.

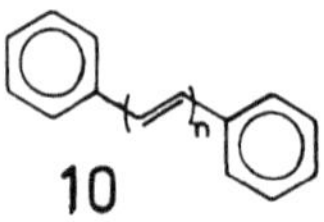

10

The spectra of both the *trans-* (*11a*) and *cis*-stilbene (*11b*) are well known and have been measured under different conditions [35]. The absorption spectrum of *trans*-stilbene in aliphatic hydrocarbons or ethanol consists of three distinct absorption bands in the near ultraviolet region. The intense long wavelength band occurs at about 290 nm and shows considerable vibrational structure. The shorter wavelength bands are located at about 230 nm and at about 200 nm (cf. Fig. VI.5).

11

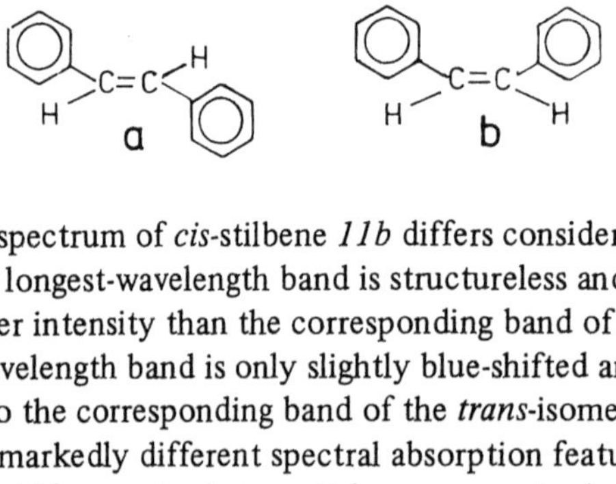

The spectrum of *cis*-stilbene *11b* differs considerably from that of the *trans*-isomer *11a*. The longest-wavelength band is structureless and appears at shorter wavelength and with lower intensity than the corresponding band of the *trans*-stilbene. The subsequent lower wavelength band is only slightly blue-shifted and exhibits higher intensity with respect to the corresponding band of the *trans*-isomer (cf. Fig. VI.5).

The markedly different spectral absorption features of the stilbene isomers result from the difference in their spatial arrangements. According to electron diffraction studies *trans*-stilbene is non-planar in the gas phase. The phenyl rings are twisted about 30° around the plane of the double bond [186]. The *cis*-isomer shows a stronger deviation from planarity [187]. The CC bonds between the benzene moieties are clearly distinguished in their lengths: 1.329 Å (*trans*) and 1.334 Å (*cis*) for the central double bond and 1.481 (*trans*) and 1.489 (*cis*) for the adjacent bonds.

Numerous theoretical studies have been made to reproduce these structural properties [188–199], but results of only a few of them agree satisfactorily with the experimental facts. Reasonable results have been obtained by methods which minimize the total energy as a sum of π and σ-contributions [35, 191–193]. Furthermore, the departure from the coplanarity of both *cis*- and *trans*-stilbene has been reproduced well by

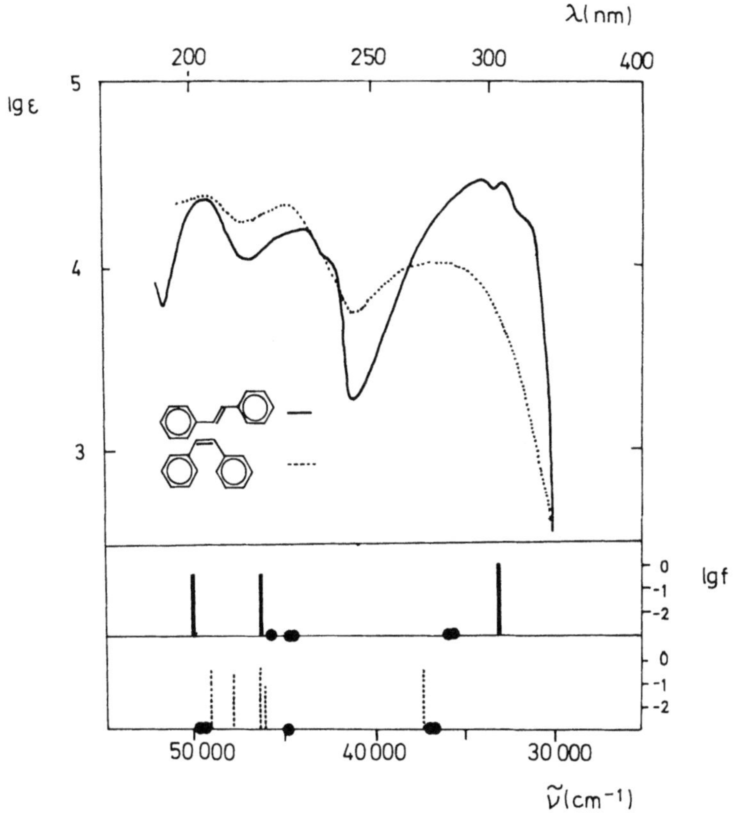

Fig. VI.5. Spectral absorption curves of *trans* and *cis* stilbene in ethanol [39] and results of PPP calculations

calculations on the semiempirical NDDO level (an all-valence-electron approximation) [200] and ab initio level [201].

The molecular geometry of *trans*-stilbene in solution is not yet known with certainty. The shape of the solution and solid state spectra have led Suzuki to suggest that the most preferred conformation of *trans* stilbene in solution would be planar [35]. PPP [35, 188, 192, 196, 197] and CNDO/S calculations [200] of the planar *trans*-stilbene gave a satisfactory description of the absorption bands in terms of allowed $\pi \rightarrow \pi^*$ transitions (cf. Fig. VI.5). Additional symmetry-allowed transitions have zero transition probability and are located beneath the strong longest-wavelength absorption band. An absorption maximum at 278 nm observed in the vapour phase has been attributed to one of these transitions [192].

The interpretation of the absorption bands of *trans*-stilbene *11a* attained by PPP-type calculations was not uniform. Whereas the longest-wavelength absorption at about 295 nm was consistently ascribed to the A → B transition, polarized along the long molecule axis, different assignments were obtained for the shorter-wavelength bands at about 230 and 200 nm (cf. Ref. [196]). The theoretical prediction is very sensitive to the adopted parametrization of the Coulomb repulsion integrals. Polarization measurements

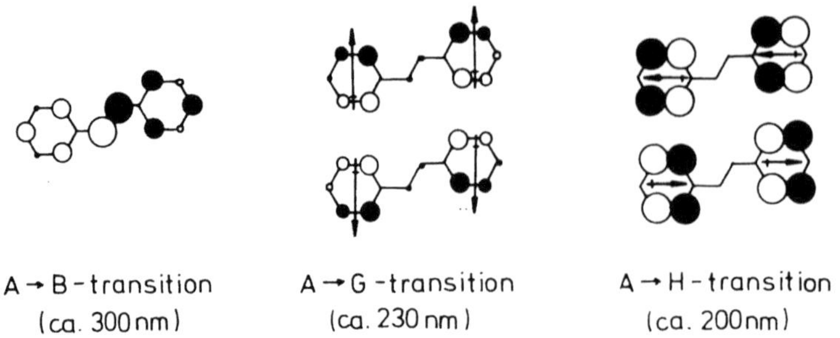

Fig. VI.6. Transition densities of the one-electron transitions responsible for the three absorption maxima of *trans* stilbene in the near UV region. The atomic transition densities result from multiplying the expansion coefficients of each atom of the delocalized MO's which are involved in the corresponding transition. The magnitude of the positive or negative numbers is illustrated by open and dashed circles, resp.

of *trans*-stilbene in stretched sheets of polymers provided contrary interpretation of the 230 nm absorption band [197, 201]. Only the results in Ref. [197] are in agreement with results of PPP calculations obtained in Mataga-Nishimoto approximation of the repulsion integrals. From the dichroic spectrum of *trans*-stilbene in stretched polyvinyl alcohol sheets the angle between the transition moments of the first two allowed electronic transitions was estimated as about 75°, which is in excellent agreement with the predicted value. According to the last-mentioned results the second absorption band at about 230 nm should originate from the partially localized $A \rightarrow G^+$ transition and the third band at about 200 nm from the completely localized $A \rightarrow H^+$ transition (labelled in Platt's notation). The presence of electronically forbidden transitions in the longest-wavelength absorption region has been recently detected by fluorescence polarization spectroscopy [201a].

The nature of these intense electronic transitions can be characterized by the transition densities, which result from the expansion coefficients of the MO's involved in the electronic excitation [197] (cf. Fig. VI.6). The $A \rightarrow B$ transition, arising from a nearly pure $1 \rightarrow 1'$ transition, is evidently delocalized over the whole π-electronic system. The following higher-energy transitions contain local excitations at each of the benzene fragments.

This interpretation of the electronic transition of *trans*-stilbene is supported by PPP configuration analysis. The analysis reveals that the L_a and B_a states of the benzene fragments mix with CT configurations to the lowest-excited B state of *11a*, whereas the H state arises from a pure excitonic combination of the L_a and B_a states (cf. Table VI.2).

Donor as well as acceptor substituents at the phenyl groups bring about a bathochromic shift of the B band. Some spectral data of substituted stilbenes *12* are assembled in Table VI.3.

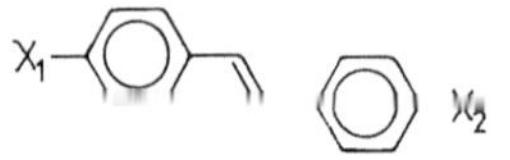

12

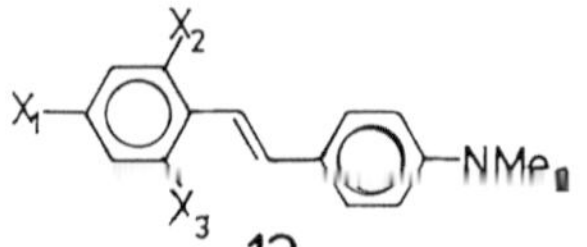

13

Table VI.2. Experimental and calculated absorption wavelengths of *trans* stilbene *11a* and nature of the relevant excited states in terms of the fragmented analogue according to the PPP configuration analysis

Transition	$\lambda_{exp.}$ (in nm)	$\lambda_{calc.}$ (in nm)	Dominant Configurations[a]	$r_M \cdot 100$ (%)
A → H	201.5[b]	196	41% L_a, 33% B_a	82
A → G	228.5	213	24% B_b, 16% CT (*B* → *E*) 16% CT (*E* → *B*), 18% CT (*B* → *B*)	81
A → B	294.1	301	18% L_a, 14% B_a, 19% CT (*B* → *E*), 19% CT (*E* → *B*)	81

[a] *B* = benzene, *E* = ethylene [b] in heptane [35]

Table VI.3. Longest-wavelength absorption maxima of substituted *trans* stilbenes *12* and *13* in nm

12[a]	X_1/X_2					
	H/H	NH_2/H	NH_2/NH_2	NO_2/H	NO_2/NO_2	NO_2/NH_2
λ_{max} (lg ε)	293.5 (4.30)	318 (4.27)	329 (4.43)	340 (4.24)	341.5 (4.27)	409 (4.36)
13[b]	$X_1/X_2/X_3$					
	NO_2/H/H	NO_2/NO_2/H	$NO_2/NO_2/NO_2$			
λ_{max} (lg ε)	435 (4.5)	480 (4.4)	525 (4.3)			

[a] in ethanol [240] [b] in benzene [241]

Both types of substituents cause an increasing participation of CT configurations of the type benzene-to-ethylene (CT_1, CT'_1) or ethylene-to-benzene (CT_2, CT'_2) (Fig. VI.7).

If one of the phenyl groups is substituted by a strong donor (D) and the second one by a strong acceptor (A) substituent, an additional CT configuration CT_3 contributes to the lowest excited energy state. This CT_3 configuration consists in an electron excitation from the donor-substituted to the acceptor-substituted phenyl group. Effective configuration mixing gives rise to low-energy electronic transitions of 4-donor-4'-acceptor substituted stilbenes, which is reflected in long-wavelength absorptions (cf. Table VI.3).

Introduction of additional acceptor groups in 2'- and 6'-positions of an 4-donor-4'-acceptor substituted stilbenes of the structure *13* leads to progressive bathochromic shifts, but the interference between the *ortho*-nitro group and the ethylenic hydrogen atoms causes a deviation from the coplanarity, which is indicated by the decreasing ab-

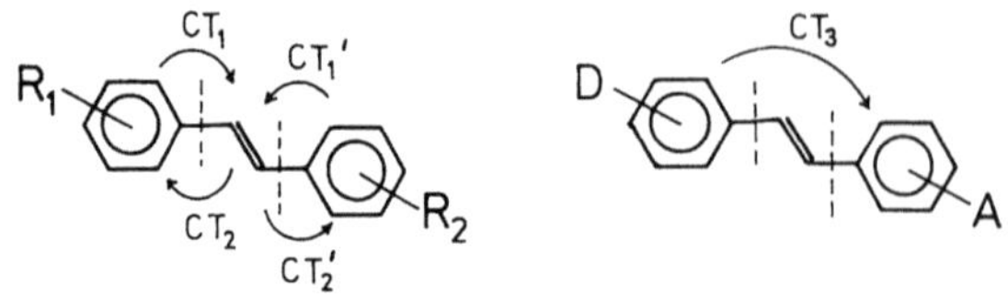

Fig. VI.7. Labelling of intramolecular CT transitions in a composite molecule description of *trans* stilbene

sorption intensity of the color band. The same effect is well-known for stilbenes which contain methyl groups in *ortho*-position of the phenyl groups or at the ethylene bridge [35, 191]. Attempts have been undertaken to correlate the wavenumbers of the first absorption band by substituent constants [202] and HMO transition energies [203].

Very strong donor and acceptor substituents bring about compounds which no longer have the typical absorption spectrum of stilbene, such as *14*. Due to the strong mixing between CT states and the NB ground state a considerable ground state stabilization occurs. This stabilization is connected with some bond equalization through the bridging ethylene structure. The electronic structure is intermediate between the two formulas *14a* and *14b*. According to formula *14b* the compound belongs to the polymethine dyes (merocyanine dyes). The compound *14* absorbs at about 620 nm in chloroform [204].

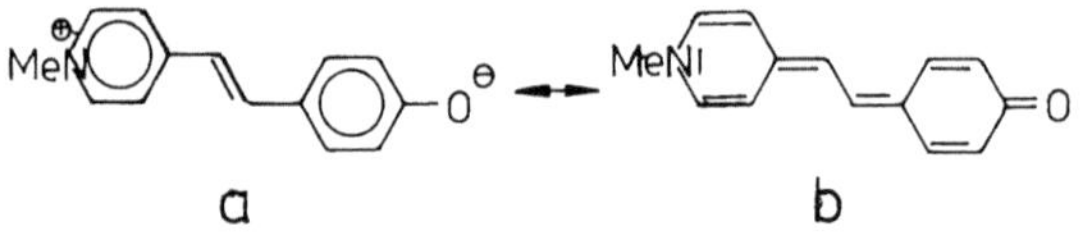

14

Spectra of various benzologues and vinylogues of *trans*-stilbene are known [205–208]. A few of them were theoretically studied [196]. The longest-wavelength absorptions of *15–20* are compiled in Table VI.4.

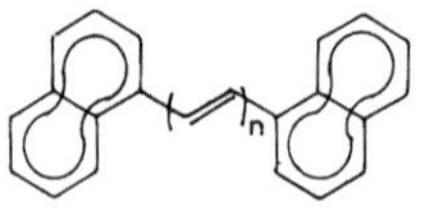
15

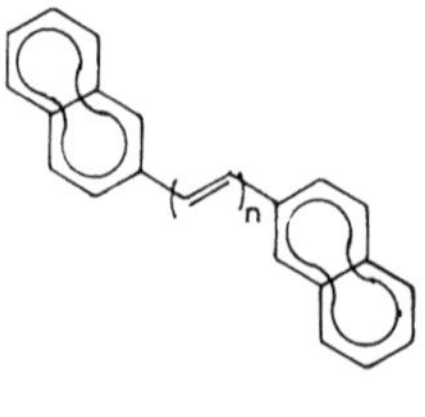
16

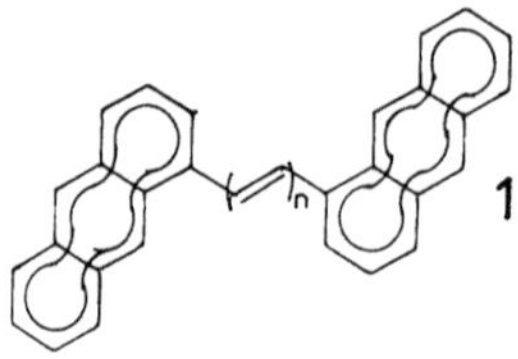
17

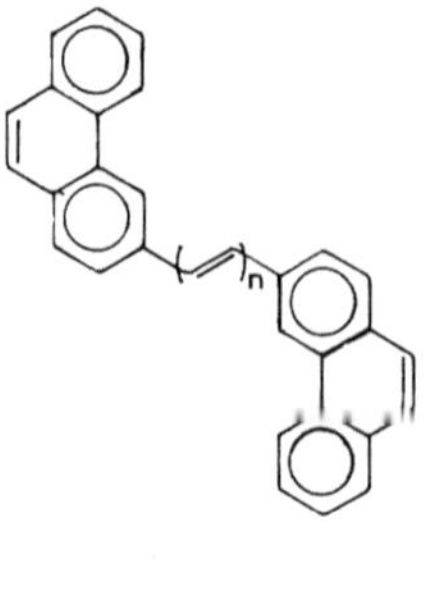
18

Table VI.4. Longest-wavelength absorption maxima of diaryl polyenes in nm (lg ε)

n	*10*[a]	*15*[b]	*16*[b]	*17*[c]	*18*[d]	*19*[e]	*20*[d]
1	306(4.38)	341(4.31)	333(4.71)	406(4.31)	348(4.89)	340(4.30)	357(4.75)
2	334(4.60)	362.5(4.58)	356(4.88)	423(4.51)	371(5.01)	367(4.56)	378(4.91)
3	358(4.88)	381(4.72)	379(4.95)	433(4.65)	390(5.08)	384(4.67)	396(5.02)
4	384(4.93)	406.5(4.72)	399(5.09)	440(4.73)	408(5.15)	403(4.83)	414(5.09)
5	403(4.97)	–	415(5.05)	452(4.87)	426(5.18)	420(4.81)	428(5.07)
6	420(5.05)	436.5(5.03)	433(4.94)	456(4.76)	441(5.16)	434(5.17)	443(5.13)
7	435(5.13)	–	–	–	–	–	–

[a] in benzene [184]
[b] in tetrahydrofuran [207]
[c] in tetrahydrofuran [206]
[d] in tetrahydrofuran [208]
[e] in benzene [208]

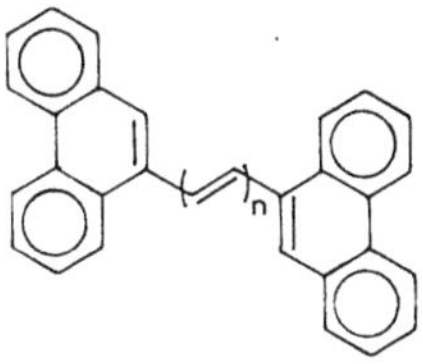

19

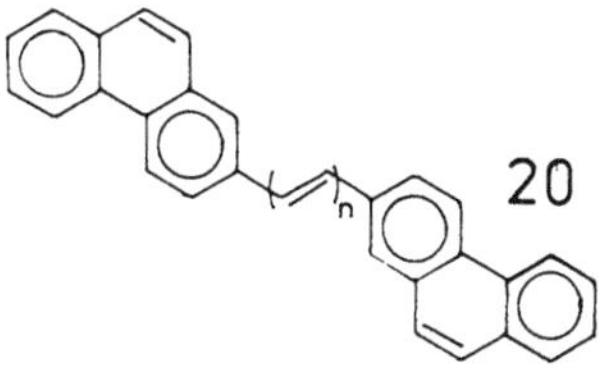

20

Absorption curves with a well-defined fine structure have been observed for *14, 18* and *20*, but not for the other three compounds *15*, *17* and *19*. These differences have been attributed to a violation of coplanarity caused by steric interference between the hydrogen atoms of the ethylene bonds and those of the peri-positions of the hydrocarbon nuclei [206, 208]. The comparison of absorption maxima reveals, moreover, that the absorption maxima of the corresponding iso-π-electromic systems differ only slightly. Series *17* represents an exception. The longer-wavelength absorptions of *17* with respect to *18–20* are mainly due to the low-energy A → L_a transition of the terminal anthracene group which effectively contributes to the delocalized electronic excitation of the polyene. Whereas the color band of *17* suffers a progressive red shift with the increase in length of polyene chain, a shorter wavelength absorption band at about 260 nm is nearly independent of the chain length (Fig. VI.8). The location of this band is similar to that of the B_b band of anthracene. Obviously, a higher-energy electronic transition of *17* is strongly localized as discussed for stilbene.

A more recent study has shown that long chain anthrylopolyenes do not exist in the all-*trans* configuration [209].

1,8-Diphenyl-tetraene (*10*, n = 4) received much attention in connection with a postulated low-energy ππ* state, to which electronic excitation is forbidden for symmetry reasons [79, 210, 211, 211a]. Several experimentalists could not support the existence of this electronic state [212, 213], but according to more recent studies [214, 215] their experimental results are not inconsistent with the aforementioned assumption. In this context it is worth mentioning that the existence of such an electronic state has been recently confirmed for stilbene (*10*, n = 1) by two-photon excitation (maximum absorp-

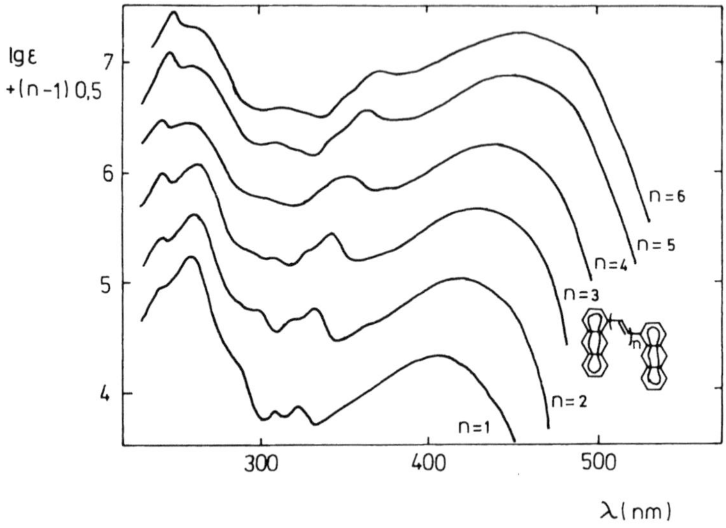

Fig. VI.8. Spectral absorption curves of di(1-anthryl)polyenes [206]. The curves with the exception of n = 1 have been displaced upwards as indicated on the ordinate axis

tion at about 244 nm). A strongly allowed transition is observed under this excitation condition [216].

The absorption spectrum of *cis*-stilbene *11b* is satisfactorily reproduced by PPP calculations, if the distortion is appropriately taken into account [188]. The twisting vibrational mode around the ethylene-phenyl bond is mainly responsible for the blurring of the fine structure [35, 93]. This out-of-plane rotation of the phenyl rings also proves to be the most convenient molecular deformation to relieve the strain in sterically hindered diphenylbutadienes [217].

The configuration of branched-conjugated carotenoids has been studied recently, among them the triaryl carotenoid *21* [218]. The spatial arrangement indicated by the formula led to agreement between the experimentally and theoretically derived polarization directions. Whereas the A → B transition is directed along the longest partial chain, the A → C transition is polarized along the shorter one.

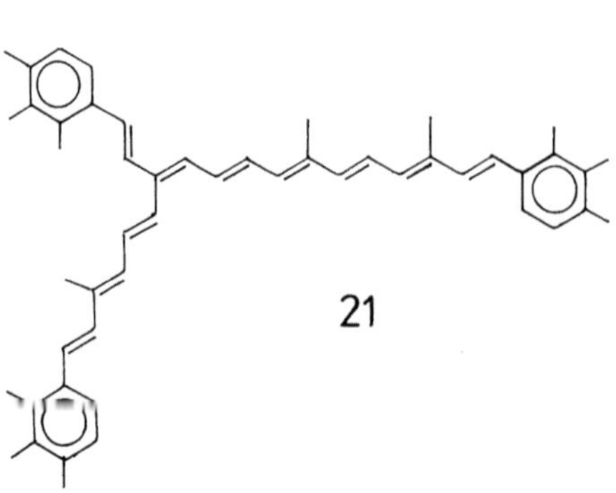

21

Table VI.5. Effect of internal aza substitution of polyenes on the longest-wavelength absorption maxima in nm (lg ε)

$H_5C_6-(CH{=}CH)_n-CH{=}X-X{=}CH-(CH{=}CH)_m-C_6H_5$				$[CH_3-(CH{=}CH)_n-CH{=}X-]_2$		
n	m	X = N[a]	X = CH[a]	n	X = N[b]	X = CH[c]
0	0	304 (4.56)	328 (4.61)	1	275 (4.60)	299
1	0	339 (4.72)	349 (4.83)	2	342 (4.80)	352
1	1	352 (4.81)	375 (4.95)	3	372 (4.94)	395.5
2	1	378 (4.88)	403 (4.97)			
2	2	385 (4.68)	420 (5.06)			

[a] in ethanol [242] [b] in dioxane [239] [c] in ether or petroleum ether [243]

From the derivatives of the diarylpolyenes only the diaza heteroanalogues should be mentioned. Spectral data of some azines are collected in Table VI.5 along with data of the corresponding iso-π-electronic hydrocarbons. The longest-wavelength absorptions of azines are hypochromically shifted in comparison with the parent hydrocarbons. The same has been observed for purely aliphatic polyenes (cf. Table VI.5).

For reasons mentioned in Sect. VI.1 replacement of CH by N should not affect the transition energy appreciably. However, the N–N π bond in azine-type heteropolyenes is probably weaker than the CH–CH π bond in polyenes. Since the corresponding bond order increases upon excitation, the HOMO is less destabilized with reducing bond strength than the LUMO. Consequently, a slight upturn in transition energy is observed, which gives rise to the observed hypsochromic shift.

5. Polyenes Bridged by Heteroatoms

Some naturally occurring dyes contain polyene chains which are singly or multiply bridged by heteroatoms. The polythienyls *22* belong to them.

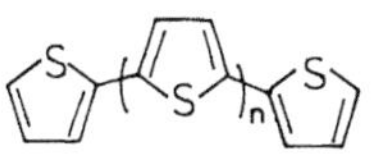

22

n = 0 305 nm
n = 1 355 nm
n = 2 391 nm
n = 3 418 nm

in benzene [219]

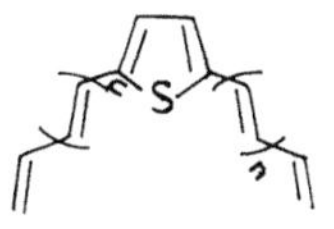

23

n = 1 328 nm
n = 2 381 nm
n = 3 422 nm

in cyclohexane [220]

They absorb in the same wavelength region as the polyenic parent compounds (*1*) of equal chain length. The correspondence between polythienyls and polyenes has been supported by polarization measurements [151].

Similarly, the absorption wavelengths of *23* nearly coincide with those of the polyenes of the same chain length. This observation has been interpreted by the fact that molecular orbitals localized over the polyene fragment are mainly involved in the electronic excitation [220]. The absence of a bathochromic substituent effect is in accordance with a FMO rationale based on the molecular symmetry [130]. However, the occurrence of the partly localized transition does not imply that a partial excitation over a molecular fragment exists.

PPP-CA calculations reveal that the lowest-energy electronic states are only poor described by the fragmented analogues. In good accordance, the former LHM calculation on *23* (n = 0) provided less satisfactorily numerical results [221].

One may be tempted to include in this section polyene derivatives, known as indophenine and pyrrole dyes. Their alternative classification to the indigoid dyes is, however, more advantageous for comparison purposes (cf. Sect. X.3).

Similarly, compounds of the structure *24* and related compounds represent borderline cases. Since the HOMO of the polyenic parent chain (10 π electron system) is symmetrical with respect to the symmetry plane perpendicular to the molecular plane, bridging by lone pair orbitals raises the HOMO energy. This leads to a bathochromic shift. However, considering the central heteroatom as terminal group of two equal polymethine chains, the compounds were alternatively assigned to the bichromophoric cyanine dyes [222].

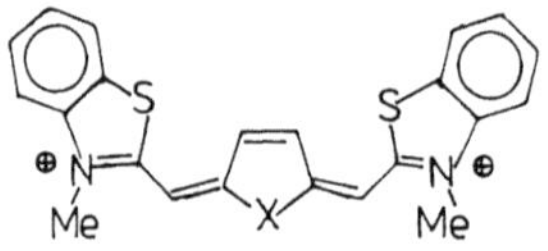

24

a X = NH 545 (4.38)
b X = O 522
c X = S 534
d X = Se 530 (4.65)

in water [222–224]

By N–H deprotonation *24a* is converted to a true polymethine dye, whose color (664 nm) is well understood by taking into account both the effect of aza substitution and ethylene bridging [224].

6. Medium Effect

In spite of the fact that polyenes are weakly polar compounds, appreciable wavelength shifts have been observed for carotenoids. Thus, the color band of *8* is shifted from 537 nm in cyclohexane to 577 nm in carbon disulfide [225]. The red-shift of the absorption band and the concomitant decrease of oscillator strength are correctly described by the Coulomb interaction between the electronic transition moments of the solute molecules and those of the molecules of the solvent shell [226]. In equeous solutions, carotenoids form aggregates, which absorb 60 to 80 nm blue-shifted with respect to the monomers [226a].

Considerable attention has been paid to the solvent dependence of the spectral absorption of retinal and its derivatives. If the polar retinal is dissolved in nonpolar solvents a good linear correlation is obtained by plotting wavenumber shifts versus $(n^2 - 1)$ $(2n^2 + 1)$ (McRae equation) [155]. Spectral solvent effects of polar solute in polar solvents require the change of dipole moment upon excitation to be considered [227]. According to the continuum theory spectral shifts increase with increasing change of the dipole moment upon excitation. As demonstrated for 4-nitro-4′-amino stilbene, increase of the dipole moment gives rise to bathochromic shift in passing from non-polar to polar solvents. A bathochromic shift has been observed, for example, in cyclohexane (415 nm) to chloroform (440 nm) [228]. In contrast, the merocyaninic *14* undergoes blue shift in passing from non-polar to polar solvents (cf. p. 194).

A striking bathochromic shift occurs in strong acids or by addition of Lewis acids [229]. Thus, a deep blue coloration has been observed when vitamin A, dissolved in chloroform, is mixed with a saturated solution of $SbCl_3$ in the same solvent (Carr-Price reaction). Addition of sulphuric acid or trichloroacetic acid to solutions of carotenoids causes similar colour changes. The exact nature of the colored compounds is not yet known in the most cases, but protonation undoubtedly leads to the formation of polyenyl carbonium ions [230–232]. Thus, vitamin A acetate *25* undergoes protonation to the retinylic cation *26*, which absorbs at 583 nm (in H_2SO_4/CH_3OH). The observed absorption wavelength of *26* is sufficiently in line with the absorption wavelength of corresponding symmetrical carbenium ion of the series *27* (n = 3) (λ_{max} = 626 nm) [233]. The formation of vinylogous carbenium ions is also supported by the finding that the absorption wavelengths of protonated carotenoids increase linearly with the number of effective double bonds ($\lambda_{max} = 300.5 + 65.5$ n). Similarly α,ω-diphenyl polyenes undergo protonation in concentrated sulphuric acid with formation of colored carbenium cations [234].

Retinal and its Schiff bases are protonated at the oxygen and nitrogen, respectively. The thereby observed bathochromic effect is considered to be important for the understanding of color vision (cf. Sect. VI.3).

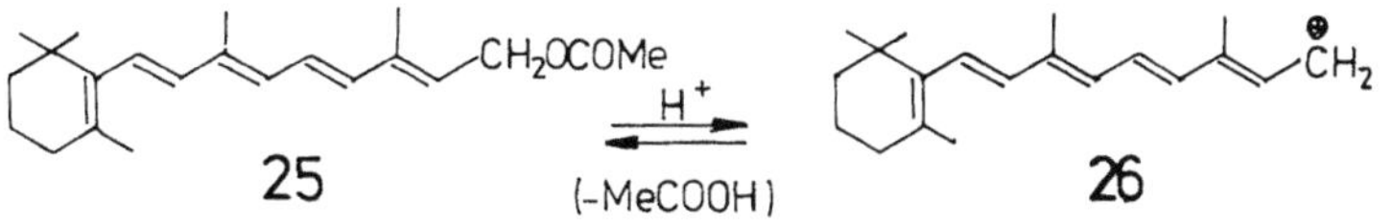

25 26

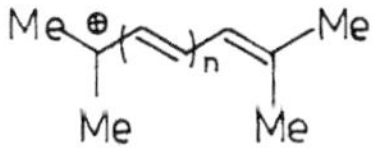

27

VII. Azo Dyes

The importance of the azo colorants is shown by the fact that they embrace over 60% of the total number of dye structures known to be manufactured. The application involves acid dyes for wool and polyamide, basic dyes for acrylic fibres, direct dyes for cotton and viscose rayon, disperse dyes for polyester and reactive dyes for cellulose [31, 244–246]. Further areas of their application are dying of paper and leather and their use as pigments. The azo dyes comprise practically every shade from yellow to black.

1. Chromophoric System

Azo dyes are characterized by the presence of one or more azo groups –N=N– which form bridges between organic residues. In the most stable configuration the residues are *trans* positioned. Simple azo compounds have the general formula *1* in which the residues R_1 and R_2 are saturated, unsaturated or aromatic groups. Dyes of practical importance contain carbocyclic or heterocyclic π electron systems.

$$R_1-\underline{N}=\underline{N}-R_2$$

1

In dependence on the intramolecular bonding azo compounds can be considered from different points of view. Some models are illustrated in Fig. VII.1.

Firstly the π electron structure may be considered as a whole according to model *A*. This description permits one to reveal similarities between azo and stilbene dyes which are iso-π-electronic to each other.

Secondly, an alternative approach for describing the π electron system of azo compounds consists in applying a composite molecule model. Proper subunits are then the fragments R_1 and R_2 and the central azo group, as schematically indicated in Fig. VII.1 by three rectangles (model *B*).

Thirdly, one of the residues R_1 or R_2 may either display marked donor (R_D) or marked acceptor properties (R_A). Then the azo compound can be modelled according

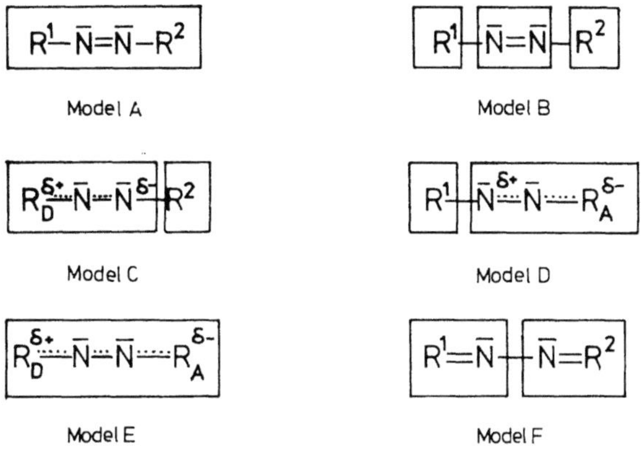

Fig. VII.1. Different fragmentation modes for describing azoic structures

to *C* or *D* of Fig. VII.1. Due to the acceptor property of the azo group considerations according to model *C* may become particularly advantageous.

Fourthly, one moiety R_1 may be a strong donor, whereas another one R_2 is a strong acceptor group. In that case the azo group more or less loses its individuality and the compound can be viewed as heteroanalogous polymethine dye rather than as an azo dye (model *E*). Therefore, dyes of the model type *E* are considered, in part, in Sect. XII.

Fifth, the electronic structure may finally modelled according to model *F*. Structures of this type are azines which are heteroanalogous polyenes (cf. Sect. VI.4).

2. Weak Long-Wavelength Absorptions

Before we review the spectral absorptions of azo dyes a striking difference between the azo and stilbene dyes should be pointed out. In contrast to stilbene dyes some simple azo dyes display a weak long-wavelength absorption which results from the azo group itself.

In place of the two C–H bonds of the vinylene group the azo group has two lone-pair orbitals n which are appropriately sp^2 hybrid orbitals. Owing to the overlap between both orbitals they form a symmetrical and an antisymmetrical combination $n_+(n_s)$ and $n_-(n_a)$, respectively, as illustrated in Fig. VII.2 for the *trans* structure.

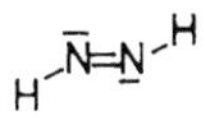

2

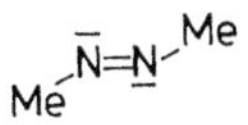

3

In the *trans* isomer the antisymmetrical combination is of lower, and in the *cis* isomer of higher energy.

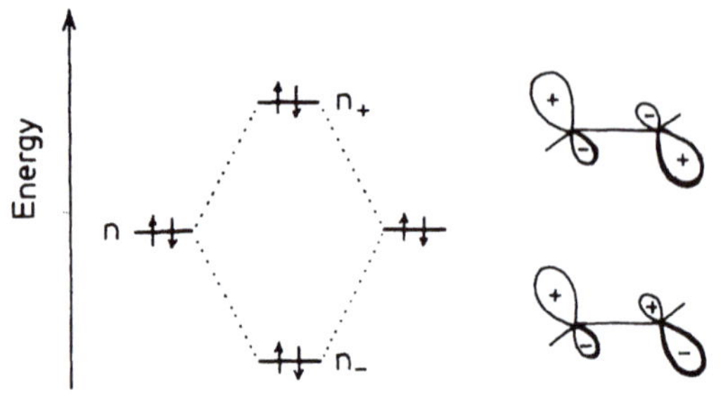

Fig. VII.2. Splitting of the lone-pair n-orbitals of the azo group

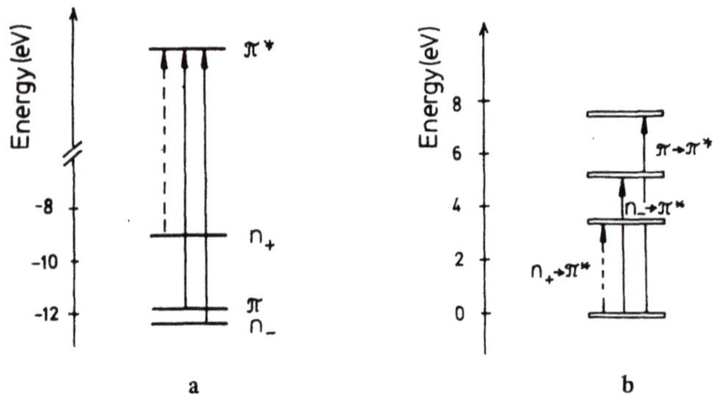

Fig. VII.3. MO energy level (a) and term level diagram (b) of *trans*-azomethane *3* according to data given in Ref. [255] and [261]

Theoretical studies on *trans*-diimide (*2*) [247–253] and *trans*-azomethane (*3*) [253, 254] predicted a large energy splitting between n_+ and n_-. This prediction was established by the photoelectron spectra of *3* [255, 256]. Assuming the validity of Koopmans theorem, the energy levels n_+ and n_- differ by more than 3 eV[1]. The energy of the π MO is intermediate between the two lone pair energy levels (cf. Fig. VII.3a).

Accordingly, two $n \rightarrow \pi^*$ transitions are anticipated: the lower energy $n_+ \rightarrow \pi^*$ transition, and the higher energy $n_- \rightarrow \pi^*$ transition. The $n_+ \rightarrow \pi^*$ transition is symmetry forbidden in the *trans*-azo isomer. Vibronic interactions account for the occurrence of a weak and structureless long-wavelength absorption band which causes the yellow color of azomethane and related simple azo compounds [258–260]. A linear correlation has been found between $\Delta E_{(n_+ \rightarrow \pi^*)}$ and the vertical ionization energy $I_v(n_+)$ [256].

The simple MO picture (Fig. 3a) suggests, however, that the $\pi \rightarrow \pi^*$ transition of *1* should be interposed between the two $n \rightarrow \pi^*$ transitions. Experimental and theoretical studies indicate, however, that the correct energy sequence of the transitions is: $\Delta E_{(n_+ \rightarrow \pi^*)} < \Delta E_{(n_- \rightarrow \pi^*)} < \Delta E_{(\pi \rightarrow \pi^*)}$ [254, 261] (cf. Fig. VII.3b).

1 A splitting in nearly the some order of magnitude probably also occurs for *trans*-azobenzene [257]. According to the photoelectron spectrum its first ionization potential I_{n_+} is 0.52 eV lower than that of azomethane.

This term level ordering is due to the different contributions of electron-electron interaction terms to the n → π and π → π* transition energies [254]. The assignment of higher-energy transitions, however, has not yet been settled [262]. Additional information about the electronic term levels of *3* was recently obtained by electron impact excitation [263].

Low n → π* excitation energies are also expected for simple *cis* azo compounds. In this case the antisymmetrical combination n_- of the lone pair orbitals posseses the higher energy. The n_- → π* transition is allowed. Aliphatic *cis*-azo compounds therefore show more intense n → π* absorption bands [264].

The location of the n → π* absorption band depends strongly on the electronic structure of the residues linked to the azo group. As demonstrated by Bock [265], the n → π* absorption ranges from 300 nm up to 600 nm for simple azo compounds. The wavelength varies in the order : $F < OR < NR_2 < CH_3 < SO_2R < COR < C_6H_4R < POR_2 < SiR_3$. Some examples are given in Table VII.1.

Table VII.1. Spectral data of the low-intensity $n_+ \rightarrow \pi^*$ absorption band of trans-R–N=N–R

R	λ_{max} in nm (lg ε)	Solvent	Ref.
H	350	hexane	[388]
CH_3	357 (1.11)	hexane	[261]
Bu^+	368 (1.08)	gas phase	[266]
($Bu^+CH=CH$)	440 (2.37)	hexane	[389]
C_6H_5	444 (2.67)	hexane	[261]
CN	445 (1.26)	acetonitrile	[390]
COC_6H_5	469 (1.62)	dioxane	[265]
$PO(C_6H_5)_2$	568 (1.36)	dioxane	[265]
$Si(CH_3)_3$	784 (0.70)	aliph. hydrocarb.	[266]

Table VII.2. Longest-wavelength absorption maxima of aliphatic and aromatic phenylazo compounds *1*

R_1 (R_2 = Ph)	λ_{max} in nm (lg ε)		Solvent	Ref.
	n → π*	π → π*		
H	417.5 (1.87)	260 (3.87)		[391]
Me	398 (2.04)	262 (3.94)	ethanol	[279]
CH=CH–Ph	444 (2.80)	344 (4.58)	hexane	[392]
C≡C–Ph		360	ethanol	[393]
phenyl	444 (2.65)	316 (4.36)	cyclohexane	[205]
1-naphthyl	454 (2.95)	371 (4.08)	cyclohexane	[205]
2-naphthyl	446 (2.96)	329 (4.35)	cyclohexane	[205]
1-azulyl		608 (2.76) 415 (4.47)	cyclohexane	[394]

The spectral shifts generated by different residues of the azo group are approximately additive. The unique strong substituent effect of the $SiMe_3$ group has been interpreted by the low ionization potential of the substituted azo compounds [266]. Bathochromic shift has been generally observed in passing from the alkyl substituted to conjugated residues (cf. Table VII.2). However, only a few azobenzenes show a well-resolved $n \rightarrow \pi^*$ absorption maximum in non-polar solvents [267–268]. In the spectra of the most commercial azo dyes the $n \rightarrow \pi^*$ absorption is buried underneath the intense longest-wavelength $\pi \rightarrow \pi^*$ absorption band. Nothing is known about its exact location in these cases.

The $n \rightarrow \pi^*$ absorption band undergoes blue shift in passing from non-polar to polar solvent and mostly disappears when the azo compound is dissolved in acids.

Attempts have been undertaken to calculate $n \rightarrow \pi^*$ transition energies of azobenzenes within the framework of the simple HMO [35, 268a] and PPP procedure [269]. The low-energy $n \rightarrow \pi^*$ transition energies were well reproduced in the last mentioned study. A second $n \rightarrow \pi^*$ transition was predicted in the near ultraviolet region. The remarkably intense $n \rightarrow \pi^*$ absorption band of *cis*-azobenzene (lg ϵ 3.1) in comparison to that of aliphatic azo compounds of *cis*-structure reflects its non-coplanarity (see below).

3. Azo Compounds with Carbocyclic Groups

Apart from the weak $n \rightarrow \pi^*$ absorption the spectral absorption curve of *trans*-azobenzene *4* is quite analogous to that of *trans*-stilbene (Fig. VII.4).

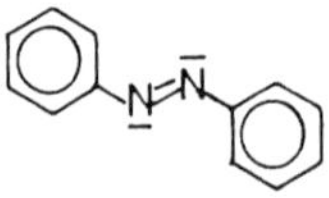

4

Only the first $\pi \rightarrow \pi^*$ absorption band of azobenzene is slightly shifted to longer wavelengths. The close resemblance suggests a very similar molecular and electronic structure of both molecules. *Trans*-azobenzene was found to have a nearly planar structure in the solid state [270, 271], while a non-planar structure with phenyl groups twisted approximately 30° around the N-phenyl bonds exists in the gas phase [272]. The latter result is satisfactorily in line with the calculated geometry [158]. With bond lengths of about 1,24 Å (1.26 Å) for N=N and about 1.45 Å (1.42 Å) for C–N a pronounced bond alternation has been found in the central part of the molecule in the X-ray [270, 271] and gas electron diffraction study [272], resp. Judging from the fact thath the solution is very similar to the KCl-disk spectrum, Suzuki inferred that the most probable conformation of *trans*-azobenzene in solution is nearly planar [35].

The electronic spectrum of *4* was interpreted with the aid of PPP-type calculations [269, 273–276]. The intense long-wavelength absorption band at about 315 nm comprises two allowed electronic transitions [276]. The high intensity of this band is determined by a long-axis polarized transition, which corresponds to the A → B transi-

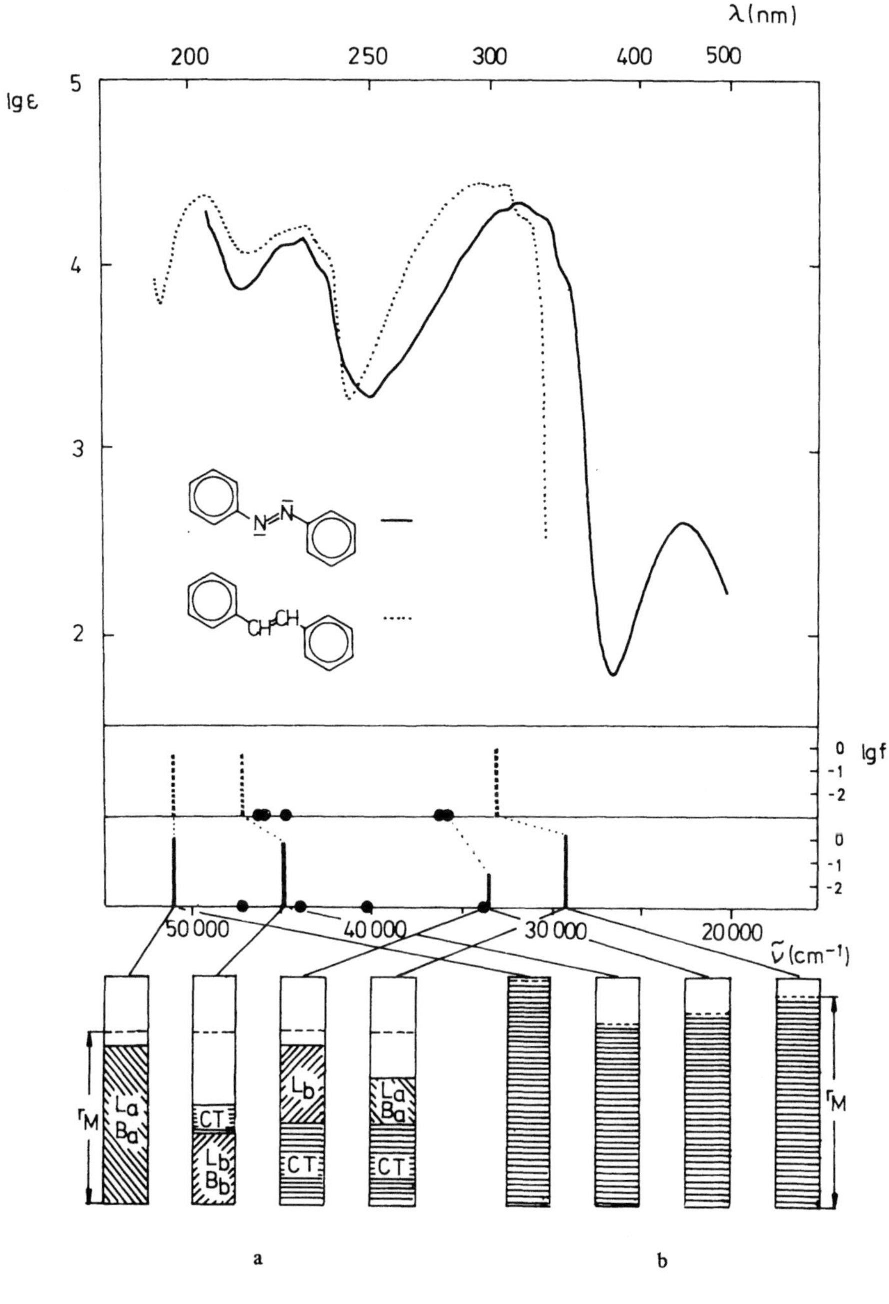

Fig. VII.4. Spectral absorption curves of *trans*-azobenzene in hexane [39] and of *trans*-stilbene in ethanol [39] and results of PPP calculations. The rectangles illustrate the PPP-CA interpretation of the most important excited states of azobenzene with respect to molecular fragments (on the left) or with respect to the corresponding excited states of stilbene (on the right)

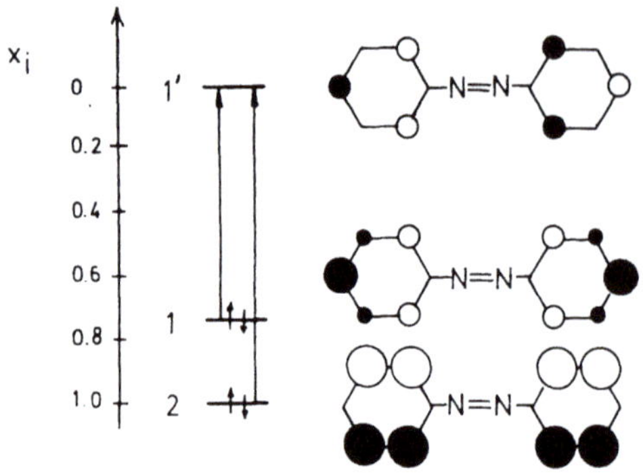

Fig. VII.5. Lowest-energy one-electron transitions of azobenzene and pictorial representation of the expansion coefficients of the MO's involved therein. The open (shaded) circle signifies that the positive (negative) lobe of the $2p_\pi$ orbital is above the molecular plane. Their areas indicate the magnitude of the squares of the expansion coefficients

tion of stilbene. The nearby higher-energy A → C transition, polarized in the short molecular axis, is of low intensity. In a rough approximation the A → B transition corresponds to an one-electron transition from the highest filled MO to the NFMO (1 → 1′), whereas the transition from the subjacent MO (2 → 1′) contributes the A → C transition (cf. Fig. VII.5).

The absorption bands at nearly 225 nm and 200 nm arise from the short-axis polarized A → G and the long-axis polarized A → H transition, respectively. The assignment corresponds closely to that given for *trans*-stilbene (see p. 33). Polarization measurements, however, have not yet provided an unequivocal confirmation of this assignment [201, 277, 278].

The close correspondence between the state wave functions of azobenzene and stilbene is supported by PPP configuration analysis. The results are depicted in Fig. VII.4. The nearly identical band position of the 200 nm absorption reflects the strong localization of the relevant electronic state at the benzene moieties. Only the intense longest-wavelength absorption band of azobenzene is more strongly affected by the heteroatomic substitution. Replacement of CH=CH by N=N increases the contribution of the charge-transfer states, arising from electron excitation from the phenyl rings to the N=N-acceptor group. This fact is documented by 35% CT contribution to the lowest energy $\pi \rightarrow \pi^*$ state of azobenzene compared to 19% in the case of stilbene.

Benzylideneaniline *5* is iso-π-electronic with stilbene and azobenzene. Its absorption spectrum, however, differs considerably from the spectrum of the latter compounds. Haselbach and coworkers [190, 279] supported a former assumption that the molecule is twisted around the C–N single bond, thus subdividing the conjugated system of *5* into two weakly interacting π systems. One of these π systems is localized essentially at the benzylideneimine fragment and the other one at the benzene moiety, comprising the lone pair orbital of the adjacent nitrogen (aniline fragment). Theoretical studies also suggested torsion around the C–N single bond [280]. The non-coplanarity of *5* has been

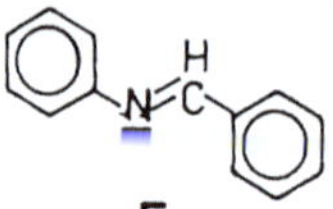

5

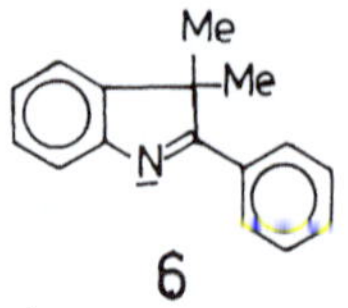

6

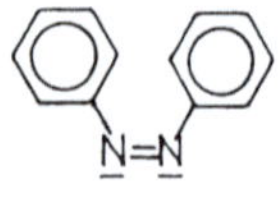

7

confirmed by X-ray analysis [281] and electron diffraction studies [282]. In good accordance the planar 3,3′-dimethyl-2-phenyl-indolenine (*6*) has a spectrum strikingly different from *5*, but intermediate between that of *trans*-azobenzene *4* and that of *trans*-stilbene [283].

A satisfactory theoretical reproduction of the electronic spectra of the planar and non-planar benzylideneaniline was achieved by Hohlneicher and coworkers, who took into account the nitrogen lone pair orbitals in the PPP approximation [269]. As the

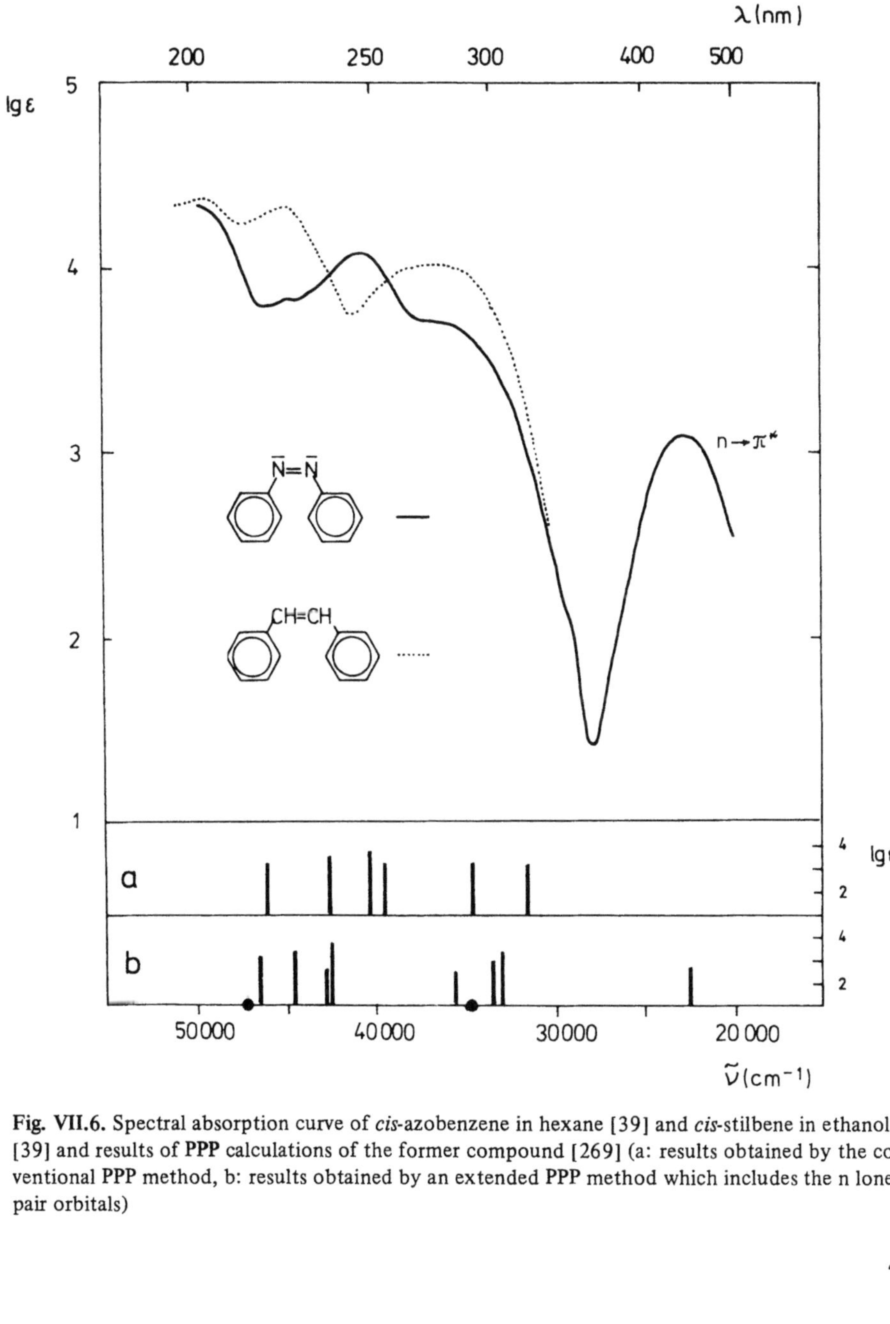

Fig. VII.6. Spectral absorption curve of *cis*-azobenzene in hexane [39] and *cis*-stilbene in ethanol [39] and results of PPP calculations of the former compound [269] (a: results obtained by the conventional PPP method, b: results obtained by an extended PPP method which includes the n lone pair orbitals)

calculated orientation of the transition moments was not reported in Ref. [269], the theoretical results cannot be compared with the more recent experimental finding that the two lowest-energy transitions are long-axis polarized [277, 284].

The close correspondence of the spectral curves of some substituted *5* and *6* suggests that, in contrast to the parent compounds, derivatives may exhibit a planar or nearly planar structure.

Cis-azobenzene *7* is less stable than its *trans*-isomer, in good agreement with calculations by the refined all-valence-electron method PCILO [285]. Due to steric hindrance of the benzene rings *7* was found to be symmetrically twisted around the double bond about 56° in the crystalline state [286]. Due to this molecular structure the absorption spectrum of *cis* azobenzene differs markedly from that of the *trans* isomer. The spectral differences account for the photochromic properties of numerous azobenzenes which are brought about by *trans* → *cis* isomerization [287].

On the other hand, the spectrum of *cis*-azobenzene also deviates remarkably from that of *cis*-stilbene (cf. Fig. VII.6).

Hohlneicher and coworker [269] attempted to reproduce the spectral curve of *cis*-azobenzene by PPP-type calculations without and with explicit consideration of the n-electrons. Due to the mixing of n-orbitals with π-orbitals occurring in the non-planar conformation, the long-wavelength n → π^* transition gains significantly intensity from the allowed $\pi \rightarrow \pi^*$ transitions. But also the position of the higher energy transitions are markedly affected by mixing between the nπ^* and $\pi\pi^*$ wave functions. This accounts for the pronounced spectral differences mentioned above.

Benzologues of *trans*-azobenzene exhibit the first $\pi \rightarrow \pi^*$ absorption band at longer wavelengths than the parent compound [205]. Examples are listed in Table VII.2. The longest-wavelength absorption of phenylazoazulene at 608 nm arises from an electron excitation mainly localized within the azulene fragment. This is in accord with a recent PPP-CA calculation with respect to the fragmented analogue [288].

4. Substituted Azobenzenes

Both donor and acceptor substituents produce red shift of the intense 320 nm absorption band of the parent compound. This is exemplified in Table VII.3 with substituted azobenzenes *8*. The spectral shifts correspond closely to those of other substituted benzenes.

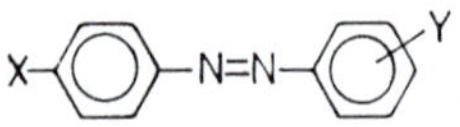

8

Donor substituents, such as hydroxy-, mercapto- or amino-groups give rise to a stronger bathochromic shift (up to 400 nm) than acceptor substituents. The shift values correlate fairly well with Bloor's spectroscopic substituent constants σ_s which, for comparison purposes, are included in Table VII.3. For 4-donor-substituted azobenzenes, the spectral shifts also correlate with the ionization potentials of the correspondingly substituted benzenes [276].

Table VII.3. Longest-wavelength $\pi \rightarrow \pi^*$ absorption maxima of para-monosubstituted and p,p'-disubstituted azobenzenes *8*

X	λ_{max} in nm (lg ϵ)		σ_S^{+} [h]	σ_S^{-} [h]
	8 (Y = H)	*8* (X = Y)		
H	317 (4.34)[a]	320 (4.24)[d]	0.00	
Me	324 (4.36)	330 (4.29)[d]	0.19	
F	319 (4.36)	325 (4.23)[d]	0.10	
Cl	324 (4.40)	331 (4.42)[d]	0.22	
Br	326 (4.40)	326 (4.39)[d]	0.26	
J	331 (4.32)	–	0.53	
OH	339 (4.39)	–	0.47	
OMe	342 (4.41)	355 (4.59)[e]	0.55	
SMe	362 (4.44)	380 (4.39)[f]	0.90	
NH_2	364 (4.45)	399 (4.53)[g]	0.89	
NHMe	380 (4.47)	–		
NMe_2	400 (4.49)	460 (4.52)[e]	1.04	
NHCOMe	346 (4.46)	–	0.64	
COOMe	325 (4.34)[b]	–		–
CN	324 (4.59)[c]	–		–
NO_2	331 (4.42)	338 (4.36)[e]		1.20
COMe	326 (4.30)	–		0.77

[a] in cyclohexane [267, 395]
[b] in ethanol [396]
[c] in ethanol [397]
[d] in ethanol [398]
[e] in ethanol [399]
[f] in ethanol [400]
[g] Ref. [401]
[h] Ref. [402], σ_S^{+} in heptane, σ_S^{-} in hexane

The hydroxy- and amino-group exert a still higher bathochromic shift when they undergo deprotonization. Red shift up to 140 nm has been observed in amino-substituted azo dyes which are dissolved in alkaline dimethylsulfoxide or dimethylformamide [289].

The color band is also red-shifted when acceptor substituents are introduced in the unsubstituted benzene ring of donor-substituted azobenzenes. Thus, 4-dimethylamino-4'-nitro azobenzene absorbs at about 475 nm. Further examples of the structure *9* (R_1 to R_4) are tabulated in Table VII.4.

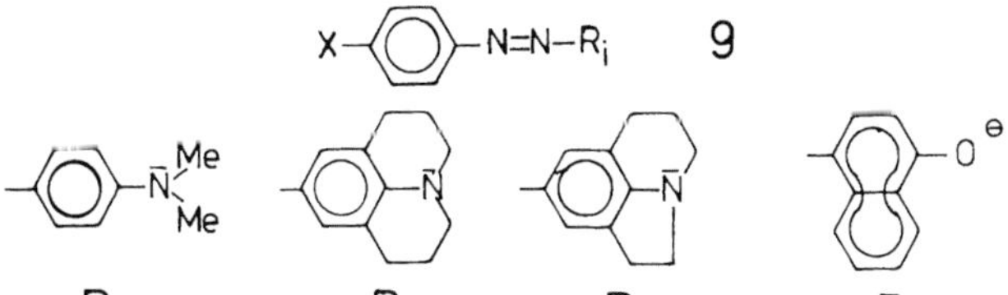

9

R_1 R_2 R_3 R_4

PPP calculations provide a satisfactory description of the absorption spectra of simple substituted azobenzenes, e.g. of 4-nitro azobenzene, 4-dimethylamino azoben-

Table VII.4. Longest-wavelength absorption maxima of para-substituted aminoazo and hydroxyazo compounds *9*

9	λ_{max} in nm (lg ϵ)				σ_P
X	R_1	R_2	R_3	R_4	
NO_2	475[a]	518 (4.55)[b]	515 (4.46)[c]	604[d]	0.78[e]
CN	–	–	487 (4.44)	556	0.66
SOMe	447	–	482 (4.43)	559	0.50
CF_3	427	471 (4.49)	461 (4.39)	–	0.54
I	422	467 (4.51)	453 (4.38)	–	0.18
Br	419	464 (4.50)	448 (4.40)	–	0.23
Cl	417	461 (4.49)	444 (4.38)	512	0.23
F	407	450 (4.44)	426 (4.33)	499	0.06
H	*407*	*442* (4.43)	*425* (4.34)	*500*	0
Me	407	446 (4.43)	422 (4.31)	498	−0.17
OMe	407	452 (4.44)	417 (4.35)	496	−0.27
OH		449 (4.44)	414 (4.36)		−0.37
NH_2	410		–		−0.66
SMe	420				
NMe_2	450				

[a] in ethanol [375]
[b] in ethanol [290]
[c] in ethanol [291]
[d] in 0.1 N KOH solution in ethanol [297]
[e] taken from Ref. [291]

zene or 4-dimethylamino-4′-nitro azobenzene, although the calculated absorption wavelengths of the A → B transitions are too low in the last mentioned two cases (Fig. VII.7). [288].

Considered as substituted azobenzenes, a different participation of CT configurations have been found. Whereas the lowest energy state of 4-nitro azobenzene is well described by the locally excited B state of the parent compound only, amino-to-azobenzene CT configurations strongly contribute to the lowest excited state of the amino-substituted compounds. This results from the effective mixing between the donor p_π-orbital and the HOMO of azobenzene, which has a high expansion coefficient in 4-position of the benzene fragment (cf. Fig. VII.5).

Since the donor orbital interacts the more strongly the higher the donor orbital energy, donors of low ionization potential are highly efficient. From this point of view the large bathochromic effect of the mercapto group is easily understandable and does not require invocation of d-orbital effects of the sulphur.

A direct donor-to-acceptor CT configuration additionally contributes to the lowest excited state in azobenzenes which bear a donor substituent group at one benzene fragment and an acceptor substituent group at the other one. This CT configuration mixes with the B state of azobenzene via amino-to-azobenzene CT configurations, thus, causing an additional diminution of the excitation energy.

But the CT configurations also interact with the NB ground state leading to some ground state stabilization. This effect is accompanied by a decrease of the appropriate-

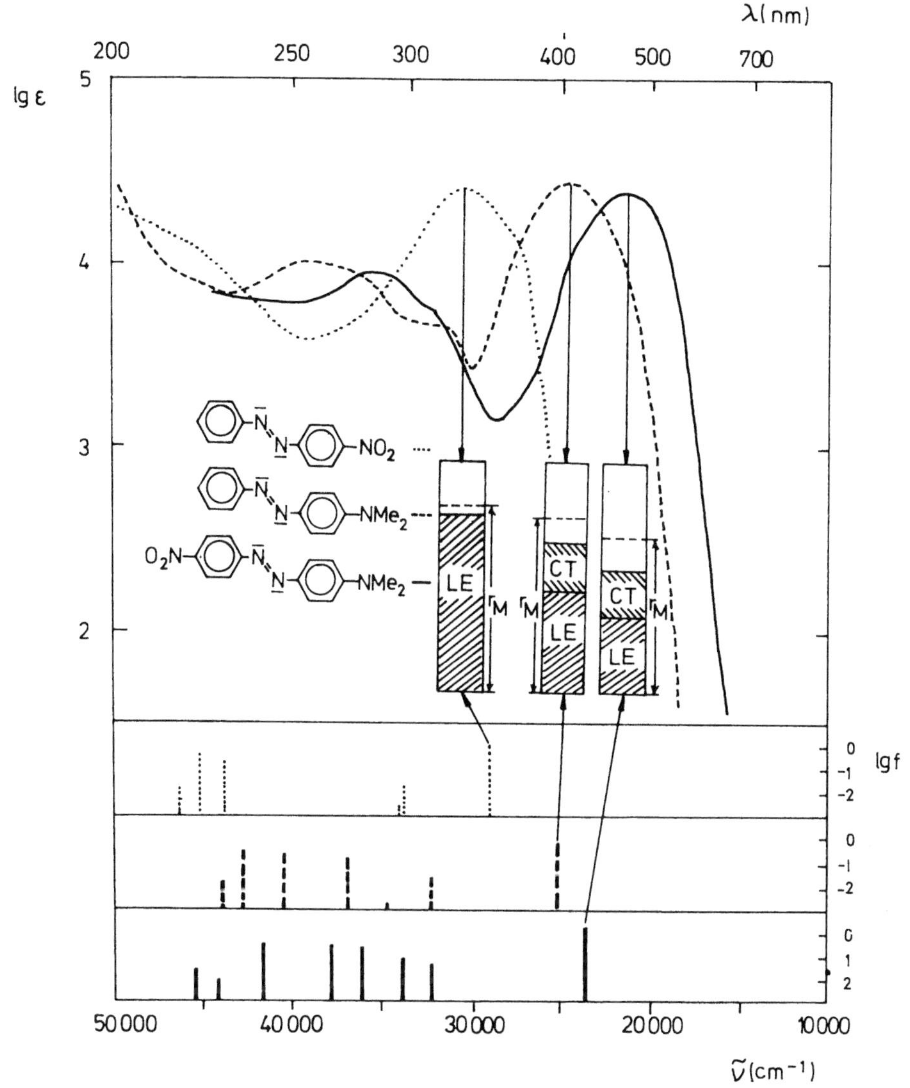

Fig. VII.7. Spectral absorption curves of substituted *trans*-azobenzenes [39, 395] and results of PPP calculations. The rectangles display the weights of the lowest-energy LE- and CT-configurations to the color-determining electronic states according to PPP-CA calculations

ness of the composite molecule model, as reflected in the Γ_M values of the amino-substituted compounds (cf. graphical illustration of the CA configuration mixing in Fig. VII.7.)

The spectral properties of substituted azobenzenes can be similarly rationalized by referring to smaller molecular fragments considering substituents of the azo group as separate substructures. However, as demonstrated in Fig. VII.8, model *B* is less suited than the above discussed model *A*. Due to the stronger partitioning, CT contributions dominate in describing the lowest-excited states.

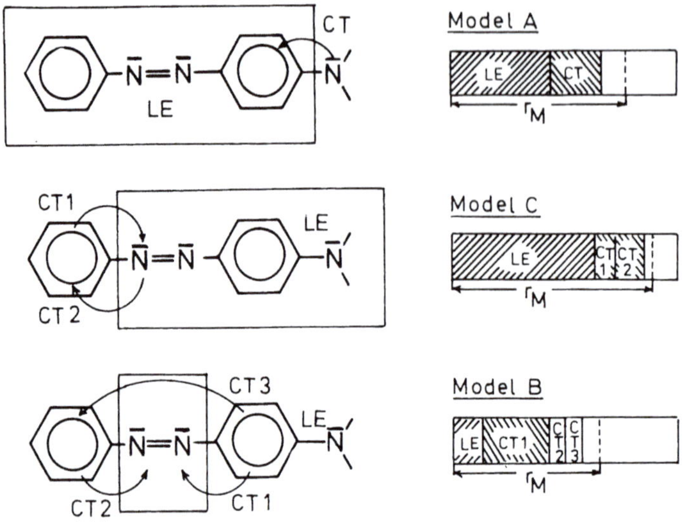

Fig. VII.8. PPP-CA interpretation of the lowest-excited states of aminoazobenzene with respect to different fragmentation modes

A more adequate fragmentation model is, however, that of model *C*, where the aminophenylazo subunit is considered as separate subchromophore. This reference takes properly into consideration the ground state stabilizing aminophenyl-azo interaction. Within this model the phenyl group appears as an perturbation of the aminophenylazo subchromophore. This approach makes easily understandable the substituent effect observed with substitution at the phenyl group (*9*, $R_i = R_1$): The absorption wavenumbers of these compounds correlate with Hammett-type substituent constants σ_p. Electron-withdrawing substituents give rise to a bathochromic shift and conversely electron donating substituents give rise to a hypsochromic shift of the color band [290, 291]. Examples are listed in Table VII.4 along with substituent constants. A similar correlation has been found for series of differently substituted disperse azo dyes [292].

Obviously, the substituted phenyl group in *9* exerts a predominantly inductive effect on the aminophenylazo subchromophore. This picture, however is not longer relevant when the substituent group X is strongly involved into the chromophoric system. This situation arises in passing from the weak donor groups OH and OCH_3 to the strong donor groups NH_2, SMe or NMe_2 (cf. Table VII.4).

The spectral data in Table VII.4 also show that the absorption wavelength depends on the structure of the amino group. Substitution of dimethylaminophenyl (R_1 of *9*) by julolidyl (R_2) or by lilolidyl (R_3) causes pronounced bathochromic shifts [290, 291]. This is connected with a concomitant increase in intensity. The spectral effect is due to the improved conjugation rather than to the *ortho*-alkyl effect [290, 293]. Ion selective crown ether dyes have been prepared recently [293a]. The amino group is involved in the crown ether structure. Effective complexation of metal cations brings about up to 120 nm hypsochromic shift of the color band relative to that of the parent dye.

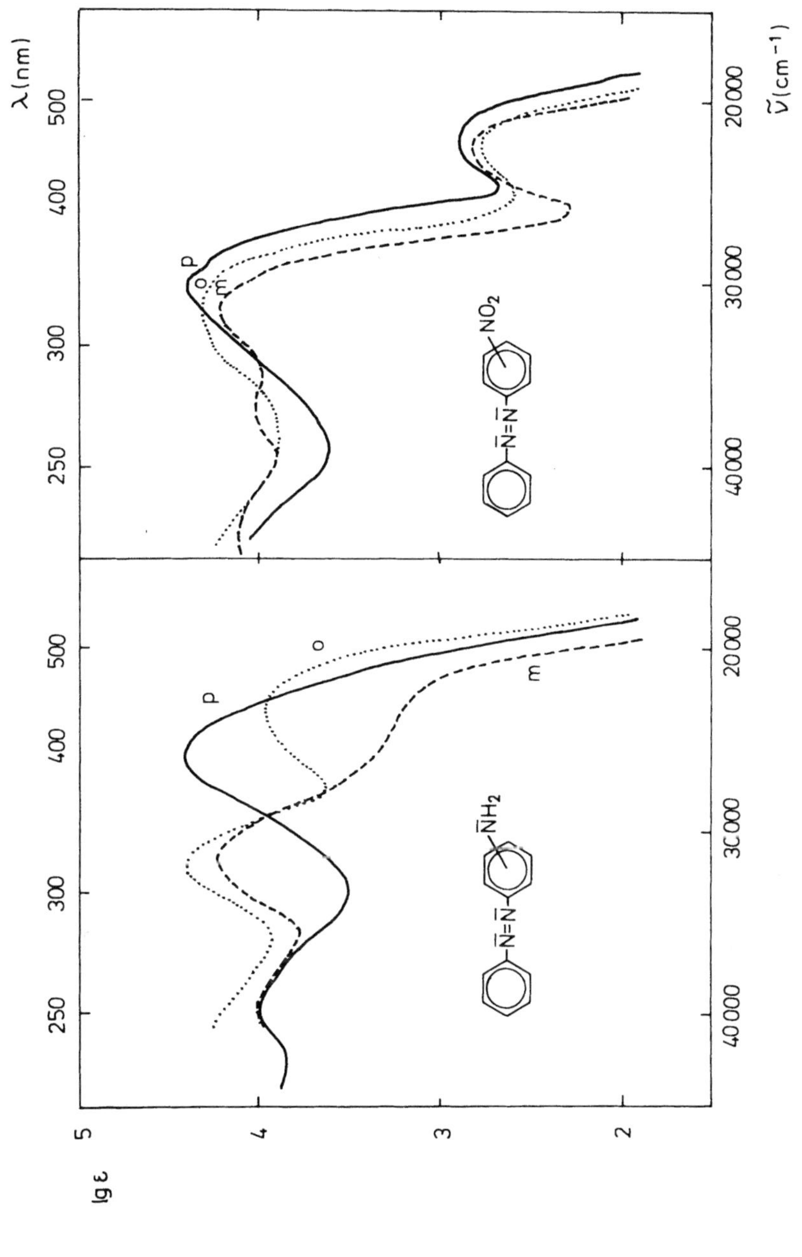

Fig. VII.9. Spectral absorption curves of *ortho-*, *meta-* and *para*-aminoazobenzenes and *ortho-*, *meta-* and *para*-nitroazobenzenes (redrawn from Ref. [403])

Correlations with Hammett-type σ-values were obtained also for phenylazo-naphthols [294, 295]. Deprotonated 4-phenylazonaphthols (*9*, $R_i = R_4$) display significant substituent shifts over more than 100 nm [294]. Pronounced spectral substituent effects have also been reported for 5-phenylazo-tropolones [296].

Only few spectral data are accessible about the full triads of mono-substituted azobenzenes substituted at one of the benzene fragments. The spectral pattern of the *ortho-*, *meta*- and *para*-aminoazobenzene corresponds to that observed generally for donor-acceptor substituted benzenes [54]. Whereas the *ortho*- and *meta*-substituted compounds exhibit only a medium intense $\pi \rightarrow \pi^*$ absorption at long wavelengths, the *para*-isomer is characterized by an intense band at shorter wavelengths (cf. Fig. VII.9).

This effect can be easily rationalized by considering the MO's of azobenzene which are involved in the lowest energy excitation. Due to the high expansion coefficients of the higher-energy occupied MO 2 in 4-position depicted in Fig. VII.5, the donor orbital raises this MO more strongly than that of the highest occupied MO 1. Consequently, the weak $2 \rightarrow 1'$ transition of azobenzene now appears at longer wavelengths.

A similar spectral behaviour has been observed for other mono-donor-substituted azobenzenes. However, in contrast to substituted azobenzenes, donor-substituted amino-azobenzenes absorb at nearly the same wavelength as the amoniazobenzene parent compounds. This is demonstrated in Table VII.5 for isomeric methoxysubstituted 4-amino-azobenzenes.

Table VII.5. Influence of the position of donor and acceptor substitution on the color band of azobenzenes of the structure *8*

X	Y	λ_{max} in nm (lg ϵ)			Ref.
		ortho	*meta*	*para*	
H	NH_2	427 (4.00)	430 (3.20)	384 (4.40)	[403]
NHC_2H_4CN	OMe	394 (4.43)	394 (4.45)	395 (4.45)	[404]
H	NO_2	320 (4.30)	315 (4.20)	333 (4.40)	[403]
$N(C_2H_5)_2$	CN	462 (4.48)	446 (4.45)	466 (4.51)	[405]
NHC_2H_4CN	NO_2	422 (4.45)	418 (4.44)	451 (4.47)	[404]

Table VII.6. Effect of multiple donor or acceptor substitution on the color band of the azobenzenes *8* in dependence on the positions of substitution (X: a single substituent, Y: two or three substituents)

X	Y:	λ_{max} in nm (lg ϵ)							Solvent	Ref.
		2',3'	2',4'	2',5'	2',6'	3',4'	3',5'	2',4',6'		
H	OMe		364 (4.19)	391 (3.58)					acetic acid	[287]
NO_2	OMe		407 (4.35)	427 (3.86)		393 (4.31)			ethanol	[297]
H	O^-		473 (4.37)	572 (–)		501 (–)			ethanol/ KOH	[297]
NO_2	O^-		574 (4.17)	655 (–)		613 (4.24)			ethanol/ KOH	[297]
NEt_2	CN	490 (4.54)	514 (4.60)	495 (4.56)	503 (4.52)	500 (4.59)	478 (4.53)	562 (4.67)	ethanol	[405]

Introduction of an additional donor substituent in the aminophenyl residue shifts the color band to longer wavelengths. This has been demonstrated with 2-donor substituted 4-N,N-diethylamino-4′-nitroazobenzene dyes [296a]. The 2-acceptor substituted compounds absorb at shorter wavelengths, however. Extensive studies of multiple donor substitution confirmed the validity of Kaufmann's "rule of the distribution of auxochromes" [297]. As exemplified in Table VII.6, the wavelengths of donor-substituted azobenzenes *8* (Y = donor group) generally increase in the sequence 2′,4′ < 3′,4′ < 2′,5′.

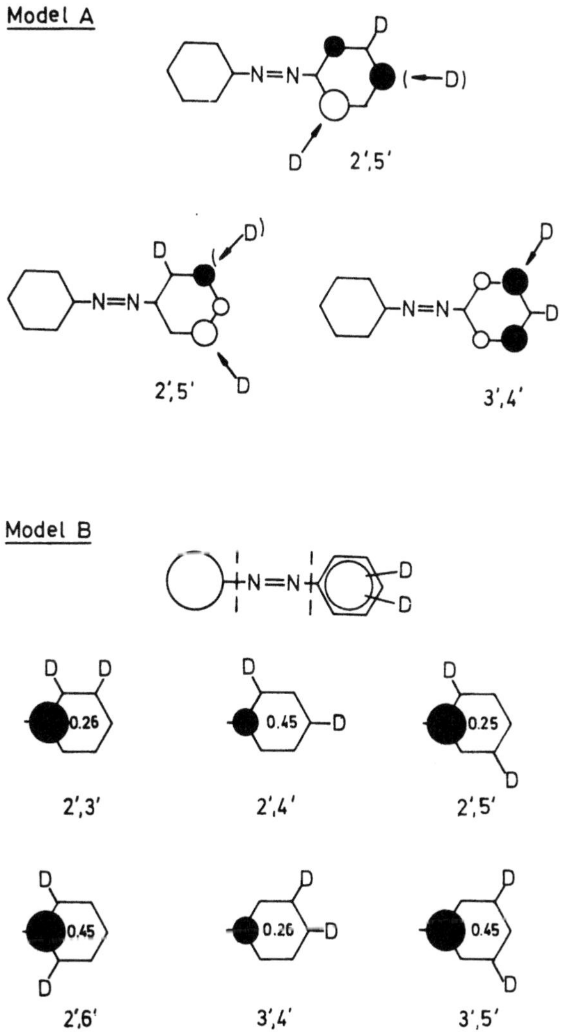

Fig. VII.10. Pictorial representation of the expansion coefficients of the HOMO's of mono-substituted azobenzenes (upper part) and of the LUMO's of di-substituted benzenes (lower part) (cf. caption to Fig. VII.5). The figures indicate the HOMO energy of the substituted benzenes

This observation can be understood by asking for the positions of the most effective HOMO-HOMO interaction occurring between a donor substituent orbital and the orbital of the singly donor-substituted azobenzenes (model *A*). According to the expansion coefficients of the HOMO, pictorially illustrated in Fig. VII.10 (upper part), introduction of a second donor group D in *ortho*(2′)-substituted compounds is more effective in producing a bathochromic shift in 5′-position than in 4′-position and corresponding substitution in *para*(4′)-substituted compounds is more effective in the 3′-position than in 2′-position. The same results is obtained by starting from the mono-substituted *meta*(3′) isomer: Particular long wavelength absorption should result, in turn, from 2′,5′ (≙ 3′,6′) and 3′,4′-substitution.

Perturbation theoretical considerations can also be based on model *B*, referring to an aryl-azophenyl composite molecule. Now the phenyl residue should be the more effective in producing long-wavelength absorptions, the higher its donor capacity (the higher the energy of the HOMO) and the better the LUMO-LUMO interaction, for the LUMO-LUMO interaction determines the strength of mixing between the LE-excited states at benzenoid residue and the phenyl-to-azo CT states. Energies and expansion coefficients of the LUMOs are given in Fig. VII.10 (lower part) for model-type structures of substituted benzene residues. Due to low HOMO energies, 2′,3′-, 2′,5′- and 3′,4′-substitution is now clearly favoured over other substitution patterns and 2′,3′- and 2′,5′-substitution should be most effective because of the higher expansion coefficient (stronger LUMO-LUMO overlap). Remarkably, the predicted strong bathochromic effect with 2′,3′-substitution has not been concluded above from considerations according to model *A* (upper part of Fig. VII.10).

Three points are additionally worth mentioning. First, the strongly bathochromic-shifting substituents need not necessarily cause strong absorption. Since the transition may have strong CT-character, the transition probability is low, if no sufficient stabilizing overlap between the subunits takes place (cf. Sect. IV.2). Second, HOMO-LUMO overlap (not depicted here) brings about ground state stabilization, which counteracts the bathochromic shift. Intense and long-wavelength absorptions are, consequently, only obtained if both a strong excited state and moderate ground state stabilization occurs. This is the case for 2′,5′-substitution. Third, if the ground state is strongly stabilized, the composite molecule model is less appropriate. This situation is encountered with 2′,4′-substitution of strong donor groups. As discussed for *para*-dimethylamino azobenzene, the benzenoid residue and the azo group merge and gives rise to a separate chromophore. This chromophoric system attains some polymethine character. Like polymethines compounds of this type may display stronger absorption at longer wavelength than is expected from the composite molecule picture.

Reasoning along these lines, the consideration can be easily extended to azobenzenes which are singly or multiply substituted by acceptor groups. The isomeric mono-nitro azobenzenes show absorption maxima only slightly shifted with respect to azobenzene and resembling each other (cf. Fig. VII.9). Contrary to aminoazobenzenes, increasing red shift has been observed in the sequence m $<$ o $<$ p, as exemplified in Table VII.5. Similarly, 2′,4′- and 2′,6′-positioning of acceptor groups is favoured in producing bathochromic shifts over 3′,5′-positioning as demonstrated with nitrilo-substituted aminoazobenzenes (cf. Table VII.6).

Table VII.7. Longest-wavelength absorption maxima of substituted 4-nitro-4′-aminoazobenzenes *10* ($Y = Y_1 = H$, $R_1 = H$, $R_2 = CN$) in methanol [298]

$X(X_1)$	λ_{max} in nm (lg ε)	
	Mono-Substitution ($X_1 = H$)	Di-Substitution ($X = X_1$)
H	453 (4.64)	453 (4.64)
Me	454 (4.62)	383 (4.38)
Cl	475 (4.60)	417 (4.49)
CN	504 (4.65)	549 (4.58)
NO_2	491 (4.58)	520 (4.60)

Steric hindrance may occur when substituents are in the 2′,6′-positions. Marked steric effects have been observed for 4-nitro-4′-amino-azobenzenes *10* and derivatives [298, 299]. Data given in Table. VII.7 demonstrate that introduction of two substituents possessing roughly spherical shape (Me, Cl) in the same ring both *ortho* to the azo group cause a hypsochromic as well as a hypochromic effect. *Ortho*-cyano groups, however, do not seem to produce steric interference, because of their rod-like shape, whereas *ortho*-nitro groups probably rather undergo a rotation out of plane than interfere with the coplanarity of the azobenzene chromophore.

In summary, introduction of substituents into azobenzene allows a considerable variation of the color. The strongest effects have been observed when donor and acceptor substituents are attached to different phenyl groups. Along these lines even azo dyes with blue and bluish-green shades have been produced. Disperse dyes of the structure *10* are blue, for example if the substituents are $X = NO_2$, $X_1 = Cl$, $Y = OMe$, $Y_1 = NHAc$ and $R_1 = R_2 = CN$ (λ_{max} 610 nm). [300].

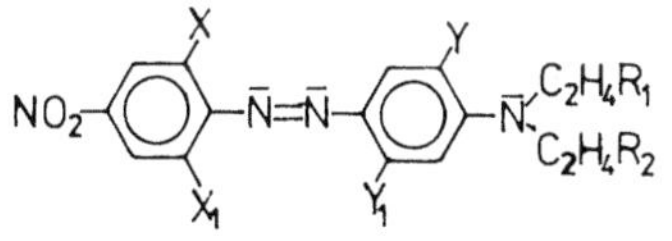

10

The practical importance of substituted azobenzenes has stimulated attempts to estimate the intense longest-wavelength absorption maxima by means of additive substituent shifts. However, as has been already shown by Skulski, systematic deviations confine the applicability of a simple additivity rule [301]. 4,4′-Bis-donor and 4,4′-bis-acceptor substituted azobenzenes absorb at significantly shorter wavelength as predicted from the shifts observed with mono-substitution. Only weak substituent shifts are nearly additive, for instance, those produced by chlorine [302]. A more general approach consists in defining different substituent increments which take into account the substitution pattern and the degree of co-planarity of the chromophoric system [303].

Only few systematic investigations on the substituent effects have been performed on the spectral properties of the benzologues of azobenzene, such as for symmetrically substituted 1,1′- and 2,2′-azonaphthalenes bearing two bromine atoms or two nitro groups in different positions [304]. A vast number of commercial azo dyes containing one or two naphthalene rings owes their long-wavelength absorption donor substituents in 2- or 4-position. The arylazo-substituted 1-naphthylamines *11a* and *11b* for example absorb at about 440 and 500 nm.

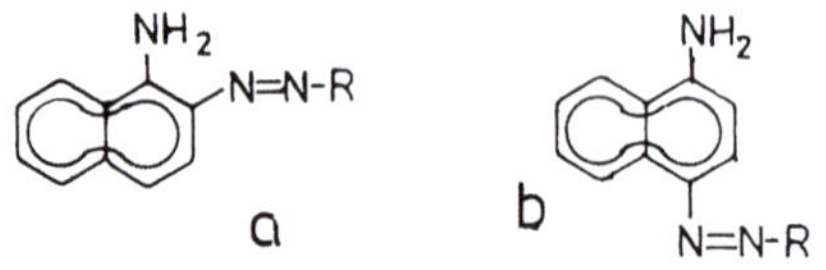

11

	a	b
R = phenyl:	468 nm (4.16) [305]	438 nm (4.36) [305]
R = 1-naphthyl:	490 nm (4.17) [306]	465 nm (4.36) [306]
		in alcohol

2-Substituted dyes are, in general, deeper in color than the 4-substituted ones.

Many hydroxy substituted azo dyes are capable of undergoing tautomerization (see Sect. VII.7).

5. Polykisazo Compounds

Bisazo and polykisazo dyes form a large group of commercial dyes the hues of which cover nearly all shades from yellow to green. Many of them are direct dyes with a high substantivity on cellulose fibres and good fastness properties to light [31].

The position of the first intense absorption maxima of polykisazo benzenes *12* and polykisazo naphthalenes *13* is shifted to longer wavelengths with extension of the π system (cf. Table VII.8).

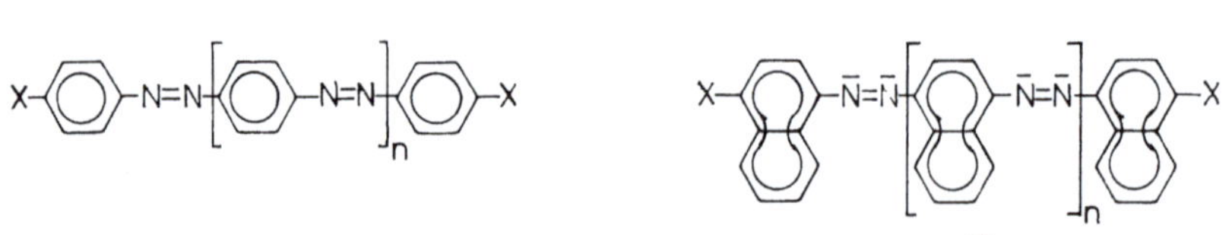

12 13

In good agreement with theoretical results [259], the intense longest-wavelength absorptions converge with increasing chain length. Semiempirical calculations performed within the framework of the molecular exciton model predict the convergence limit of unsubstituted para-polykisazo benzenes at about 435 nm.

As in the case of monoazobenzenes, donor substituents influence the color band of polykisazo compounds more effectively than acceptor substituents [307]. The longest-wavelength absorptions have been observed for compounds which bear both donor and acceptor substituents [308] (cf. Table VII.8).

Table VII.8. Longest-wavelength $\pi \rightarrow \pi^*$ absorption maxima of para-polykisazobenzenes *11* and para-polykisazonaphthalenes *12*

n	λ_{max} in nm (lg ϵ)					
	11				*12*	
	X = H[a] Y = H	X = NO_2[b] Y = NO_2	X = NH_2[c] Y = NH_2	X = NO_2 Y = NMe_2	R = H[d]	R = NH_2[d]
0	318 (4.34)	338 (4.43)	390 (4.50)	470 (4.41)	405 (4.18)	490 (4.40)
1	359 (4.64)	377 (4.68)	474 (4.78)	500 (4.50)	450 (4.31)	585 (4.63)
2	380 (4.76)	403 (4.83)	490 (4.82)		485 (4.50)	610 (4.74)
3	400 (4.80)		500 (4.85)			

[a] in ethanol or tetrahydrofuran [307]
[b] in benzene [406]
[c] in ethanol [308]
[d] in different solvents [407]

Coupling of two azobenzene moieties through an additional π system R leads to bisazo dyes according to the formula *14*. Their absorption properties are influenced both by the nature of R_i and the kind of their junction. The *ortho*- and *meta*-bisazo benzene *14* ($R_i = R_1$ and R_2, resp.) absorb at shorter wavelength than the *para*-isomer ($R_i = R_3$), and the positions of their absorption maxima hardly differ from that of the azobenzene parent compounds *9* (R_i = Ph) (cf. Table VII.9). Obviously, *ortho*- and *meta*-phenylene bridges are insulators of conjugation.

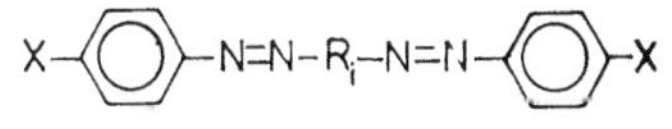

14

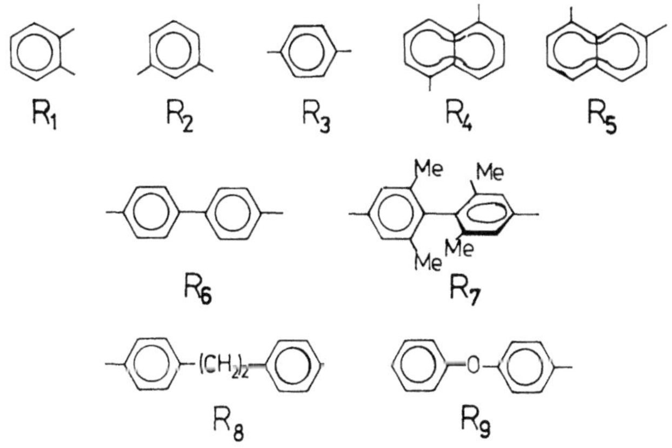

Whereas the low effect in *ortho*-position may have steric reasons, the weak *meta* effect results from the electronic structure. Considering *14* as acceptor substituted azobenzene, the LUMO-LUMO interaction should be lowest in *meta*-position, for the LUMO expansion coefficient of azobenzene is there very low (cf. Fig. VII.5).

Table VII.9. Longest-wavelength $\pi \rightarrow \pi^*$ absorption maxima of bisazo dyes of the structure *14* in comparison to related monoazo parent compounds (in ethanol)

X	R_i	Bisazo Dye *14* λ_{max} in nm (lg ϵ)	R_i	Parent Dye *9* λ_{max} in nm (lg ϵ)	Ref.
H	R_1	304 (4.45)	Ph	318 (4.34)	[307]
H	R_2	320 (4.53)	Ph	318 (4.34)	[307]
4-OH	R_2	360 (4.70)	Ph	350 (4.30)	[313]
H	R_3	359 (4.64)	Ph	318 (4.34)	[307]
4-OH	R_3	400 (4.60)	Ph	350 (4.30)	[313]
4-Me_2N	R_4	440 (4.63)	1-Naphthyl	432 (4.40)	[314]
4-OH	R_5	357 (4.50)	2-Naphthyl	362 (4.49)	[314]
2-OH-5-Me	R_6	415 (4.50)	Ph	388	[315]
2-OH-5-Me	R_7	390 (4.50)	Ph	388	[315]
4-Me_2N	R_8	410 (4.50)	p-Tolyl	400	[309]
4-Me_2N	R_9	420 (4.60)	p-MeO-phenyl	408 (4.44)	[316]

Weak bathochromic effect are also observed when two distinct phenylazo fragments are joined through *meta*-phenylene bridge [309–313]. These dyes unify the absorption feature of the two participating subchromophores. Insulators of conjugation are similarly naphthalene groups which are linked in the 1,5- or 1,7-positions (e.g. *14*, $R_i = R_4$ or R_5) [314].

Bisazo dyes derived from benzidine show the absorption feature of their chromophoric subunit (but with double the intensity) if the bridging diphenyl moiety is strongly twisted around the central essentially single bond, as exemplified by *14* with $R_i = R_7$ [311, 315]. In absence of overcrowding *ortho*-substituents the overlap between the two azobenzene fragments is no longer negligibly small and red shift and concomitant intensification of the longest-wavelength absorption band has been observed (e.g. *14*, $R_i = R_6$).

A slight coupling of the chromophoric system takes place when two azobenzene molecules are linked through methylene groups (e.g. *14*, $R_i = R_8$) [309–311] or through oxygen (e.g. *14*, $R_i = R_9$, cf. Table VII.9) [316].

Weak coupling of yellow and blue azobenzene chromophores leads to green azo dyes. These compounds may be considered as multiple chromophore dyes with structurally related subchromophores.

6. Heterocyclic Azo Compounds

Replacing one or both benzene rings of azobenzene by heterocyclic units leads to heterocyclic azo compounds which find increasing practical interest, e.g. as disperse dyes [317]. The discussion in the preceding sections facilitates basic understanding of their longest-wavelength absorption pattern. 4-Phenylazo-pyridines and 4-phenylazo-pyridinium salts display spectral curves which are closely related to those of azobenzenes (cf.

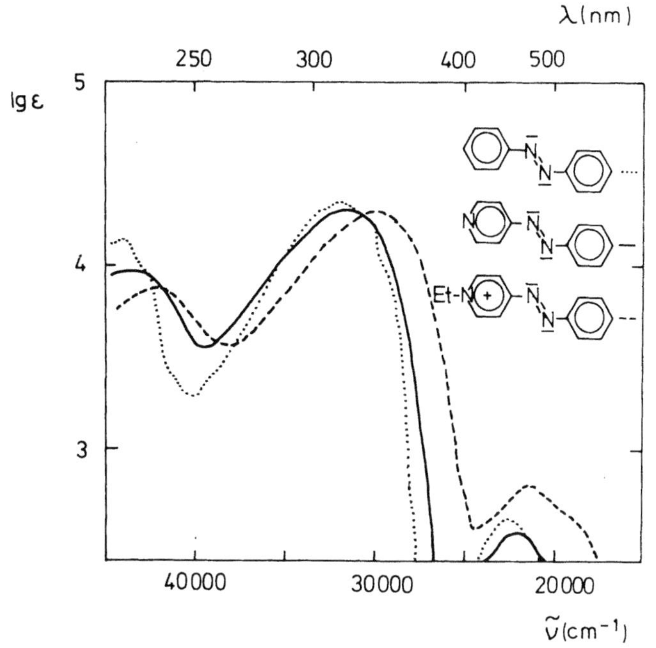

Fig. VII.11. Spectral absorption curves of azobenzene in hexane [39], 4-phenylazo-pyridines in ethanol [415], and 4-phenylazo-pyridinium tetrafluoroborate in acetonitrile [39]

Fig. VII.11). This spectral similarity has also been demonstrated by the absorption characteristics of 2-substituted 5-phenylazopyridines in comparison to substituted azobenzenes [267].

The six membered heterocycles pyridine and pyridinium ion are acceptor π systems which evidently behave like acceptor-substituted benzenes. In contrast, remarkable long-wavelength absorptions are produced by cationic, neutral or anionic heterocyclic fragments, such as R_1 to R_5, R_6 to R_9 or R_{10} in azo compounds of the structure *15*.

R_i–N̄=N̄–R_i **15**

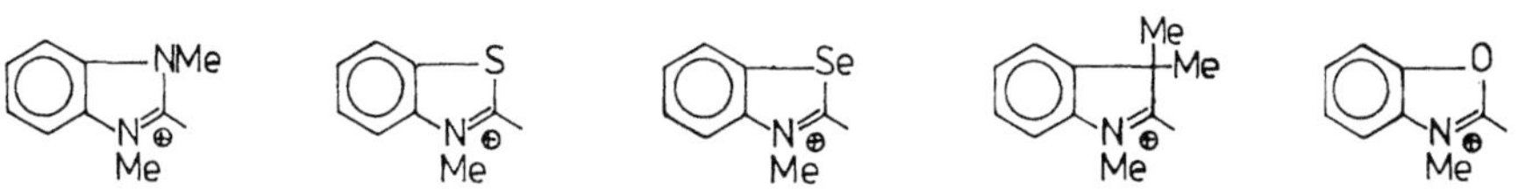

R_1: 552 (4.50) R_2: 510 (4.38) R_3: 516 (4.42) R_4: 625 (4.48) R_5: 578 (4.51)

in acetonitrile [318] in sulphuric acid [318]

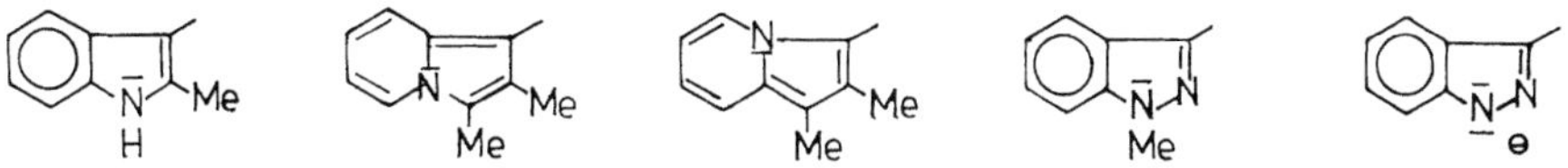

R_6: 411 (4.43) R_7: 510 (4.69) R_8: 568 (4.79) R_9: 395 (4.46) R_{10}: 552 (4.46)

in ethanol [319] in methylene chloride [318] in methanol [320] in dimethylformamide [320]

Table VII.10. Longest-wavelength absorption maxima of some neutral heterocyclic azo dyes

15		λ_{max} in nm (lg ε)	Solvent	Ref.
X	Y			
S	OMe	483 (4.97)		
S	H	482 (4.98)	in ethanol/HCl	[321]
S	NO_2	554 (5.00)		
S	NO_2	546		
O	NO_2	538	in chloroform	[322]
Se	NO_2	564		

Dimethylamino-thienyl substituted phenylazo compounds (*16*, X = S) absorb at about 70 nm longer wavelengths than the iso-π-electronic p-dimethylamino-phenyl systems (X = CH=CH) [321]. On the other hand, the absorption wavelength increases on passing from furan to selenophene [322] (cf. Table VII.10).

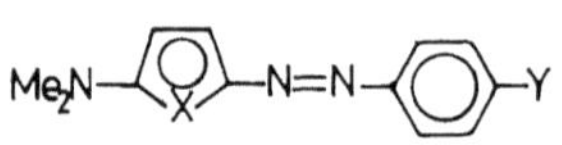

16

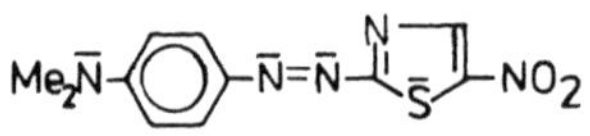

17

Whereas the dyes of the structure *16* are strongly red-shifted by acceptor substituents Y, donor substitution gives rise to deep colors in nitro-substituted thiphene, thioazole and isothiazole azo dyes [300, 317, 323], for example *17*. The dyes *17* absorb at wavelengths about 100 nm longer than the iso-π-electronic nitrobenzene derivatives, and suffer additional red shift with intramolecular rigidization of the donor amino group [300].

On the other hand, amino-substituted derivatives of the structure *18* (X = NR_2) absorb wavelengths about 200 nm longer and with considerably higher intensities than the unsubstituted compounds (X = H) as demonstrated with three series of compounds in Table VII.11.

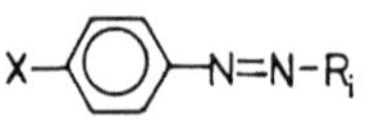

18

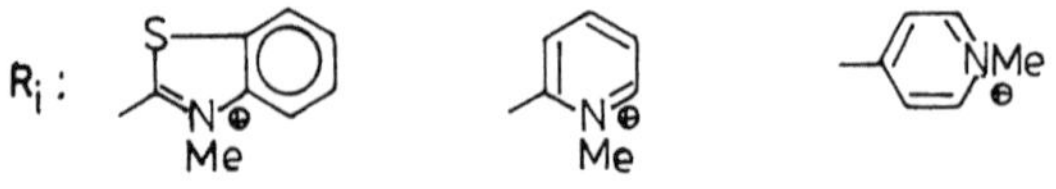

a b c

Cationic dyes such as *20* absorb at considerably longer wavelengths than the corresponding neutral parent dyes *19*. The red shift varies in the order m ≈ o < p:

Table VII.11. Longest-wavelength absorption maxima of some cationic heterocyclic azo dyes in acetonitrile [408]

X	λ_{max} in nm (lg ε)		
	18a	*18b*	*18c*
H	407 (4.39)	348 (4.36)	336 (4.31)
Cl	417 (4.43)	359 (4.41)	347 (4.36)
Me	425 (4.42)	369 (4.42)	358 (4.35)
NR_2[a]	594 (4.83)	563 (4.75)	509 (4.44)
NHPh	604 (4.74)	535 (4.68)	524 (4.62)

[a] dimethylamino in the case of *18a* and piperidino in the case of *18b* and *18c*

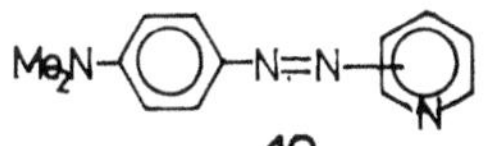

19

o: 432 (4.47)
m: 428 (4.46)
p: 444 (4.48)

in ethanol [324]

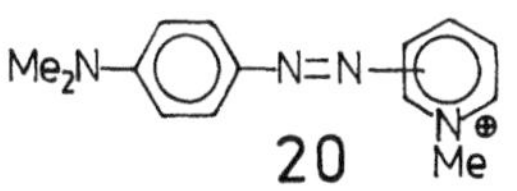

20

o: 553
m: 458 (4.47)
p: 555 (4.48)

in ethanol [325]

Heterocyclic dyes of the general formula *21* have found extensive interest as red, violet and blue basic dyes for polyacryl nitrile fibres. Spectral data were compiled for a wide range of heterocyclic cations [326–333]. Some examples follow:

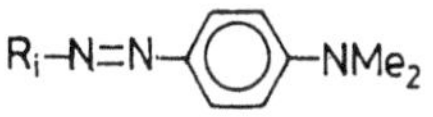

21

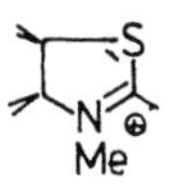

R_1: 515
in methanol [328]

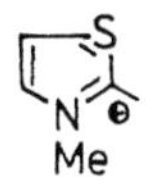

R_2: 588
in methanol [327]

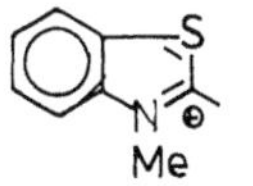

R_3: 600 (4.90)
in methanol [333]

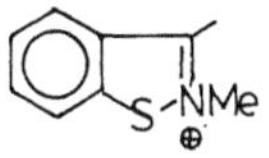

R_4: 586 (4.71)

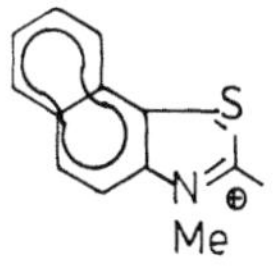

R_5: 619
in methanol [327]

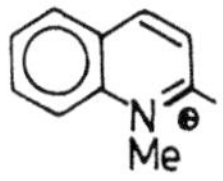

R_6: 591 (5.00)
in acetonitrile [326]

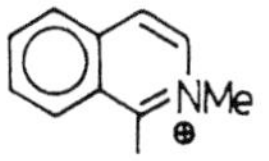

R_7: 570 (4.63)

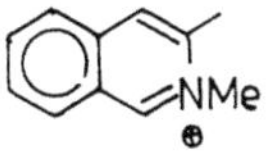

R_8: 537 (4.74)

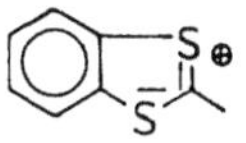

R_9: 566 (4.59)
in methanol [333]

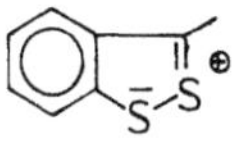

R_{10}: 625 (4.44)

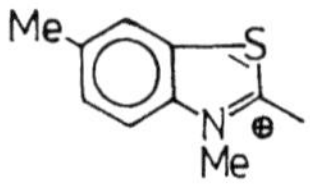

R_{11}: 600

in methanol [327]

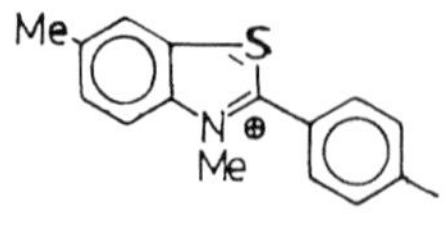

R_{12}: 481

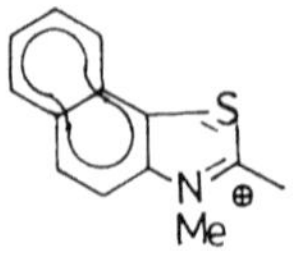

R_{13}: 619

in methanol [334]

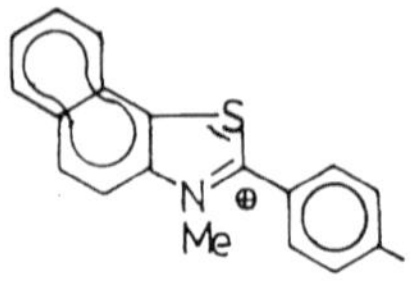

R_{14}: 554

Most of the experimental findings agree with conclusions reached from the simple MO picture. Corresponding to the MO theoretical arguments given in Sect. IV.3, large red shift is anticipated with heterocyclic residues of low ionization potential. For that reason compounds of the structure *15* undergo bathochromic shift when the indolyl residue R_6 is substituted by indolizinyl residue R_7 or R_8. On the other hand, replacement of CMe in the indolyl residue R_6 by the more electronegative aza nitrogen, giving rise to R_9, causes blue shift. Furthermore, as the ionization potential is lowered in passing from the indazenyl residue R_9 to the anionic R_{10}, red shift is observed for the corresponding azo compounds of the structure *15*.

At first glance the long-wavelength absorption of the dicationic azo compounds *15* along the series R_1 to R_6 seems to be of quite different origin. PPP-CA calculations have revealed, however, that even the cationic residues act as electron donor rather than as electron acceptor group [288]. To understand this, the enhancement of the electron affinity of the central azo group by the nearby positive charges of the heterocyclic residues must be taken into consideration. Therefore, extremly long wavelength absorptions have been observed for *15* ($R_i = R_4$) because the positive charge of R_4 is highly localized, in that case, near to the azo group.

Inspection of the key orbital interactions which are responsible for ground and excited state stabilization shows that the structure of the series *15* gain stabilization in both electronic states. If the azo group is linked to the heterocyclic rings in certain other positions than hitherto experimentally implemented, excited state stabilization should dominate and absorptions at even longer wavelengths might be found [288].

The importance of donor capacity is also supported by nonsymmetrical dyes, such as the bathochromic shift occurring in passing from the aminophenyl residue, via aminofuryl and aminothienyl to the aminoselenyl residue in *16*. But another point gains importance: the acceptor capacity of the second residue at the azo group. The higher the electron affinity of the acceptor residue, the lower the CT transition energy. Furthermore, the donor-to-acceptor CT configuration between the residue should effectively mix with the donor-to-azo CT configuration. They interact if overlap occurs between the LUMO of the donor and the LUMO of the acceptor. For example, since R_6 of *21* has a higher electron affinity than R_7 or R_8 according to theoretical calculations, the color band of *21* ($R_i = R_6$) appears at longer wavelength.

The different color properties of *21* ($R_i = R_7$) and *21* ($R_i = R_8$), on the other hand, are governed by the LUMO expansion coefficients: As the coefficient is larger in the 1-position of the isoquinolinium residue, the aminophenyl-to-isoquinolinium CT configuration of *21* ($R_i = R_7$) affects the lowest excited state more strongly than the corresponding CT configuration of *21* ($R_i = R_8$). In accordance, the first mentioned compounds absorbs at longer wavelength. Considerations along this line are useful for deriving the dye structure of desired color properties.

Dyes of the structure *21* undergo blue shift with insertion of phenylene between the azo group and the heterocyclic residue, for example, in replacing R_{11} by R_{12} or R_{13} by R_{14}. Insertion of phenylene groups separates the heterocyclic acceptor groups from the azo acceptor group. Then, the aminophenyl-to-heterocyclus CT transition energy is raised. The phenylogous dyes are therefore chromophorically more closely related to substituted aminoazobenzene dyes than to heterocyclic azo dyes.

The electron affinity of the acceptor residue should decrease if acceptor substituted benzenes are substituted by acceptor substituted five-membered aromatics. This is in striking disagreement with the experimental finding, because longer wavelength absorptions are observed in the latter case, as mentioned above for *17*. A closer examination by PPP configuration analysis revealed that besides an increased LUMO-LUMO interaction between the acceptors, the considerable lower LE excitation energy of the substituted heterocycles is important.

Instead of discussing color-structure relationships in terms of a donor-azo-acceptor composite molecule (model *B* in Fig. VII.1), the azo-acceptor part of the molecule could be alternatively considered as a unified acceptor unit (model *C*). This kind of approach is useful if only the donor part is varied, as is the case in the three series of structures *18*. The wavelengths reflect now the graduation of the ionization potentials of the benzenoid fragments like in series of carbocyclic azo compounds.

The donor-acceptor composite molecule models *B* and *C* become the more unlikely the more the CT configuration contributes to the ground state stabilization. Extreme cases are compounds with anionic donors and cationic acceptor groups, such as *22a*. Their molecular structure displays some bond equalization towards *22b*. Consequently, these compounds could be classified as diaza merocyanines (model *E*), as shown in extensive studies by Hünig and coworker [335]. They absorb intensively between 500 and 600 nm (cf. Table VII.12).

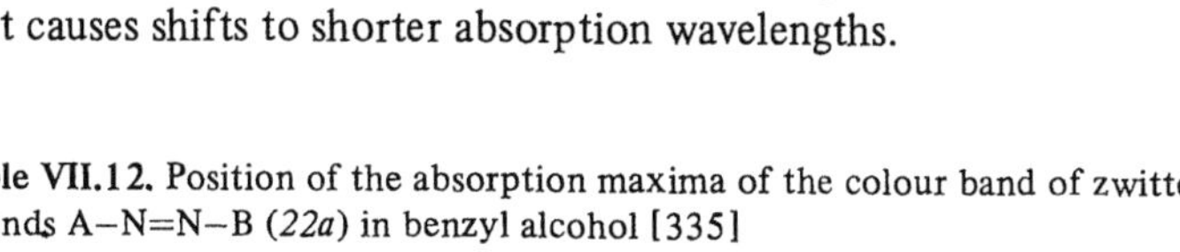

22

The benzologues of the dyes *22* show interesting regularities: Annelation of the heterocycles leads to higher absorption wavelengths, whereas annelation of the phenolic part causes shifts to shorter absorption wavelengths.

Table VII.12. Position of the absorption maxima of the colour band of zwitterionic azo compounds A–N=N–B (*22a*) in benzyl alcohol [335]

B \ A	λ_{max} in nm (lg ϵ) phenoxide	1-naphthoxide	9-anthroxide
4-pyridinium	558 (4.88)	550 (4.65)	530 (4.41)
4-chinolinium	608 (4.83)	578 (4.62)	542 (4.41)
10-acridinium	625 (4.46)	585 (4.31)	542 (4.15)

Contrary to the iso-π-electronic aza-free merocyanines (styryl dyes), the compounds *22* show positive solvatochromism throughout the series. This behaviour indicates a less bipolar character of the azo dyes compared to the merocyanine dyes. Negative solvatochromism has been observed, however, by replacing the phenoxide group by the cyano-substituted cyclopentadiene anion in passing from *23* to *24* [326]:

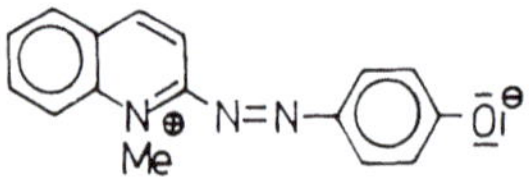
23

483 nm in xylene
499 nm in acetonitrile

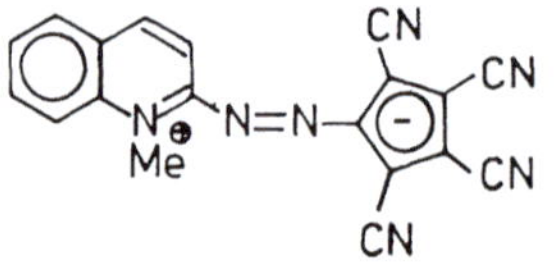
24

558 nm in xylene
486 nm in acetonitrile

We do not consider in this section the practically important "pyrazolone azo dyes", which in textbooks are mostly designated by formula *25a*. In contrast to former assumptions these dyes are not real azo dyes of the structure *25a*, however. They exist exclusively in a tautomeric phenylhydrazone form (cf. p. 73).

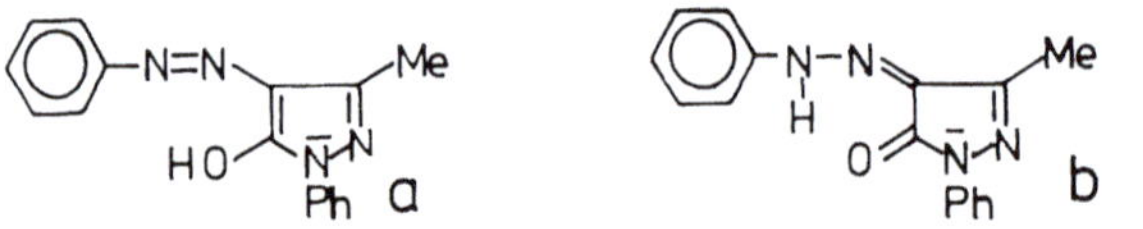
25

7. Tautomeric Azo Compounds

The elucidation of structure-color relationships becomes more complicated when the azo dyes are in tautomeric equilibrium. Since the tautomers can differ markedly in color, the phenomenon of tautomerism is very important in color chemistry.

Whereas *para*- and *ortho*-aminoarylazo compounds probably solely exist in the azo form *26* or the chelated form *27*, a broad experimental material furnished much evidence for the existence of azo-hydrazone equilibria in hydroxyazo dyes.

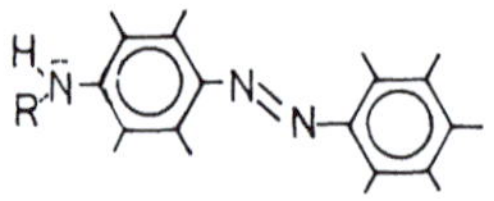
26

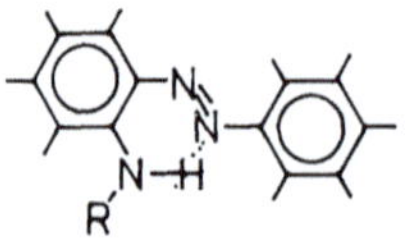
27

The voluminous literature has been reviewed elsewhere [336, 337]. Our discussion is confined to a brief outline of the spectral features.

A large number of studies has been carried out to elucidate the structure of 1-aryl-azo-2-naphthols (*28*) and 1-arylazo-4-naphthols (*29*), which exist both as true azo compounds (*a*) or as naphthoquinone hydrazones (*b*)[1].

1 According to Nurmukhametov et al. the quinoid structure should be replaced by the bipolare azo structure [337a].

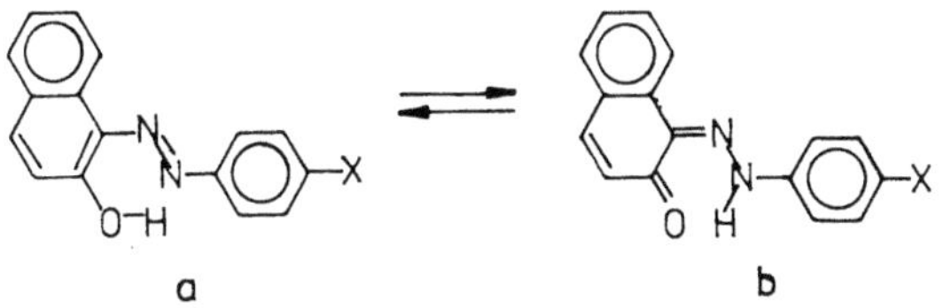

28

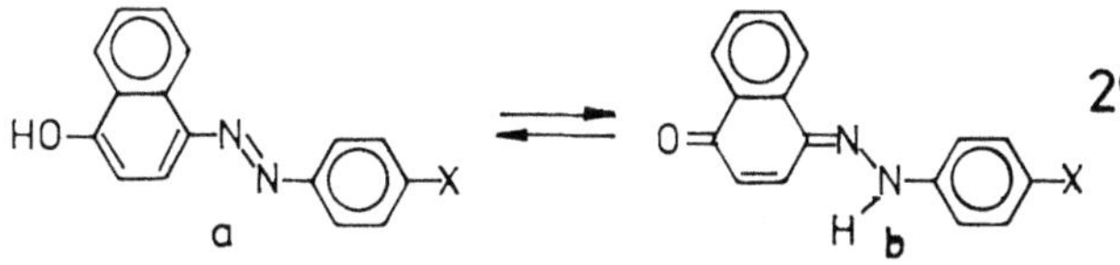

29

The presence of equilibria has been demonstrated by the solvent dependence of the solution spectra. The spectral curves obtained in different ethanol/water mixtures display two absorption bands of distinct relative intensity. They intersect each other giving rise to the occurrence of an isosbestic point, thus indicating the existence of a simple equilibrium [338].

The two bands were assigned by the aid of the spectra of O- and N-methylated derivatives which possess unequivocally azo or hydrazone structure. From that comparison it was concluded that the longest-wavelength absorption generally belongs to the hydrazone form and the following shorter wavelength absorption to the azo form, as exemplified for *29* in Fig. VII.12. The results of the PPP calculation of the methylated *29* suggest that each of the longest-wavelength absorption bands arises from a single electronic transition [276] (cf. Fig. VII.12).

However, the occurrence of two distinct absorption maxima at long wavelengths does not necessarily indicate the presence of a tautomeric equilibrium. 4-Phenylazophenols, for instance, exhibit two low-energy electronic transitions which solely arise from the pure hydrazone tautomer [339].

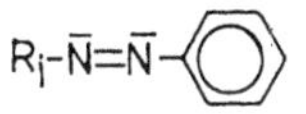

30

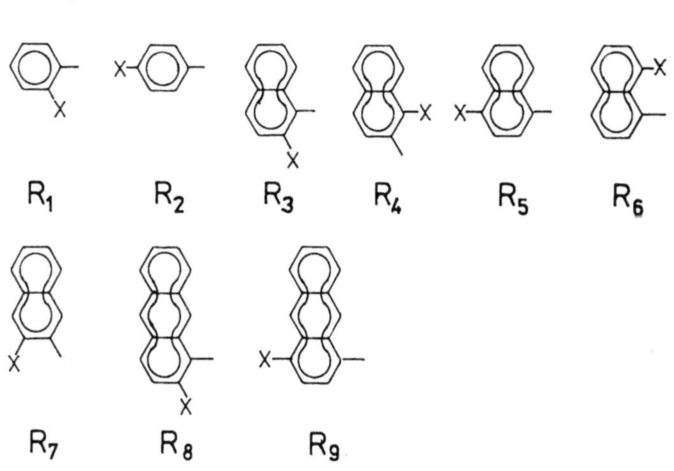

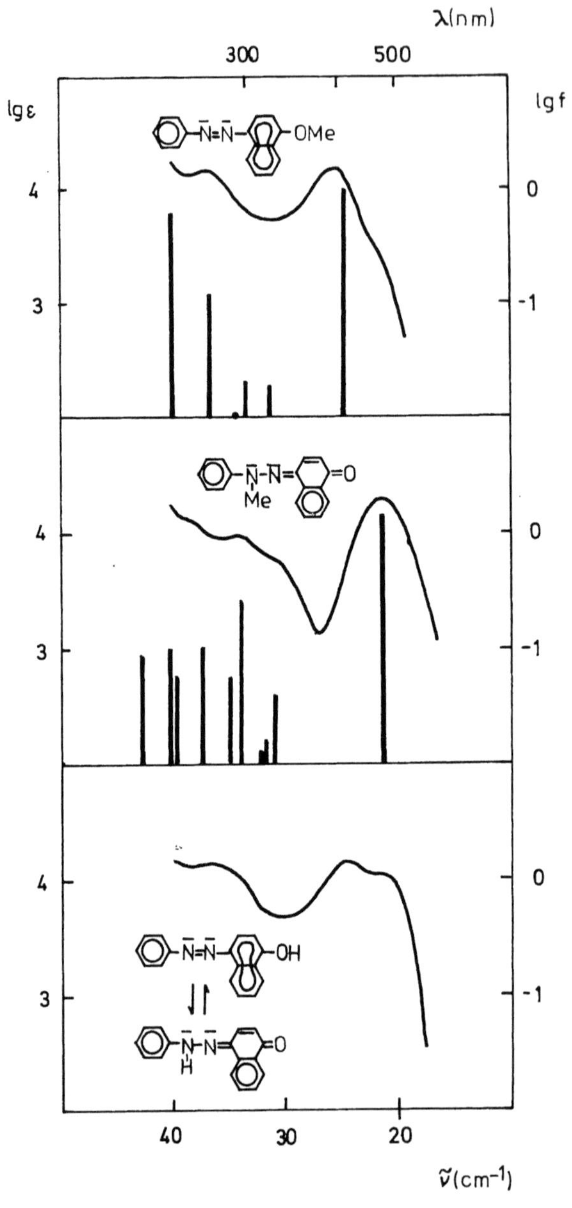

Fig. VII.12. Absorption curves of 1-arylazo-4-naphthol and its O- and N-methylated derivative, resp., and results of PPP calculations [276]

Since the true maxima and intensities of the individual tautomers are not exactly known, the quantitative analysis of the equilibria has not yet been given. Only the tautomers of *29* were said to be separable by crystallization from different solvents [340]. Despite of some uncertainty about the exact position of the equilibria of hydroxyarylazo dyes, some general rules are well established. Thus, the content of the hydrazone

Table VII.13. Spectral absorptions of unsubstituted hydroxyazo and methoxyazo compounds *30* and content of the hydrazone tautomer (the values given in brackets refer to the absorption wavelengths of the tautomeric hydrazone)

Compound 30 R_i	λ_{max} in nm[a] X = OH	X = OMe	% Hydrazone[b]	Solvent[c]
R_1	325	332	29.3[d,e]	E/W
	[373]		28.2	CH
R_2	349	343	0[a]	
	[–]			
R_3	422	380	69.5[e]	E/W
	[480]		56.3	
R_4	[490][h]		75.9[e]	E/W
R_5	410	392	62[e]	E/W
	[470]		24	CH
R_6	480[e]		0[f]	E
R_7	440[f]		0[g]	E
R_8	–	435		
	[511]		100[a]	E
R_9	–	442		
	[505]		100[a]	E

a in ethanol [409]
b estimated by electron spectroscopy, % quinone hydrazone = $100 \cdot A_H/(A_H + A_A)$, where A denotes the molar absorbance of the hydrazone (*H*) and azo (*A*) tautomer respectively
c E = ethanol, CH = cyclohexane, E/W = 50% aqueous ethanolic solution
d 2-hydroxy-5-methyl-azobenzene
e Ref. [344]
f Ref. [410]
g Ref. [411]
h Ref. [412]

form of *30* (X = OH) increases in passing from the 1,4- to 1,2-isomers and in passing from benzene derivatives to benzologous systems (cf. Table VII.13).

The effect of variation in aromatic ring size was illuminated from the quantum chemical viewpoint [341]. HMO calculations reproduce correctly the low tendency of phenylazo-phenols and of 2-phenylazo-3-naphthols to tautomerize to quinoid structures. *Ortho*-substituted compounds undergo stabilization by intramolecular hydrogen bonding in both tautomeric forms. Experimental and theoretical results indicated that the additional stabilization for the hydrazone form is more than twice for the azo form.

More recent calculations using the refined all-valence-electron method PCILO on the tautomers and conformers of 1-phenylazo-2-naphthol (*28*, X = H) [342] and 4-phenylazo-1-naphthol (*29*, X = H) [343] suggest that a better knowledge of the solvent effect is needed before quantitative predictions of the equilibrium can be made.

Besides the effect of the ring size, the attachment of various substituents is known to affect the position of the azo-hydrazone equilibrium. NMR-studies confirmed that electron-withdrawing substituents favour the hydrazone tautomer over the azo tautomer [344]. Remarkably, even nitro substituted and sterically hindered *para*-phenylazo-phenols undergo appreciable tautomerization [345, 346]. The influence of the nitro group in compounds of the formula *31* is exemplified by data in Table VII.14. Whereas acceptor substitution favours the hydrazone tautomer, donor substitution favours the azo tautomer. This has been demonstrated with 4-arylazo-1-naphthols [346a].

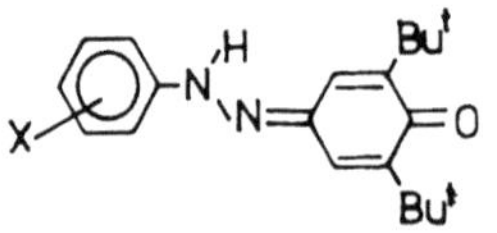

31

Systematic variation of the substituents in 4-phenylazo-phenols [346] and 4-phenylazo-1-naphthols [294] has provided only a moderate linear correlation between absorption wavenumbers of the hydrazone tautomers and Hammett-type substituent constants.

If two or more hydroxy groups are attached to the azo dyes, several tautomeric forms may be involved in the tautomeric equilibrium. Thus, the absorptions of 1-(2-hydroxyphenylazo)-2-naphthol (*32*) has been attributed to the tautomers *32a* to *32c*, assisted by the use of PPP calculations [347].

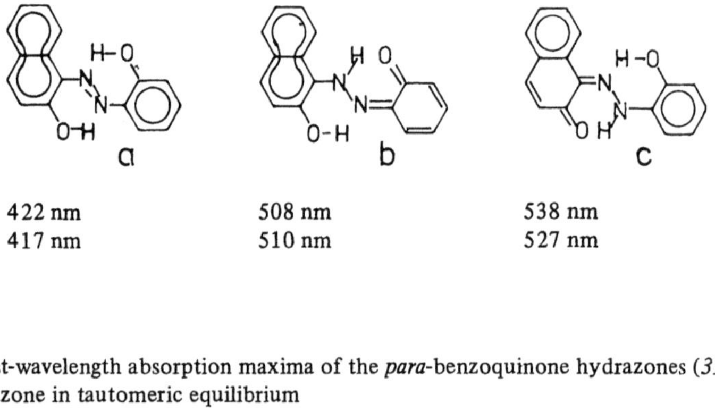

32

	a	b	c
$\lambda_{exp.}$	422 nm	508 nm	538 nm
$\lambda_{calc.}$	417 nm	510 nm	527 nm

Table VII.14. Longest-wavelength absorption maxima of the *para*-benzoquinone hydrazones (*31*) and content of hydrazone in tautomeric equilibrium

31		λ_{max} in nm (lg ϵ)			
	X:	H	4-NO_2	2,4-di-NO_2	2,4,6-tri-NO_2
λ_{max} (lg ϵ)		443 (3.38)	440 (4.05)	438 (4.55)	410 (4.60)
% hydrazone[a]		0	35	100	
$\epsilon_H/(\epsilon_H + \epsilon_A)$[b]		0.09	0.32	0.54	1.00

[a] determined by NMR spectroscopy in deuterochloroform, Ref. [413]

[b] ratio of the molar absorptivity ϵ of the hydrazone band to the total molar absorptivity of the bands of both hydrazone (H) and azo (A) tautomers, presuming that each of the two longest-wavelength absorption bands can be attributed to one isomer; solvent: carbon tetrachloride, Ref. [346]
(cf. also Ref. [345], and a critical remark in Ref. [339]

Similarly, the spectral absorption of 2,2′-bis-hydroxy azobenzene has been discussed [348].

Tautomeric equilibria may exist also in heterocyclic azo compounds which bear hydroxyl groups. The tautomers of 3-hydroxy-6-phenylazo pyridines *33* were studied experimentally as well theoretically [349, 350].

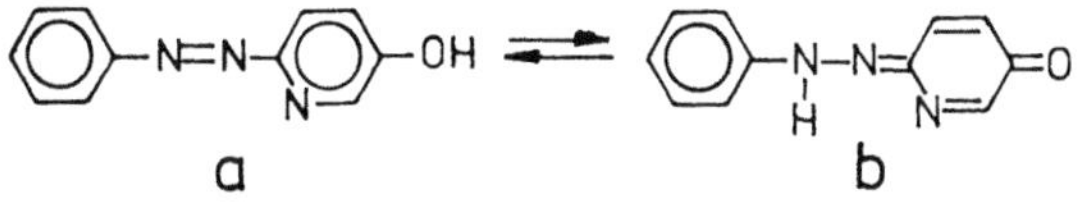

33

The yellow arylazo pyrazolones (e.g. *34*) and arylazo β-dicarbonyl compounds (e.g. *35*) have been proved to exist in the tautomeric hydrazone form [351]. They absorb near 400 nm.

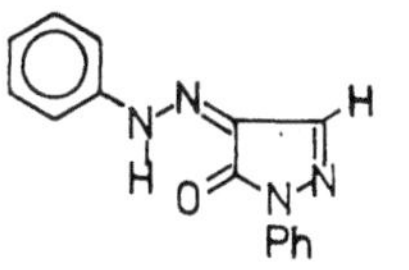

34

398 (4.34)
in ethanol [352]

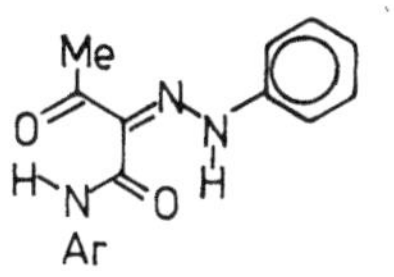

35

Ar=Ph: 375 (4.48)
in ethanol [353]

The arylazo pyrazolone [354–356] and arylazo acetanilide dyes [356a] were studied theoretically. Spectral data of various dyes of the structure *34* and *35* were reported in Ref. [352, 357].

Tautomeric equilibria have been studied also in the solid state [339, 358–360]. According to the crystal structure analyses, azo pigments involving β-naphthol are keto-hydrazone tautomers [360a]. There are also cases in which both tautomers are packed into a common crystal lattice [339]. The electronic spectra of *trans*-hydroxyazo aromatics show profound differences between the preferred species in the crystalline solids and those prevalent in solution [361]. The strong red shift observed in the solid state was attributed to intramolecularly hydrogen-bonded hydrazone-like aggregates.

The tautomeric equilibria depend strongly on the solvent molecules which surrounded the dye molecules. Polar solvents favour, in general, the hydrazone tautomer over the azo tautomer. More recent studies on the azo-hydrazone tautomerism in different solvents have shown that the effect of the solvents on the position of the equilibrium does not correlate with the dielectric constant of the solvent, but depends on the solvent structure and the microscopic environment of the dye in the solvent matrix [338].

In the presence of acids or bases both tautomers are converted to a common ion which absorbs at longer wavelengths than either of the tautomers. This phenomenon is discussed and exemplified in the following subsection.

8. Medium Effects

The solvent shift of the absorption maxima of azobenzene amounts only a few nanometers [362], but increases considerably for stronger polar azobenzenes, such as 4-nitro-4'-dimethyl-amino-azobenzene [363, 364], or the zwitterionic 4-(N-methylpyridinium-4'-azo)-phenoxide (*22*) [335, 365]. These compounds display positive solvatochromism. The solvent effect reflects the increase in dipole moment upon excitation [365]. Moreover, the electronic structure might be somewhat changed in polar solvents towards a polymethine-like structure with a high extent of bond equalization.

Under certain conditions dimerization or aggregation of azo dyes has been observed. Dimer spectra of arylazo-naphthols in water consists of an intense band hypsochromic to the monomer band and a weaker band bathochromic to the monomer band [366, 367]. According to the molecular exciton model, in-plane structural models give the most consistent results.

Striking color effects may occur in acids if they are strong enough to protonize the azo compounds (halochromism). Thus, protonated azobenzenes absorb at wavelengths shifted up to 200 nm towards red in comparison to that of the unprotonated compounds (cf. Table VII.15).

From the study of protonization equilibria Jaffé and coworker [368] concluded that symmetrical diazonium cations *36* are formed, but later experimental results supported the classical structure *37* of the protonated azobenzene [369–371].

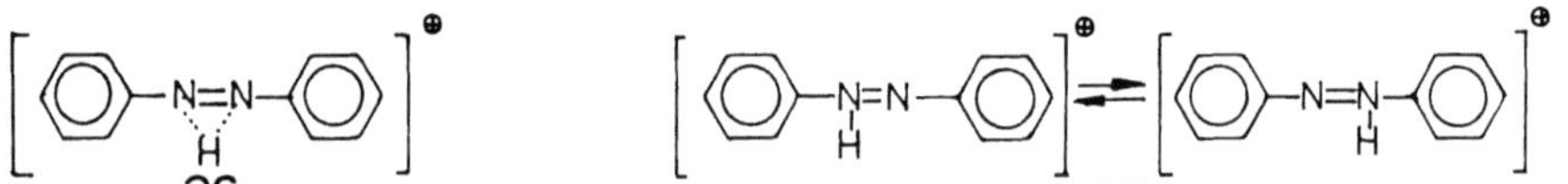

36 37

This structure is in accordance with recent X-ray photoelectron-spectroscopic [372] and theoretical studies in CNDO all-valence-electron approximation [373]. The formation of the N–H σ bond is accompanied by a significant change of the π electron structure [373, 374]. The red shift occurring by protonation could be correctly accounted for within the PPP approximation [373]. The higher electron affinity of the protonated azo group leads to lower-energy aryl-to-azo CT configurations and, consequently, to lower-

Table VII.15. Longest-wavelength absorption maxima of the *para*-substituted azobenzenes *8* in acids and spectral shift $\Delta\lambda$ in nm in relation to the respective absorption maxima in ethanol

8 X = Y	λ_{max} (lg ϵ)	$\Delta\lambda$	Ref.
H	430 (4.34)	110	[307]
OMe	500 (4.63)	146	[414]
OMe	589 (4.75)	208	[307]
NMe_2	664 (4.55)	204	[307]

energy excited states. In good agreement with this model the red shift increases significantly when azobenzene bears donor substituents or is benzoannellated [205].

Many aminoazobenzenes undergo easily protonation [290–292, 375, 376]. Some derivatives are commonly used as acid-base colourimetric indicators, for instance Methyl Orange [377].

Aminoazobenzenes have alternative protonation sites [378, 379]. Protonation may either occur at one of the azo nitrogen atoms (*38b*) or at the amino group (*38c*).

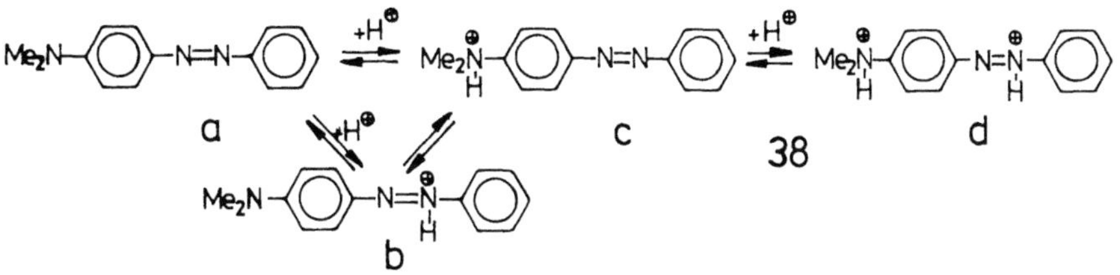

38

a	b	c	d
402 (4.44)	520 (4.79)	320 (4.34)	410 (4.45)

mixture of ethanol, water and sulphuric acid [378]

As the longest-wavelength absorption of *38c* is similar to that of azobenzene (λ_{max} 316 nm), the observed change in color can only be due to *38b*. In stronger acidic solutions blue shift occurs which was attributed to the formation of the dication *38d*.

For amino-bisazobenzenes, such as *39*, even the second protonation deepens the color, but addition of a third proton leads to a shift towards shorter wavelengths [336].

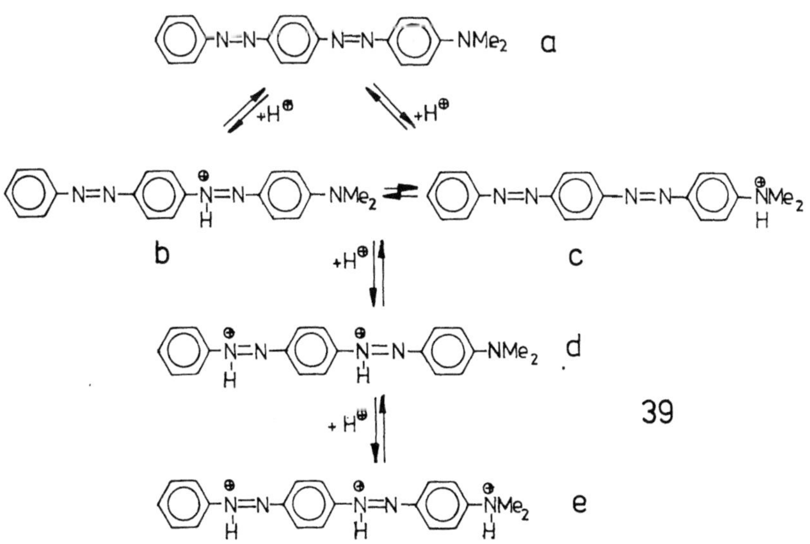

39

a	b	c	d	e
476 nm	540 nm	355 nm	630 nm	500 nm

Table VII.16. Longest-wavelength absorption maxima of protonated aminoazobenzenes and the spectral shift $\Delta\lambda$ in nm in relation to the unprotonated parent compounds

X	*9* ($R_i = R_1$)[a]		*9* ($R_i = R_3$)[b]	
	λ_{max} (lg ϵ)		λ_{max} (lg ϵ)	
NO_2	508	+ 33	506 (4.81)	− 9
CN	–		504 (4.74)	+ 17
COMe	–		515 (4.76)	+ 33
CF_3	505	+ 78	501 (4.73)	+ 40
I	528	+106	525 (4.73)	+ 72
Br	523	+104	520 (4.75)	+ 72
Cl	522	+105	520 (4.74)	+ 76
F	518	+111		
H	518	+111	518 (4.71)	+ 92
Me	533	+126	*519* (4.73)	+ 94
OMe	555	+148	532 (4.71)	+110
SMe	555	+135	553 (4.71)	+136
NMe_2	664	+241		

[a] in ethanol/HCl [375] [b] in ethanol/HCl [291]

The substituent effect on the azo-protonated aminoazobenzene of the structure *38b* differs remarkably from that of the unprotonated species. Whereas the visible absorption band of aminoazobenzenes moves towards longer wavelengths with increasing electron withdrawing strength of X (cf. Table VII.4), the reverse is true for their corresponding cations (cf. Table VII.16). This fact can be attributed to the opposite direction of charge migration accompanied with the electronic excitation (cf. Fig. VII. 13a and b). The different charge migration brings about opposite spectral shifts by the phenyl substituents, which act predominantly inductively. The distinct substituent effect in both series leads to a characteristic substituent dependence of the wavelength shift $\Delta\lambda$ observed upon pro-

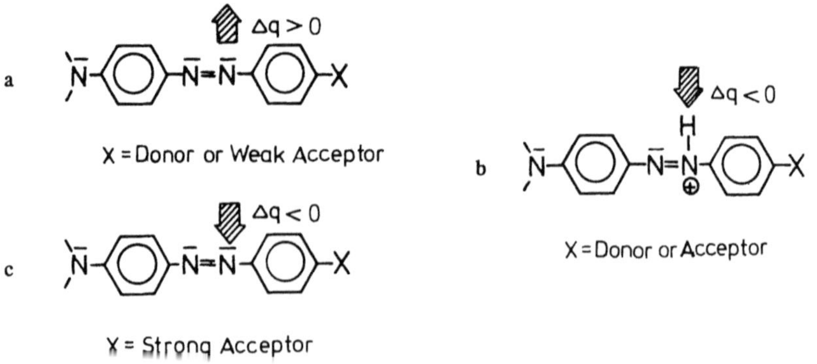

Fig. VII.13. Change of charge densities of substituted azobenzenes upon excitation (cf. text)

tonation: acceptor substituted compounds show appreciably lower shifts than donor substituted ones. The NO_2-substituent may even cause a hypsochromic shift.

Negative halochromism has also been observed with some acceptor substituted *9* [299, 375, 380] and *16* [321] (cf. Table VII.17). The subject has been reviewed recently [380a].

The halochromic effect can be easily rationalized within the simple MO model [381]. As mentioned above the change of the electron density at the corresponding azo nitrogen is, in general, characterized by the fact that the nitrogen atom gains electron density upon excitation (cf. Fig. VII. 13a). Therefore, in accord with arguments from perturbation theory normally bathochromic shift is observed with protonation. This typical electron migration, however, needs not necessarily proceed in donor-acceptor substituted azobenzenes and heteroanalogues. Owing to the more polymethine-like electronic structure, the reactive nitrogen atom may rather lose electron density upon excitation (cf. Fig. VII.13c), and protonation consequently causes hypsochromic shift.

Although hydroxyarylazo dyes are widely used as colorants, less is known about their protonated species formed in acid solutions. The color band of protonated 1-phenylazo-2-naphthol appears at slightly longer wavelengths than the tautomeric hydrazone of the unprotonated parent compound. Theoretical studies on the CNDO level furnished evidence for protonation at the lone pair oxygen of the hydrazone [382]. The strong positive solvatochromism observed in sulphuric acid (~100 nm) has been attributed to the dimerized protonated hydrazone [383].

The behaviour of hydroxyarylazo dyes in alkaline solution is simpler. Deprotonation of both tautomeric dye molecules leads to the same anion. Thus, *29c* may result from both *29a* and *29b* (X = H):

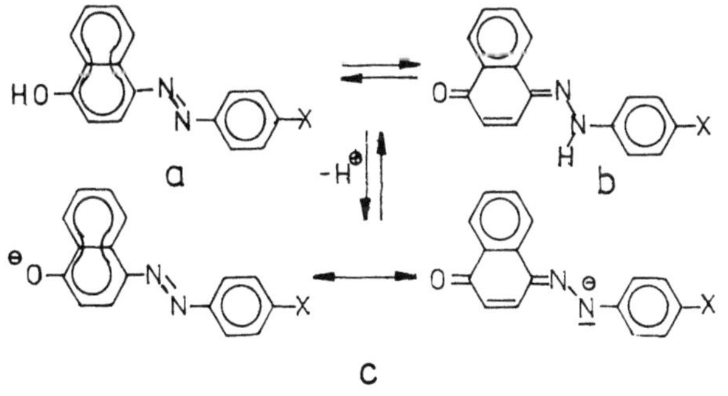

29

a	b	c
407 nm in ethanol	478 nm in acetic acid	500 nm in ethanol/KOH [294] Δλ: +93 nm and +22 nm, resp.

The strength of the positive halochromism depends on the substituents at the phenyl groups of *29*. The largest red shift is observed for *para*-nitrosubstituted 1-phenylazo-2-naphthols (Δλ = 172 nm with respect to the absorption maximum of the azo tautomer, and 139 nm with respect to the absorption maximum of the hydrazone tautomer). The shift in wavenumbers gives an excellent straight-line relationship with Hammett-type sub-

Table VII.17. Longest-wavelength absorption maxima of protonated azobenzenes *9* and *16* and the spectral shift Δλ in relation to the unprotonated parent compounds

Dye	X	Protonated Dye	
		λ_{max} (nm) (lg ε)	Δλ (nm)
9	$-\overset{+}{N}\equiv N$	506	− 90[a]
9	$-C(CN)=C(CN)_2$[b]	586 (4.86)	− 6[c]
16[d]	O	442 (4.60)	− 96[e]
16[d]	S	460 (4.70)	− 86[e]
16[d]	Se	464 (4.68)	−100[c]

[a] in ethanol/HCl [375]
[b] Et_2N- instead of Me_2N-
[c] in benzene/HCl [380]
[d] $Y = NO_2$
[e] in methanol/HCl [321]

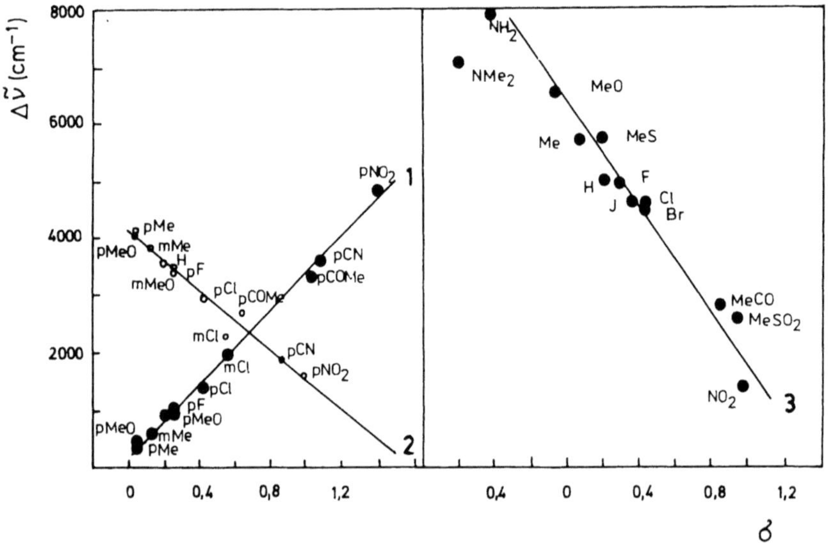

Fig. VII.14. Spectral wavenumber shifts observed with deprotonation (left hand) or protonation (right hand) plotted against Hammett-type substituent constants (curve 1: *29c* referred to *29a*, curve 2: *29c* referred to *29b* [294], curve 3: protonated *9* ($R_i = R_1$) referred to unprotonated *9* [375]

stituent constants, as exemplified in Fig. VII.14. The slope of the line represents a measure of the halochromicity of the respective series.

In this connection it should be mentioned that similar straight-line correlations are found for shifts obtained in protonation experiments of substituted aminoazobenzenes [375, 379] and phenylazo azulenes [384].

The linear relationship with reactive substituent constants results from the fact that the wavenumber shifts closely reflect the underlying chemical equilibria [294]. Hammett-

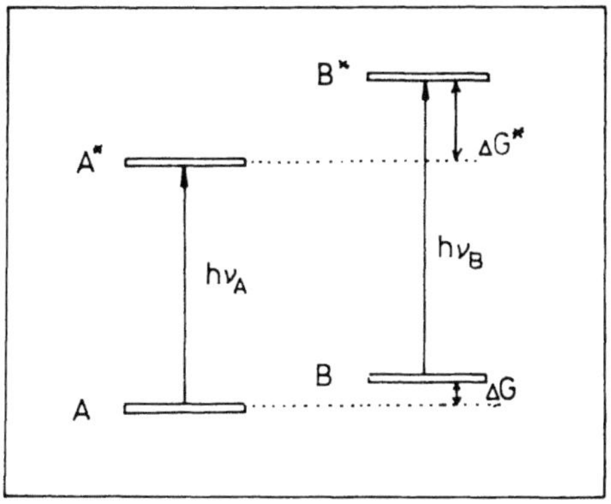

Fig. VII.15. Energy term levels of two equilibrating species *A* and *B* (cf. text)

type substituent constants describe the equilibrium constants between the unprotonated and protonated species *A* and *B* in the ground state and are, therefore, proportional to the free energy difference ΔG between *A* and *B*, as indicated in Fig. VII.15. The same kind of relationship may be true for the free energy difference ΔG* between the excited states.

It is apparent from Fig. VII.12 that the following equation holds:

$$h \cdot c/\lambda_A + \Delta G^* = h \cdot c/\lambda_B + \Delta G \qquad \text{(VII.1)}$$

where λ_A and λ_B are the absorption wavelength of *A* and *B* respectively. Reformulation of Eq. (VII.1) and introduction of the wavenumber $\tilde{\nu}$ leads to:

$$h \cdot c/\lambda_A - h \cdot c/\lambda_B = h \cdot c(\tilde{\nu}_A - \tilde{\nu}_B) = \Delta G - \Delta G^* \qquad \text{(VII.2)}$$

Since the right hand expression of Eq. (VII.2) may be assumed to be related to the σ-constants, we observe the same type of relationship for the wavenumber shifts, too.

Remarkable spectral shifts also occur by addition of Lewis acids to azobenzenes [385]. Less is known about the nature of the bonding, but the wavelength shifts suggest that the same sites of the dye molecule are involved which undergo protonation.

Treating of hydroxyazobenzenes with chromium or other transition metal salts gives rise to metal complexes which are of current use as pre-metallized or mordant dyes for wool, silk and leather [386]. This reaction is also applied in analytical chemistry [387]. The longest-wavelength absorption maxima of the complex are similar to those of the deprotonated ligands [13].

VIII. Carboximide, Nitro and Quinacridone Dyes

Besides azo dyes there are many organic colorants which contain aromatic cycles, substituted by both donor and acceptor substituents. Carboximide, nitro and quinacridone dyes are among these.

Carboximide dyes contain the carboximide group, which is a weak acceptor group. Colored compounds are only produced by introducing strong donors. 1.8-Naphthalimide *1a* is the parent compound of these dyes. It has been extensively studied and a strong absorption band has been found in the 300 nm region [416].

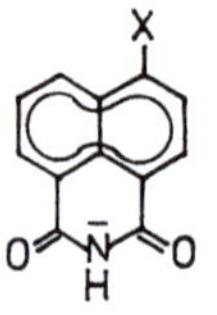

1

1a	X = H:	334 (4.06)
1b	X = $NHCOCH_3$:	372 (4.11)
1c	X = NH_2	

in dimethylformamide [416]

Whereas replacement of the imide hydrogen by alkyl or donor groups hardly influences the absorption wavelength of *1a*, marked bathochromic shift occurs by donor substitution in 4-position. 4-Acetamido-1,8-naphthalimide *1b* is slightly colored. Owing to its ability to fluoresce, this compound, and likewise 4-alkoxy-1,8-naphthalimide, are used as fluorescent brightening agents. Amino and alkylamino substituent groups produce more significant bathochromic shifts and the resulting yellow dyes absorb strongly in the visible region [417].

According to PPP calculations [116], two electronic transitions cause the longest-wavelength absorption of *1a*. A weak transition is polarized perpendicular to the two-fold symmetry axis, a strong one along this axis. In the simple MO picture they are approximately described by the $1 \rightarrow 1'$ and $2 \rightarrow 1'$ transition, resp. Inspection of the MO expansion coefficients in Fig. VIII.1 shows that the HOMO but not the subjacent MO is affected by donor substitution in the 4-position. Thus only the $1 \rightarrow 1'$ excitation energy is lowered and a single electronic transition determines the longest-wavelength absorption of the donor-substituted *1c*.

Deepening of the color is also observed in passing from *1* to the benzothioxanthene derivative *2* or in passing from *1* to the perylene derivative *3*. The dyes *2* and *3* produce deep yellow and yellowish red shades. Compound *3* is used both as vat dye and pigment dye.

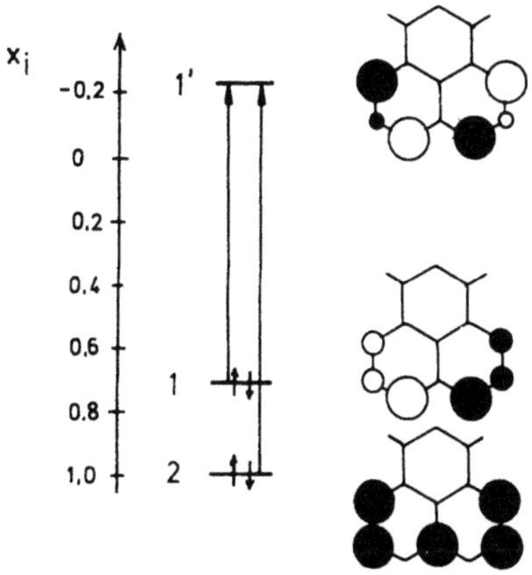

Fig. VIII.1. Lowest-energy one-electron transitions of 1,8-naphthalimides and pictorial representation of the MO expansion coefficients of the MO's involved therein (cf. caption to Fig. VII.5)

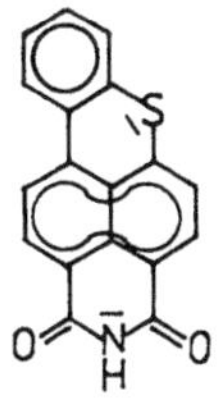

2

467 (4.29)
in chlorobenzene [418]

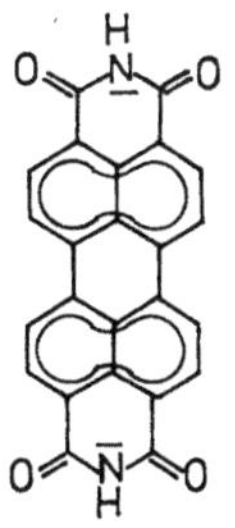

3

Structure *3* can be considered as containing in the chromophore of *1a* twice. The calculated lowest energy $\pi \rightarrow \pi^*$ transition, which is polarized in the long molecular axis can, therefore, be interpreted as in-phase combination of the local lowest-energy transitions of the subchromophores [116]. Since the local transition moments are in head-to-tail arrangement, a bathochromic and hyperchromic shift calculated in passing from *1a* to *3* is understandable. As found by LHM calculations, additional red shift results from CT contributions, that means, from electronic delocalization between both subchromophores.

Whereas only a small red shift has been observed by replacing of one carbonyl group of *1a* by the imino group or by N-phenyl substitution to *4* and *5*, resp., pronounced bathochromicity occurs by the formation of the imidazole and the perinaphthalene nuclei. This is exemplified with *6* and *7*, resp.

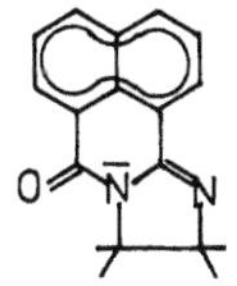

4

357 (3.95)
in methanol [419]

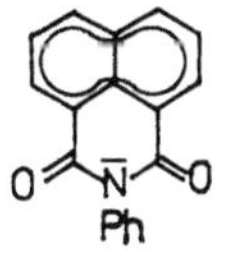

5

335 (4.11)
in dimethylformamide [416]

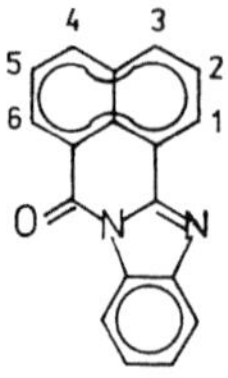

6

382 (4.07)
in dimethylformamide [416]

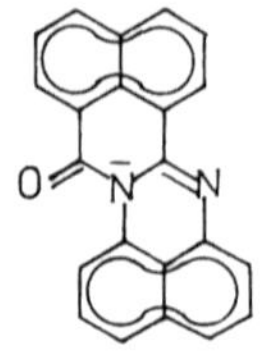

7

473 (3.99)
in dimethylformamide [416]

The absorption as well as the polarization spectrum have been recorded for *6* [420].

1-Nitro-substituted 1,2-naphthoylene-benzimidazole absorbs near 390 nm like the parent compound *6*, but 1-amino substitution shifts the absorption maximum towards about 450 nm [417]. The relations between constitution and colouring properties of various 1,2-naphthoylene-benzimidazole disperse dyes on different fibres has been extensively discussed in Ref. [421].

The isomeric aroylidene-imidazole vat and disperse dyes *8* and *9*, and *10* and *11*, do not differ greatly in their longest wavelength absorptions.

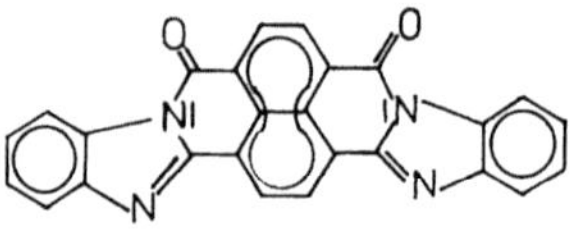

8

513 (4.10)
in methanol [419]

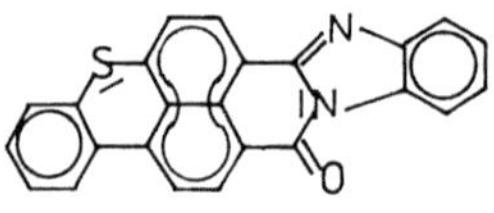

10

500 (4.33), 521s (4.24)
in chlorbenzene [422]

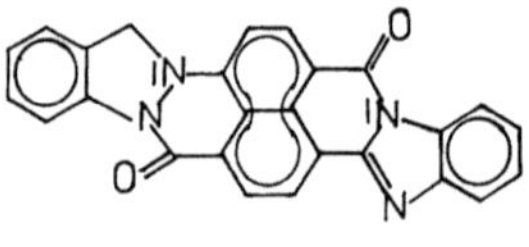

9

493 (4.06)
in methanol [419]

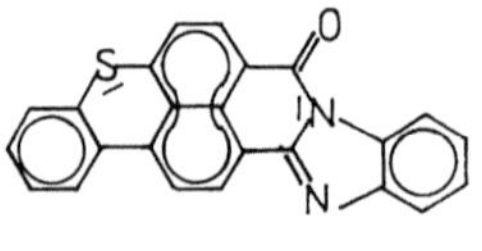

11

476 (4.18), 504 (4.17)
in chlorobenzene [422]

The isomeric mixtures *8*/*9* and *10*/*11* and their derivatives are excellent dyes which cover the range from yellow to red [423].

In comparison with the carboximide group the nitro group is a stronger electron acceptor. Thus the nitro-substituted aromatics are, in general, more bathochromic than compounds bearing the carboximide group. The deeply colored *nitro dyes* are substituted by amino groups.

Early nitro dyes were 2,4,6-trinitro-phenol (Picric Acid) and 2,4-dinitro-1-naphthol (Martius Yellow). These have lost their practical importance. Nitro-diphenylamines *12*, however, are widely used as disperse dyes. They produce yellow or reddish-yellow shades on acetate, polyester or polyamide fibres [31].

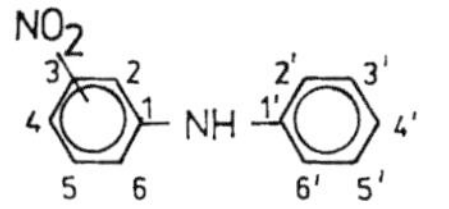

12

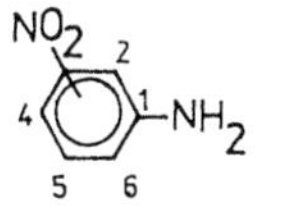

13

Replacement of one of the benzene hydrogen atoms in diphenylamine by a nitro group causes a significant spectral shift from about 290 nm to 400 nm (Table VIII.1).

The position of the color bands of the nitro-diphenylamines *12* is similar to that of the nitroanilines *13*. Consequently, the unsubstituted phenyl ring is not essential for the long-wavelength absorptions of *12*. In good agreement with this conclusion, the spectral effect of the nitro substituents depends similarly on the position of substitution in both series and decreases in the sequence o > m > p [424]. As shown formerly by LHM calculations of *13* [425, 426], the more intense color band of 4-NO_2-*12* results mainly from the L_a, and the less intense color bands of 2- and 3-NO_2-*12* from the L_b excited state of benzene. The benzene bands are strongly red-shifted by the mixing of the locally excited benzene states and low-energy charge transfer excited states. The color band of 2,4-dinitro-diphenylamines contains both the L_b- and L_a-type transition. This assignment has been confirmed by the results of PPP-CA calculations [116].

As shown in Fig. VIII.2, a single electronic transition accounts for the color bands of 2-nitro- and 4-nitro-diphenylamine in each case, whereas two electronic transitions are involved in 2,4-dinitro- and, moreover, in 2,4'-dinitro-diphenylamine. The latter case is interesting in that it has two clearly separated long-wavelength absorption bands. Based on Iz'mailskii's conception of "subautonomous molecular regions" in diphenyl-

Table VIII.1. Absorption data of nitro-substituted diphenyl amines (*12*) and nitroanilines (*13*) in ethanol

Substituent	λ_{max} in nm (lg ϵ)			
	12	Ref.	*13*	Ref.
2-nitro	425 (3.93)	[424]	403 (3.77)	[424]
3-nitro	395 (3.24)	[424]	373 (3.24)	[424]
4-nitro	393 (4.34)	[424]	375 (4.19)	[424]
2,4-dinitro	354 (4.32)	[424]	335 (4.23)	[424]
2,6-dinitro	423 (3.90)	[428]	–	
4,4'-dinitro	402 (4.58)	[439]		
2,2'-dinitro	417–423 (3.97)	[439]		
2,4'-dinitro	408 (4.18) 351 (4.13)	[424]		

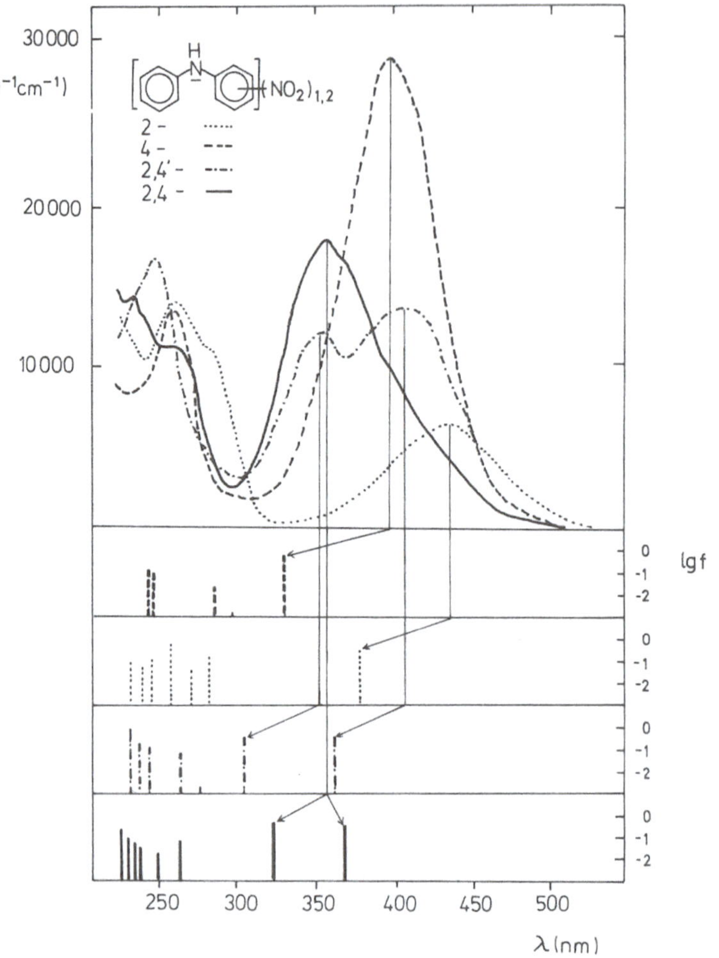

Fig. VIII.2. Spectral curves of nitro-diphenylamines in ethanol [427] and results of PPP calculations

amines [427], Asquith and coworkers [424] attributed the longer-wavelength component to the 2-nitrophenylamine moiety and the shorter-wavelength component the 4-nitrophenylamine moiety.

PPP-CA calculations [116] confirmed a distinct degree of localization of the respective excited states over the molecular parts, but they also showed that the localization consists in a localization of the electronic transitions rather than in a localization of the electronic states involved into the transitions. The localization of excitation in 2,4′-dinitro-diphenylamines is, therefore, best illustrated by the OKK analysis of the electronic transitions. The results of such calculations are illustrated in Fig. VIII.3. Dissection of 2,4′-dinitro-diphenylamine into alternative nitrophenylamine and nitrophenyl fragments documents a high degree of localization upon excitation for different nitrophenylamine fragments for the first two electronic transitions. The 2-nitro-substituted fragment is apparently more strongly involved in the lower-energy transition and the 4-nitro-substi-

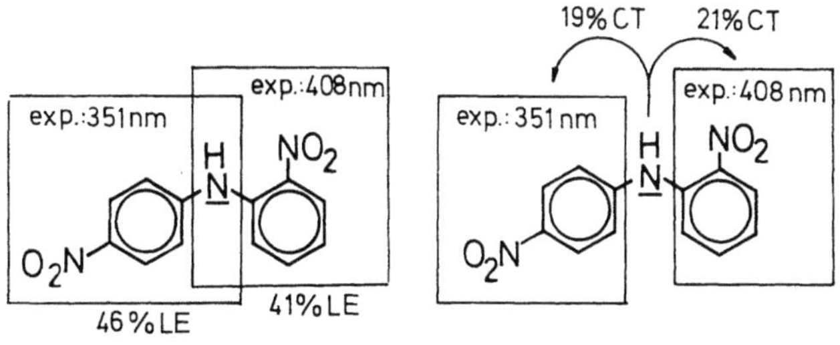

Fig. VIII.3. Two fragmentation modes of 2,4'-dinitro-diphenylamine and results of the OKK analysis of the two lowest energy $\pi \rightarrow \pi^*$ transitions

tuted fragment in the higher-energy transition. Similarly, analysis of the electronic transitions with respect to nitrogen and nitrobenzene subunits provided a predominant nitrogen-to-(2-nitro-phenylamine) CT excitation in the lower-energy transition and a predominant nitrogen-to-(4-nitro-phenylamine) CT transition in the higher-energy transition.

The partial localization of the lowest-energy electronic excitation within the nitraniline fragments accounts for the fact that substitution in the unnitrated benzene ring of mono-nitrodiphenylamines *12* has relatively little effect on the position of the color band [428]. The substituent effect may be additionally weakened by the lack of complete conjugation between the two aromatic rings [429], as suggested by the non-coplanarity of N-picryl-para-jodaniline [430].

Only a moderate shift has also been observed when donor or acceptor substituents are introduced in that benzene ring of diphenylamines which bears the nitro group (cf. Table VIII.2).

Obviously, the substitution can be considered, in that case, as perturbation of the lowest energy transition in the parent nitro-diphenylamine dyes. Attempts have been made, therefore, to correlate the spectral shifts with Hammett-type substituent constants [429, 431].

The extensively studied 1-amino-2-nitrobenzene-4-sulphonamides (such as *14*) absorb in the visible region between 412 and 419 nm (in ethanol). Variation of the shades was achieved in these series by substitution at the phenyl group rather than by variation of R within the sulphonamide substituent [432]. Absorption data of 2'-, 3'- and 4'-substituted *14* are tabulated in Ref. [429].

Table VIII.2. Color bands of nitro-diphenylamines *12* substituted in the nitrated ring (in ethanol)

Position of the Substituents	λ_{max} in nm									
	X									
	CN	$COOC_2H_5$	CF_3	H	F	Cl	Br	CH_3	OCH_3	Ref.
$2\text{-}NO_2$, 4 – X	368	367	366	393	385	382	376	396	–	[429]
$4\text{-}NO_2$, 2 – X	412	412	410	425	444	437	436	438	465	[429]
2,6-di-NO_2, 4 – X	399	402	407	423	–	–	–	–	–	[431]

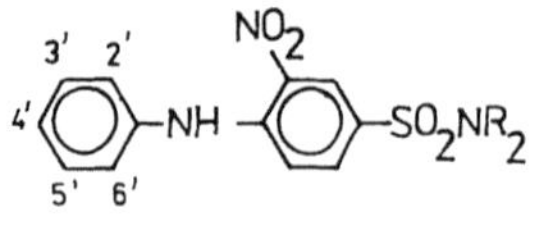

14

R = H   413 (3.77)
in ethanol [429]

More recent work on nitro-diphenylamine dyes has drawn attention to the interrelationship of lightfastness, structure and spectra [433]. From conductivity and flash-photolysis studies Asquith and coworkers inferred that the lowest excited state is of partially ionized nature. The ionization is assumed to arise from the labile hydrogen of phototautomeric quinoid forms, such as *15* and *16*.

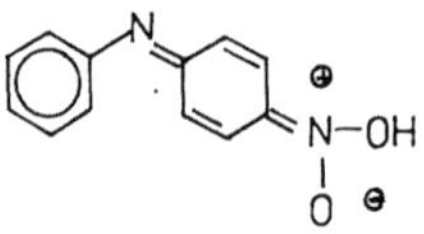

15

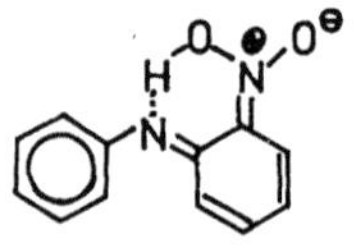

16

Comparative studies of the photochemical stability of derivatives of 2-nitro-diphenylamine and 4-nitro-diphenylamine showed the enhanced stability of the former. This was interpreted in terms of the preferential stabilization of the photoexcited states of 2-nitro-diphenylamines by intramolecular hydrogen bonding. In agreement with this assumption the light fastness decreases in passing from 2-nitro-diphenylamines to 2,6-dinitrodiphenylamines [431], in which the second nitro group cannot gain stabilization by intramolecular hydrogen bonding.

Nitro-diphenylamine dyes show, in general, positive solvatochromism. The shift of the longest-wavelength absorption maxima ranges from 10 to up 50 nm [428, 429]. Multiply nitro-substituted diphenylamines may undergo deprotonation. The ammonium salt of *17* (Aurantia) exhibits a dark red color [434]. It represents an aza analogue of a diphenylmethane dye (cf. Sect. XI.4).

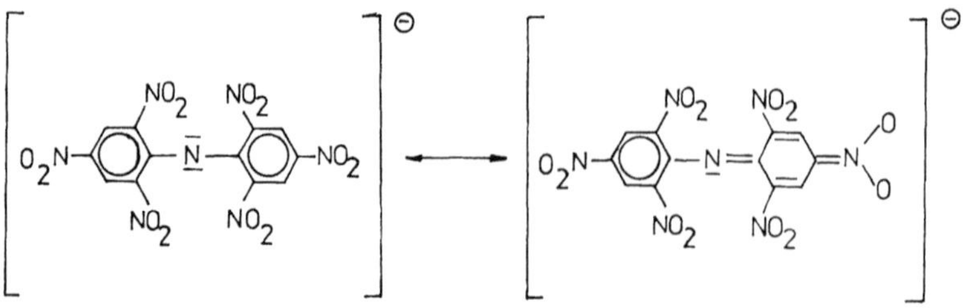

a   b

17

Quinacridone dyes contain the chromophoric systems *18* or *19*.

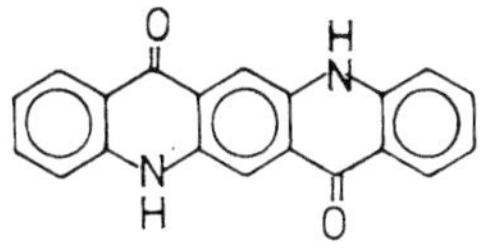

18

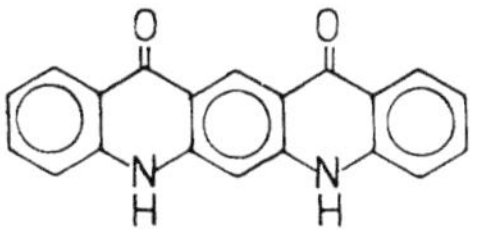

19

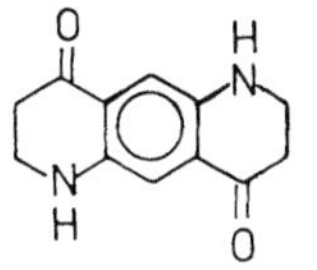

20

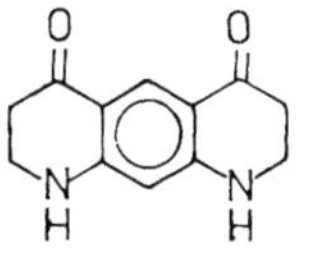

21

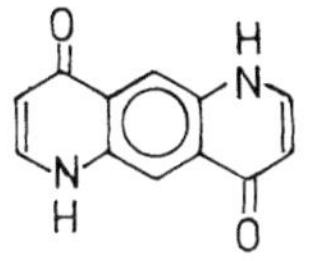

22

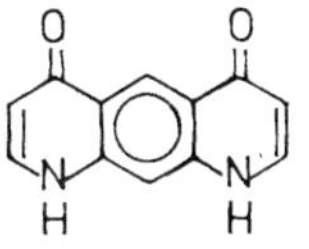

23

Trans-quinacridones absorb in the 500 nm region. The ketonic form predominates if *18* is dissolved in polar solvents [435, 436]. A slight hypsochromic shift occurs when the NH group is replaced by the NMe group. A more pronounced blue shift has been observed in passing from *trans*-quinacridone *18* to the isomeric *cis* compound *19*. The shift is a characteristic consequence of the different substitution pattern of benzene in *18* and *19* and is reproduced in LHM [54] and PPP calculations [437]. Engelhardt [437] reduced the chromophoric system of *18* and *19* in order to define the parent chromophore. He found the structures *20* and *21* still absorbing at long wavelengths, whereas the peripherically conjugated compounds *22* and *23* absorb at considerably shorter wavelengths (cf. Table VIII.3).

This hypsochromic shift can be readily explained by perturbation theoretical considerations. According to the expansion coefficient of the frontier orbitals (vgl. Fig. VIII.4), effective HOMO-LUMO overlap occurs, which results in a ground state stabilization. Thus the ground state is more stabilized than the excited state and the transition energy, is, therefore, increased. The strong coupling between the subchromophores has also been established by PPP configuration analysis [116].

Table VIII.3. Color bands of quinacridones and related compounds in dimethylsulfoxide [437]

Trans isomer		*Cis* isomer	
Compound	λ in nm	Compound	λ in nm
18	523	*19*	436
20	528	*21*	382
22	412	*23*	377

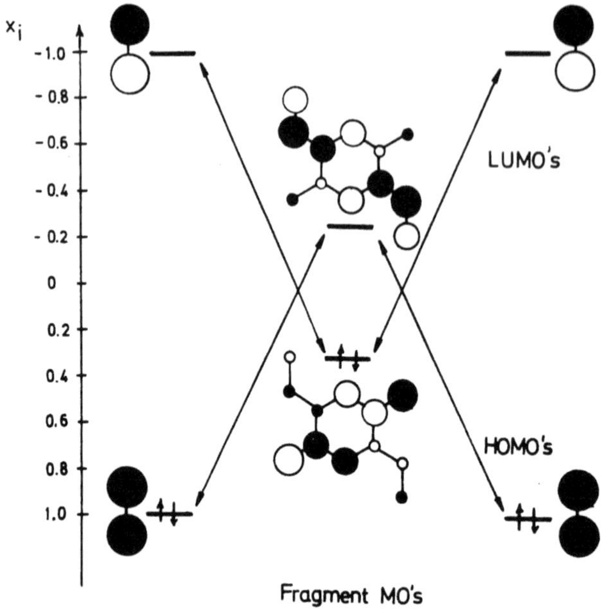

Fig. VIII.4. *Trans*-quinacridone composed of molecular fragments. The arrows indicate the predominantly interacting orbitals according to the pictorially represented MO expansion coefficients of the fragments

Due to the poor solubility of quinacridones, less is known about spectral solvent effects. The dyes undergo aggregation under certain conditions [438]. A pronounced red shift of *18* observed in sulphuric acid solutions has been attributed to the diprotonation at the oxygen atoms [436, 438]. The visible spectrum shows remarkable similarity to the visible spectrum of pentacene, which has, therefore, been considered as the parent structure of the dication [435]. A still stronger red shift has been observed with alkali which converts *18* to a dianionic species.

IX. Quinoid Dyes

The class of quinoid dyes comprises a great variety of compounds [31, 245, 246]. Applied as vat dyes they are related to indigoid colorants discussed in the next section. That means, they can be reduced to leuco compounds, which are reoxidated then to the dye on substrates such as cellulosic fibres. Quinoid compounds also belong to the acid, mordant, disperse and reactive application classes. They cover all shades between yellow and black. Also a large number of natural dyes contains quinoid structures [440, 441].

1. Chromophoric System

Quinoid dyes possess at least two carbonyl groups which are involved into a cyclic structure. They participate mostly in a six-membered ring (e.g. *ortho-* and *para*-benzoquinones) but they can also be situated in different rings. A few examples of the first type of compounds are given in Fig. IX.1. This figure also contains data about the bond lengths.

The molecular structure of these quinones is evidently characterized by the fact that the molecules consist of more or less localized C=O groups and of ethenoid, polyenoid or benzenoid subunits. Therefore, the π-electron structure of the quinone ground state can be interpreted in terms of molecular subunits according to model *A* or *B*, illustrated in Fig. IX.2. This kind of consideration has also been used for excited states of quinones.

PPP configuration analyses provided some justification for using these approaches in interpreting simple quinoid structures [442]. However, since the number of impor-

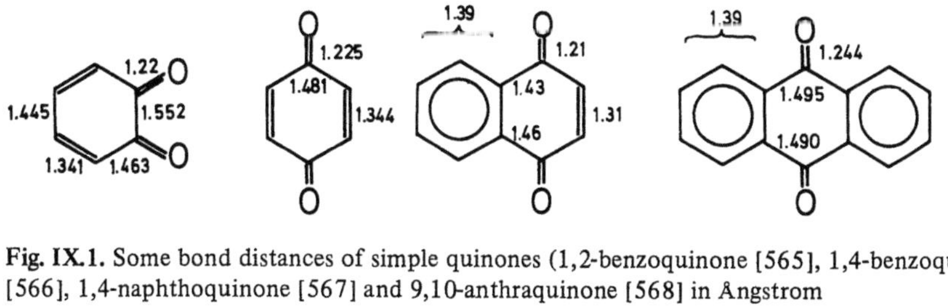

Fig. IX.1. Some bond distances of simple quinones (1,2-benzoquinone [565], 1,4-benzoquinone [566], 1,4-naphthoquinone [567] and 9,10-anthraquinone [568] in Angstrom

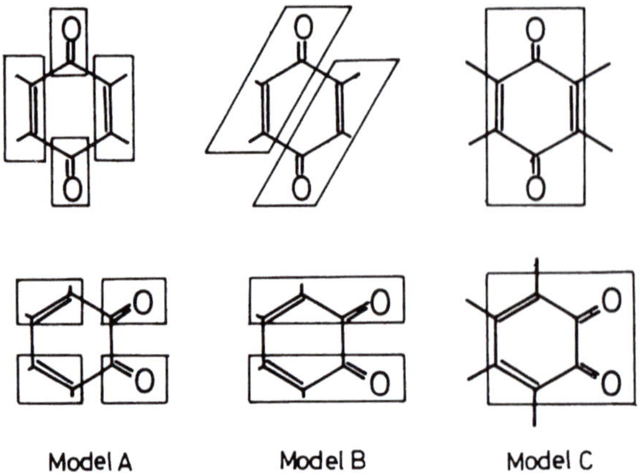

Fig. IX.2. Different fragmentation modes for describing quinoid structures

tant configurations increases strongly with the size of individual subunits, the description of the state functions of more extended quinones becomes relatively complex. In these cases the composite molecule models *A* and *B* lose their heuristic value.

For that reason it may often be more advantageous to consider some quinoid systems as particular chromophores, that means to characterize them by the whole π-electron system. In the simplest cases the benzoquinones represent reference systems according to model *C*, but in more complicated cases even the reference to more extended quinones is more favourable.

The models *A* to *C* have been used to examine the spectral features of annelated as well as substituted quinones. It should be mentioned that the suitability of the distinct fragmentation modes may be significantly depended on the substituents. Thus, in accordance with the prediction of Dähne [24], 2,5-donor substituted *para*-benzoquinones are best understood by means of model *B*, giving rise to two merocyanine subunits (see below). In fact, more recent X-ray studies have established the existence of two strongly separated merocyanine units spaced by more than 1.5 Å [27].

Before we discuss the absorption behaviour of quinoid dyes, going from simple to more complicated quinones, a brief outline should be given on the occurrence of weak $n \rightarrow \pi^*$ absorptions.

2. Weak Long-Wavelength Absorptions

As similarly described for the azo chromophore (Sect. VII.2), the carbonyl groups generate $n \rightarrow \pi^*$ transitions which in some cases give rise to appearance of a weak well-structured long-wavelength absorption band. The wavelength position of this absorption band differs between the isomeric *ortho-* and *para*-quinones and shows a characteristic hypsochromic shift in going from the *ortho*-quinones *1–3* to the isomeric *para*-quinones *4–6* [13].

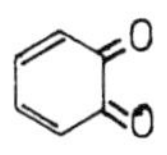

1

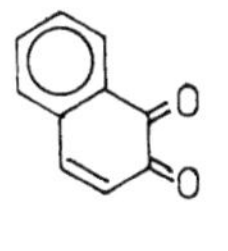

2

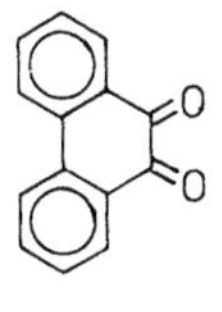

3

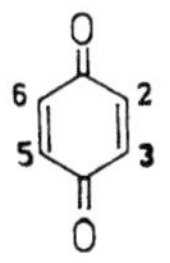

4

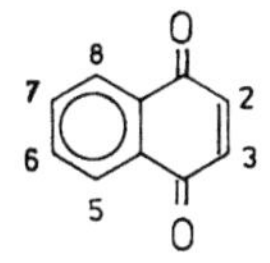

5

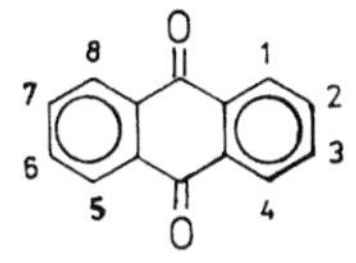

6

These differences in wavelength mainly reflect the change of the LUMO energy, which according to Koopmans' theorem corresponds to the electron affinity of the quinones. From this point of view it is worth mentioning that (in good agreement with theoretical conclusions) polarographic reduction potentials of *ortho*-quinones are generally less negative than the potentials of the *para*-isomers [443]. Moreover, in accordance with the decreasing $n \rightarrow \pi^*$ absorption wavelengths, polarographic reduction is rendered more difficult with progressive annelation.

The deeper understanding of the $n \rightarrow \pi^*$ absorption of quinones is complicated by the presence of two carbonyl groups. From all-valence-electron calculations Hoffmann and coworkers [444] concluded that the interaction between non-bonding orbitals of the two carbonyl groups of dicarbonyl compounds causes an appreciable splitting of the lone-pair orbital energies. Although the direct overlap between the lone-pair orbitals ("through-space" interaction) might be low, a remarkable splitting can nevertheless occur when they are involved into delocalized σ-wavefunctions [444, 445]. This interaction mechanism is then of the so-called "through-bond" type [446]. According to photoelectron spectra the "lone pair splitting" amounts to about 0.3 eV [447, 448]. Owing to electronic interaction terms the resulting splitting of the $n \rightarrow \pi^*$ transition energies is lower than the n-orbital splitting [448, 449] and amounts to only 225 cm^{-1} [450]. A splitting of about 1,000 cm^{-1} has been calculated in CNDO/2 approximation, however [451]. The large difference between the splitting of the n-orbital energies and the splitting of the electron state energies has been attributed to the diffuse nature of the lone-pair orbitals [452]. The results of the experimental and theoretical studies suggest, therefore, that the longest-wavelength absorption of *para*-benzoquinone comprises two $n \rightarrow \pi^*$ transitions.

Among the higher acene quinones $n \rightarrow \pi^*$ absorption bands have been found for *7* and *8*, but not for *9* and *10* [453].

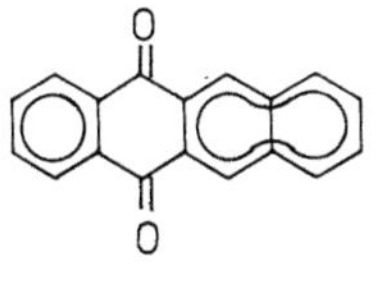

7

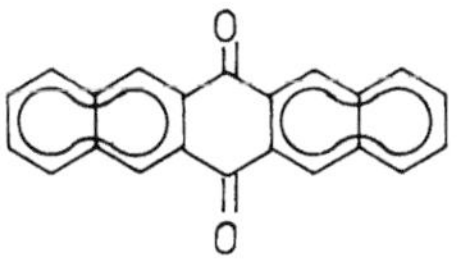

8

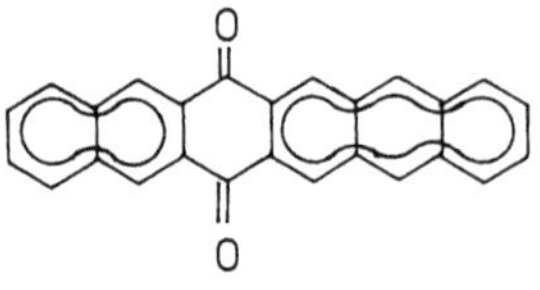

9

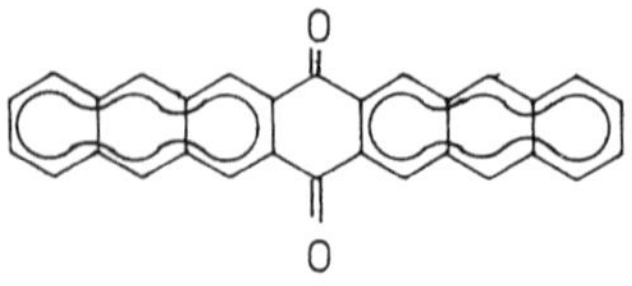

10

In the latter cases the n → π* absorptions are obviously obscured under the envelope of intense π → π* absorptions which undergo strong red shift with progressive annelation. The change of the nature of the lowest energy term levels is accompanied by an essential alteration of the photophysical properties [453–455].

The weak n → π* absorption of quinones can also be masked by the intense π → π* absorption band when red-shifted in polar solvents or by introduction of substituents [456]. The investigation of anthraquinoid vat dyes [457] showed that n → π* absorptions are still observed for various yellow or orange dyes, but not for red, blue or green ones. The occurrence of n → π* absorptions is closely related to the photochemical behaviour of the quinones and is considered to be responsible for the fibre-damaging activity of quinoid dyes [458].

3. Benzoquinones and Derivatives

The spectra of *ortho-* and *para*-benzoquinone are contrasted in Fig. IX.3. The quinones display absorption bands in the near ultraviolet region and, as found recently [459], intense shorter wavelength absorptions appear at 200 and 175 nm, resp.

The absorption curves of the isomeric *1* and *4* are strikingly different both in band position and relative intensity. The differences are reproduced, in essence, by PPP- [459–467], CNDO/S [454, 462–464] and INDO/S calculations [465–467]. *Para*-benzoquinone has also been calculated on an ab initio level [468]. The calculations confirmed that apart from the weak long-wavelength absorption π → π* transitions determine the more intense UV absorption maxima. They correspond approximately to the 1 → 1′ and 2 → 1′ transition in the one electron picture (cf. Fig. IX.4). The longest-wavelength π → π* transition of *para*-benzoquinone is forbidden for reasons of symmetry.

Configuration analyses provided a better understanding of the distinct spectral absorption feature of *1* and *4*. Whereas the interpretation according to model A [469] was less suited, a simple explanation has been obtained by model *B*, assuming enone subunits in both cases [467]. Along these lines the state wave functions of the quinones are well described by those of the molecular fragments ($r_M > 85\%$). Although *1* and *4* are built up from the same fragments, the resulting wave functions differ considerably. Whereas the excitonic-like splitting between the LE states is comparable, a distinct mixing occurs between the LE and CT states. Due to the failing symmetry in the enone fragments, the HOMO-HOMO overlap is more effective with junction to *ortho*-benzoquinone than with junction to *para*-benzoquinone. The resulting stronger excited state stabilization makes the *ortho*-benzoquinone more bathochromic than the *para*-isomer. The spectral

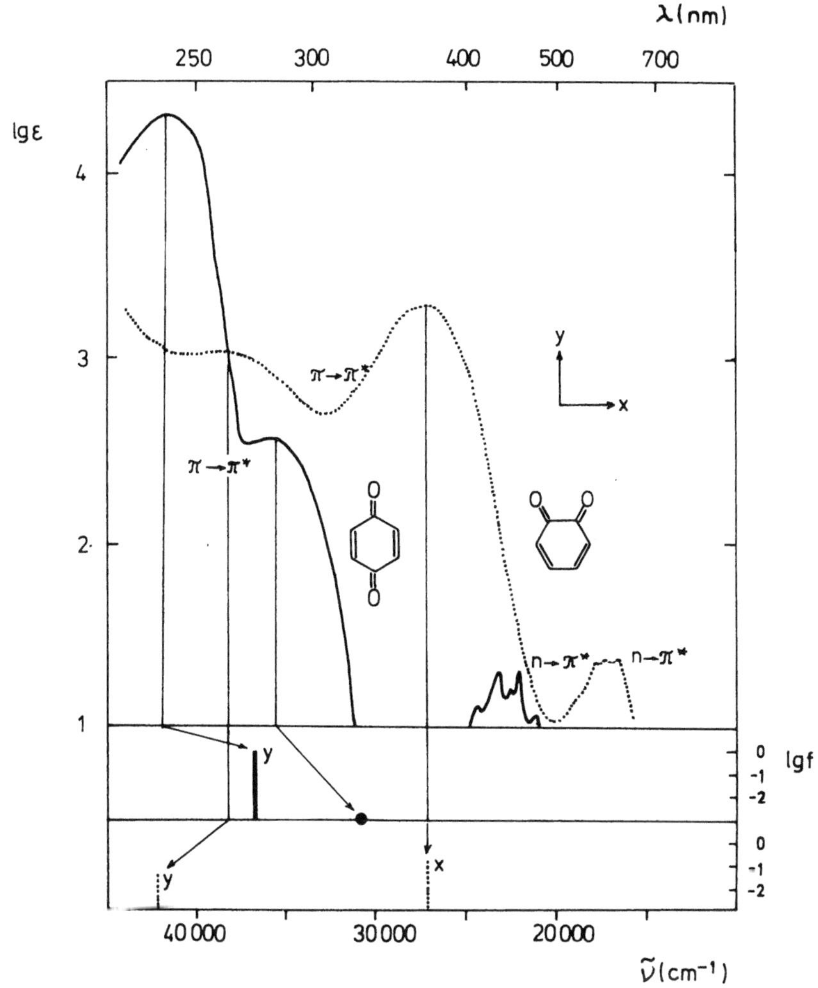

Fig. IX.3. Spectral absorption curves of *ortho*-benzoquinone in ether and *para*-benzoquinone in hexane [39] and results of PPP calculations of the $\pi \rightarrow \pi^*$ transitions

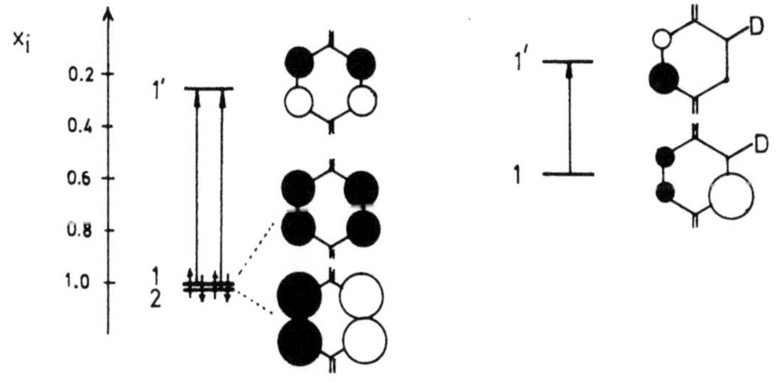

Fig. IX.4. One-electron excitations of unsubstituted and 2-donor-substituted *para*-benzoquinone. The MO's are depicted as explained in the caption to Fig. VII.5

Table IX.1. Longest-wavelength $\pi \rightarrow \pi^*$ absorption maxima of mono- and 2,5-disubstituted *para*-benzoquinones *4*

Substituents		λ in nm (lg ε)	Solvent	Ref.
2-position	5-position			
H	H	281 (2.65)	hexane	[553]
CH_3	H	308 (2.88)	hexane	[553]
Cl	H	315 (2.99)	hexane	[553]
Br	H	329 (3.08)	hexane	[553]
J	H	377 (3.10)	ethanol	[554]
C_6H_5	H	369 (3.39)	ethanol	[554]
COMe	H	370 (3.42)	ethanol	[555]
OMe	H	346 (2.99)	ethanol	[554]
SMe	H	435 (3.59)	chloroform	[556]
NMe_2	H	500 (3.75)	ethanol	[554]
NMe_2	NMe_2	495 (2.61)	dioxane	[480]
$N(CH_2)_5$	$N(CH_2)_5$	528 (2.76)	glycol mono-methylether	[479]
$N(CH_2)_4$	$N(CH_2)_4$	507 (2.63)	glycol mono-methylether	[479]
$N(CH_2)_3$	$N(CH_2)_3$	507 (2.65)	glycol mono-methylether	[479]
$N(CH_2)_2$	$N(CH_2)_2$	417 (2.79)	glycol mono-methylether	[479]

differences are expected to be smaller when the subunits become more symmetrical. In fact, as discussed on p. 96 appropriate junctions of two enaminones to isomeric diamino-benzoquinones give rise to absorption spectra which are closely similar.

The spectral absorption curves of many substituted *para*-benzoquinones resemble that of the parent compound. A progressive bathochromic shift has been observed in the sequences Cl < Br < J and OMe < SMe < NMe_2 (cf. Table IX.1). The effect of multiple substitution has been studied with methoxy-substituted *para*-benzoquinone [470]. The red shift of the longest-wavelength absorptions increases in the order 2,5 < 2,6 < 2,3. The sequence corresponds qualitatively to that derived from the size of the HOMO and LUMO expansion coefficients, which are responsible for the HOMO-HOMO and HOMO-LUMO interaction between the substituent and the parent compound (cf. Fig. IX.4, right hand part).

The relationship between the position of the absorption band and the substitution pattern of methoxy-substituted *4* has been studied within the perturbation theory approach [471, 472]. Based on model *C*, the observed absorption maximum of *para*-benzoquinone (*4*) was taken as the solution to the unperturbed problem, and the substituents were represented as perturbations to the excited states. According to this treatment, the substituent effect is divided into two components: a conjugative effect where excited states of the substituent mix with the excited state of *4*, and the inductive effect where the substituents cause the states of *4* to mix with each other. The conjugative effect depends only on the number and types of substituents, while the inductive effect depends on the relative positions of substitution as well as on the numbers and types of substituents. Simple formulae for the frequency and intensity shift of the long-wavelength $\pi \rightarrow \pi^*$ transitions of *4* and its derivatives have been derived, and inductive and

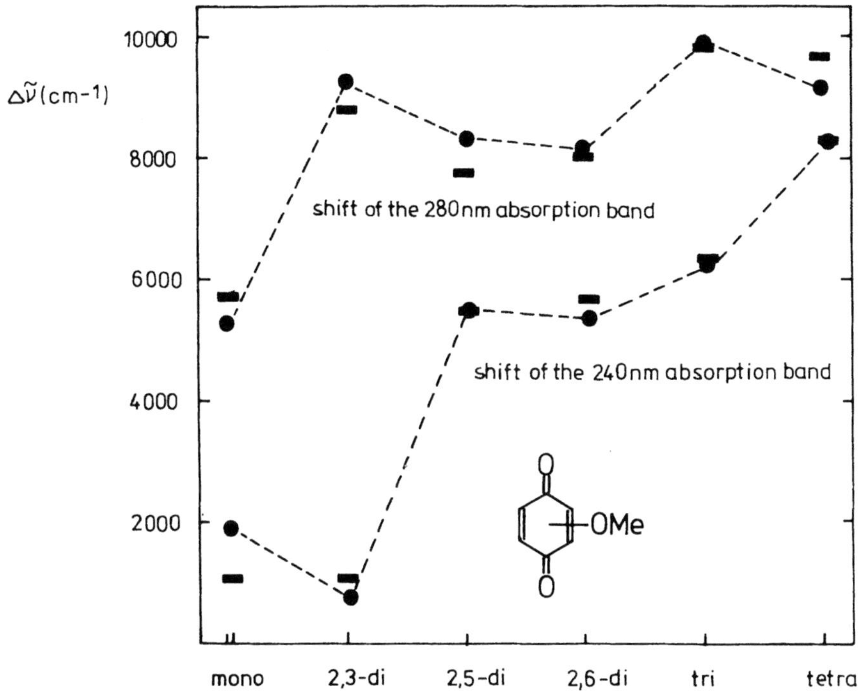

Fig. IX.5. Comparison between predicted and observed wavenumber shifts of methoxy substituents on the two longest-wavelength $\pi \rightarrow \pi^*$ absorptions of *para*-benzoquinone (exp. values are indicated by full circles) [472]

conjugative parameters for a number of substituents have been obtained from the experimental spectra [472]. The good agreement between the calculated and experimentally found data is demonstrated in Fig. IX.5.

The perturbation approach works well only for weak substituents such as CH_3, Cl and OCH_3 and breaks down for stronger substituents such as NO_2 and NMe_2.

The strongly bathochromic effect of amino substituents is reproduced by theoretical methods which involve the substituent orbitals into the delocalized π-electron system. Thus, the spectra of 2,5-diamino-*para*-benzoquinones were successfully interpreted by PPP-type calculations [105, 134, 473]. According to these calculations the medium intense absorption at long wavelengths arises from a symmetry-forbidden $\pi \rightarrow \pi^*$ transition (cf. Fig. IX.6).

2,5-Diamino-*para*-benzoquinone has been considered as substituted quinone according to model *C*, but this approach is less suited for deriving numerical data satisfactorily within the framework of the LHM method [134].

The longest-wavelength absorption of 2,5-diamino-*para*-benzoquinones undergoes blue shift when the p_π-orbitals of the amino-groups do not efficiently overlap with the molecular orbitals of *para*-benzoquinone. The striking hypsochromic shift observed in passing from the penta-, tetra-, and trimethylene-substituted 2,5-diamino-*para*-benzoquinones to the dimethylene-substituted compound (cf. Table IX.1) is caused by the fact

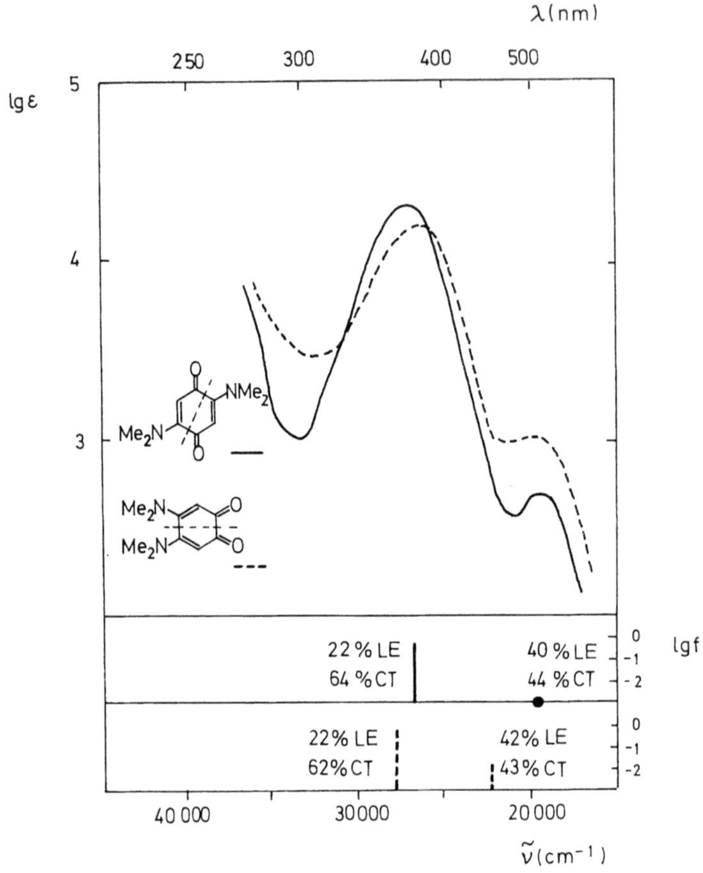

Fig. IX.6. Absorption curves of 2,5-bis(dimethylamino)-*para*-benzoquinone and 3,4-bis(dimethylamino)-*ortho*-benzoquinone in dimethylsulfoxide [477] and results of PPP-CA calculations

that the aziridine ring prevents the nitrogen p_π-orbital from overlapping efficiently with the π-electron system of quinone.

Less is known about the substituent effect on *ortho*-benzoquinone (*1*). Spectral data about alkyl-substituted *1* have been presented in Ref. [474], [475] and [476]. A recent CNDO/S study of the electronic transitions examined substituted compounds as well as the parent compound [464]. The spectral absorption behaviour of the 3,4-bis(dimethylamino)-*ortho*-benzoquinone is remarkably similar to that of the isomeric *para*-benzoquinone [477] (cf. Fig. IX.6). Although these molecules differ in symmetry, in both cases weak $\pi \rightarrow \pi^*$ absorptions appear at about 530 nm (lg $\epsilon < 3$) and strong $\pi \rightarrow \pi^*$ absorptions at about 375 nm (lg $\epsilon \sim 4$).

This can readily be understood considering the molecule as composed of two merotrimethine (enaminone) subunits (model *B*) [134, 478]. Whereas the different intensities of the two longest-wavelength absorptions reflect the out of phase and in-phase combinations of the local transition moments, which have an approximately parallel alignment to each other, the lower excitation energies result from strong admixtures of CT configurations. Results of the configuration analyses, which are illustrated in Fig. IX.6, support this interpretation.

Reasoning along these lines is consistent with the observation that strong red shift occurs by lengthening of the meropolymethinic subunits in passing from *11a* to *11b*.

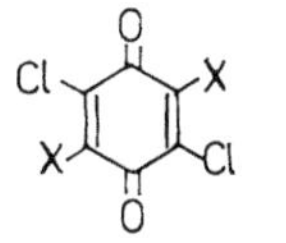

11

a: X = NMe_2
550 (2.53), 417.9 (4.03)
in glycol monoethylether [479]

b: X = CH=CH–NEt_2
645 (3.60), 537 (4.44)
in dioxane [480]

Another class of colored *para*-benzoquinone derivatives results from replacing one (or two) of the carbonyl groups with azomethine groups [481–483]. In this context indophenol and indamine dyes, which are derived from the mono and dianil of *para*-quinones resp. (*12* and *13*), should be mentioned. They are applied in color photography [13]. Both parent compounds *12* and *13* absorb in the 450 nm region, but undergo considerable red shift with *para*-donor substitution (cf. Table IX.2). The long wavelength absorption band has been classified as intramolecular CT-band [483a]. As shown for Phenol Blue (*12*, X_1 = H, Y = OH), the red shift is underestimated in PPP calculations [484].

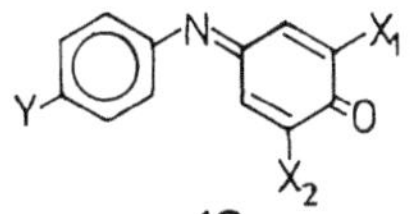

12

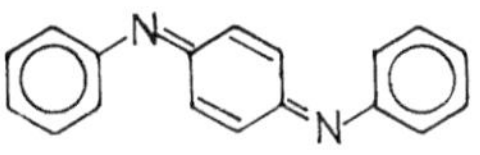

13

Table IX.2. Longest-wavelength absorption of *para*-benzoquinone N-phenylimines *12*

Compound	X_1	X_2	Y	λ in nm (lg ε or f[a])	Solvent	Ref.
12a	H	H	H	448 (3.52)	methanol	[39]
12b	H	H	OH	496 (3.83)	water pH 5.5	[557]
12c	H	H	NMe_2	552 (4.20)	cyclohexane	[558]
12d	CMe_3	CMe_3	H	439 (0.075)[a]	tetrachloromethane	[542]
12e	CMe_3	CMe_3	OH	468 (0.134)[a]	tetrachloromethane	[542]
12f	CMe_3	CMe_3	OMe	477 (0.142)[a]	tetrachloromethane	[542]
12g	CMe_3	CMe_3	NHCOMe	472 (0.148)[a]	tetrachloromethane	[542]
12h	CMe_3	CMe_3	NH_2	508 (0.214)[a]	tetrachloromethane	[542]
12i	CMe_3	CMe_3	NMe_2	549 (0.258)[a]	tetrachloromethane	[542]
12c	H	H	NMe_2	552 (4.20)	cyclohexane	[559]
12j	Me	H	NMe_2	545 (4.14)	cyclohexane	[559]
12k	OMe	H	NMe_2	536 (4.06)	cyclohexane	[559]

[a] oscillator strength f

The substituent effect strongly depends on the positions of substitution. Whereas donor substituents Y in the benzenoid fragment cause a pronounced bathochromic and hyperchromic effect (*12d–12h*), hypsochromic shift occurs with donor substitution in the quinoid part (*12j–12k*). (cf. Table IX.2) Similar regularities have been found for benzologues [485]. If the amino group in the position Y is constrained to planarity, additional red shift has been observed. Thus introduction of the julolidine ring moves the absorption up to 700 nm in polar solvents [486].

Distortion around the CN-bonds of indophenol is accompanied by distinct spectral effects. Thus, o,o′-dimethyl substitution in the benzene ring gives rise to a hypsochromic shift of 20 nm, whereas a bathochromic shift of 10 nm occurs with methyl substitution in the o,o′-position of the quinone fragment. The molar absorptivity decreases, however, in both cases. Since distortion occurs around an essential single bond C–N in the first case and around an essential double bond C=N in the second case, the spectral effects are in harmony with perturbation theoretical predictions.

Phenoxazone dyes are also derivatives of *para*-benzoquinone anil. Various naturally ocurring dyes belong to this group [487, 488]. Spectral data were given in the literature [489], but only the spectrum of Actinomycin D has been studied in more detail [490]. The basic phenoxazone-3 chromophore *14* (X = O, Y = H) results from introducing oxygen between the o,o′-positions of the anil. Since the bridging oxygen atom acts as donor on both rings, the color bands are hardly shifted in comparison to the open chain structures (cf. Table IX.3).

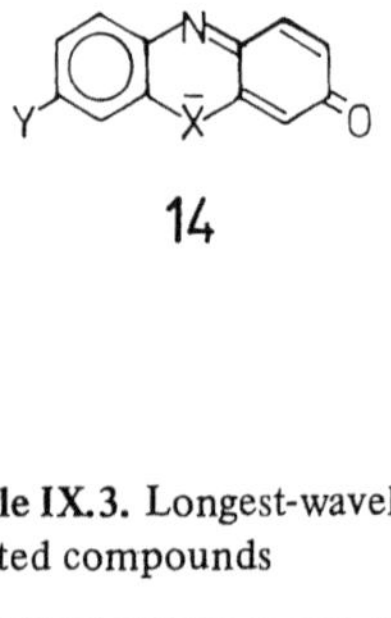

14

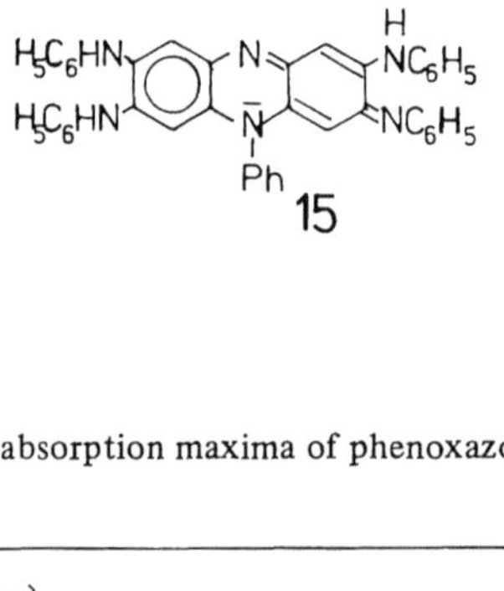

15

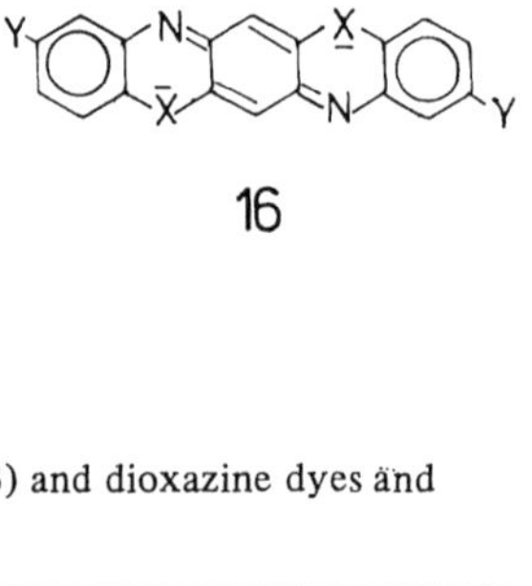

16

Table IX.3. Longest-wavelength absorption maxima of phenoxazon-(3) and dioxazine dyes and related compounds

Compound	λ_{max} in nm (lg ε)			
	Y	X: H, H	O	S
14	H	448 (3.52)[a]	442–447 (4.03)[b]	505[c]
	OH	496 (3.83)[a]	476–480 (4.00)[b]	518[c]
	NMe_2	552 (4.20)[a]	545 (4.18)[d]	690 (5.25)[e]
16	H	443 (3.90)[f]	512 (4.30)[g]	
	OEt		552 (4.90)[g]	

a cf. Table IX.2
b in methanol [487]
c in methanol [560]
d in ethanol [561], NH_2 instead of NMe_2
e in ethanol [562]
f in ethanol [563]
g in benzene [493]

However, the iso-π-electronic sulphur compounds (*14*, X = S), which are the basic chromophore of various sulphur dyes [491], absorb at longer wavelengths than the respective anils.

In alkaline medium hydroxy compounds of the structure *14* (Y = OH) undergo deprotonation which gives rise to a strong deepening of the color. The resultant symmetrical chromophores are heteroanalogous diphenylmethane dyes.

Related to the phenoxazone chromophore is also the chromophore system of indulin dyes. Indulin 6B (*15*) absorb at 500 nm (lg ϵ 4.78) in cyclohexane (605 nm in methanol) [492]. It is converted to a symmetrical ion upon protonation.

In dioxazine dyes *16* (X = O) *para*-benzoquinone dianil is doubly bridged by oxygen atoms. They absorb at longer wavelengths both with respect to the dianil parent chromophore and to the corresponding phenoxazones-3 (cf. Table IX.3). Since dioxazines are used as pigment, these compounds also received theoretical interest [493, 494]. Pronounced shifts to wavelength beyond 600 nm occur when additional donor substituents are introduced [495]. However, substituents without any influence on the light absorption in the dissolved state can also be of strong influence on the visual color in the crystalline form [495].

Spectroscopical studies showed that Emeraldine and Aniline Black, which are colored oxydations products of aniline, are quinone derivatives rather than phenazines [496]. Other dyes containing the quinone methide fragment are the fuchsone *17* and the merocyanine *18*. They are more strongly related to the triphenylmethane and polymethine dyes than to quinone dyes, however.

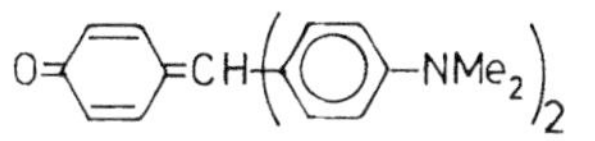

17

559 in CH_3OH [497]

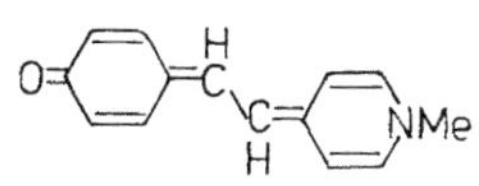

18

482 in CH_3OH [498]

Joining several quinone methide fragments leads to the compounds *19–21*. These compounds are deeply colored.

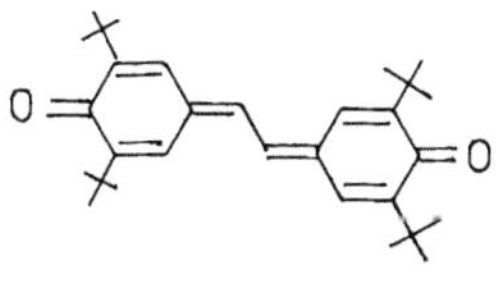

19

R = H orange [499, 500]

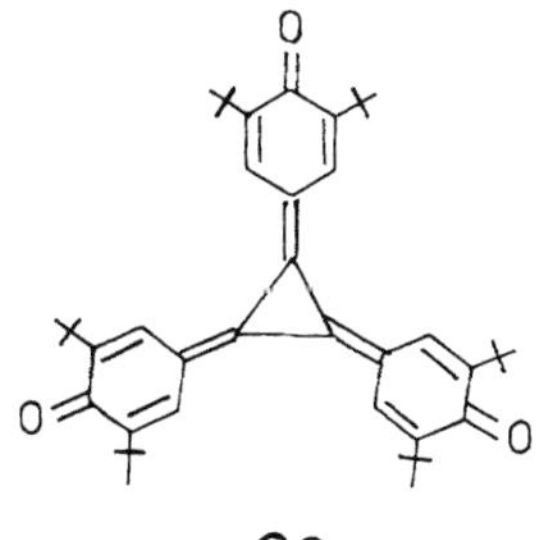

20

769 (4.71)
in benzene [501]

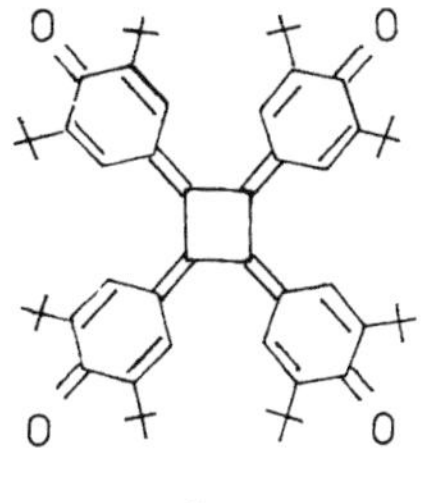

21

570 (4.50)
in cyclohexane [502]

Compound *19* represents a vinylogous electron system of *para*-diphenoquinone (*22*) which, in contrast to the *ortho*-isomer *23*, is known and has been studied experimentally. Compound *23* is expected to absorb at longer wavelength than *22*.

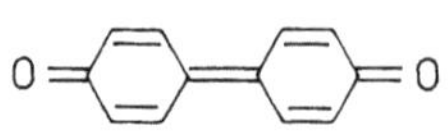

22

$\lambda_{exp.}$: 386 (4.5), 470 (2.8) [503]
$\lambda_{calc.}$: 424

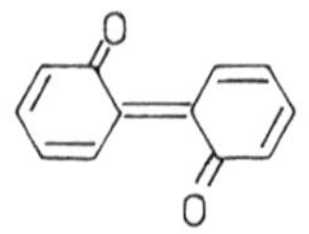

23

$\lambda_{calc.}$: 490 [116]

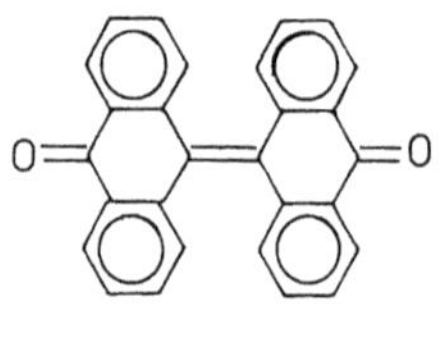

24

$\lambda_{exp.}$: 383 (4.18) in decalene [39]
$\lambda_{calc.}$: 364 [504]

The results of the PPP calculation suggest that the weak longest-wavelength absorption of *22* results from a n → π* transition. Due to the observed thermochromism, piezochromism as well as photochromism the non-planar tetrabenzo annelated *22*, the dehydrodianthrone *24*, have attracted extensive experimental and theoretical interest [504, 505].

4. Annelated *para*-Benzoquinones and Derivatives

Similar to *para*-benzoquinone the annelated derivatives show several absorption maxima in the near ultraviolet region with increasing absorption intensity towards shorter wavelengths. This is demonstrated with 1,4-naphthoquinone *5* and 9,10-anthraquinone *6* in Fig. IX.7.

Two regularities have been derived by inspection of the absorption spectra of annelated *para*-benzoquinones:

First, the wavelength of the intense long-wavelength π → π* absorption band of non-symmetrical quinones is determined by that quinoid moiety which is fused with the larger hydrocarbon fragment [453] (Hartmann's rule [506]). In the case of symmetrical annelated quinones the absorption intensity is approximately twice that of the corresponding non-symmetrical quinones.

Secondly, the spectra of the annelated quinones may be accounted for by summation of the spectral absorption maxima arising from the two aryl-carbonyl subunits

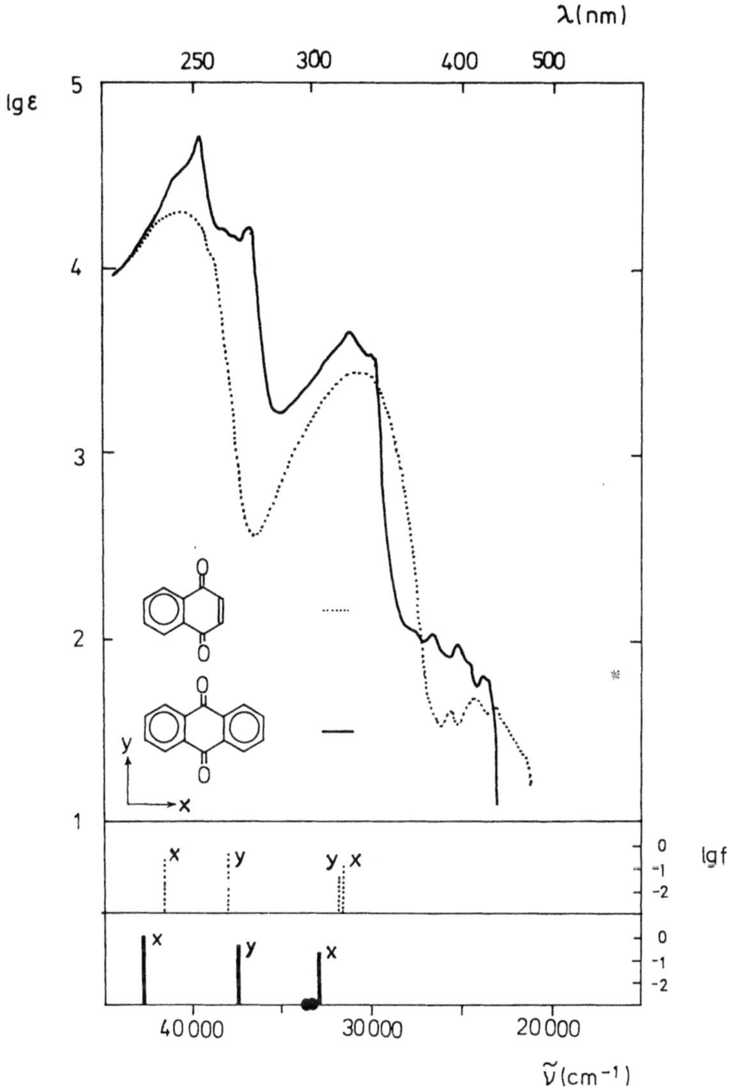

Fig. IX.7. Absorption curves of 1,4-naphthoquinone [461] and 9,10-anthraquinone in cyclohexane [39] and results of PPP calculations

("benzenoid chromophore") and those arising from the quinone fragment ("quinoid chromophore") [440, 507]. If two equal arylcarbonyl subunits are involved, the absorption intensity is twice that of the mono-annelated quinones, similarly as concluded above.

The above-mentioned regularities follow from different models in defining subchromophores. In the first case the molecule is dissected in such a way that benzenoid fragments (model *A*) or quinoid fragments (model *C*) result. In the second case two moieties are thought. Each of them includes one of the carbonyl groups (model *B*).

According to the PPP [461] and INDO/S calculations [466] the long-wavelength absorption band of 1,4-naphthoquinone *5* at about 330 nm comprises two $\pi \rightarrow \pi^*$ transitions, which are polarized perpendicular to each other. The absorption intensity is determined by the component polarized perpendicular to the carbonyl bonds (cf. Fig. IX.7). The shorter wavelength band at about 250 nm likewise consists of two electronic transitions.

Whereas shorter wavelength absorption is bathochromically shifted with substitution in the quinoid part, the longer wavelength absorption is strongly affected by substitution in the 5(8)-position of the benzenoid moiety [508, 509, 509a]. The long-wavelength benzenoid absorption band is shifted from 335 nm to 524 nm if hydroxy groups are introduced in the 5- and 8-position of 1,4-naphthoquinone (Naphthazarin, multi-banded absorption in chloroform). It is worth mentioning that this substitution pattern corresponds to that of other deeply colored benzenes bearing an acceptor substituent in 1-position and donor substituents in the 2,5-positions (cf. p. 57).

The strong effect of donor substituents in the 5(8)-position on the color band can be derived from the expansion coefficients of the occupied MO. Since the more intense $\pi \rightarrow \pi^*$ transition involved in the longest-wavelength absorption of 1,4-naphthoquinone (*5*) is approximately the $2 \rightarrow 1'$ transition, the second occupied MO has to be considered. In good harmony with the observed spectral effect, the expansion coefficients are large in the 5- and 8-position (cf. Fig. IX.8, left-hand part).

A more general interpretation of the electronic transition has been achieved by PPP configuration analyses [116]. Analyses of the excited states of *5* both with respect to model *A* and model *C* provided results which are compiled in Fig. IX.9. Obviously, two of lowest excited states of *5* correlate with excited states of *para*-benzoquinone, and two of them with those of benzene. This may be considered as a theoretical justification of the phenomenological classification in terms of benzenoid and quinoid bands. Some

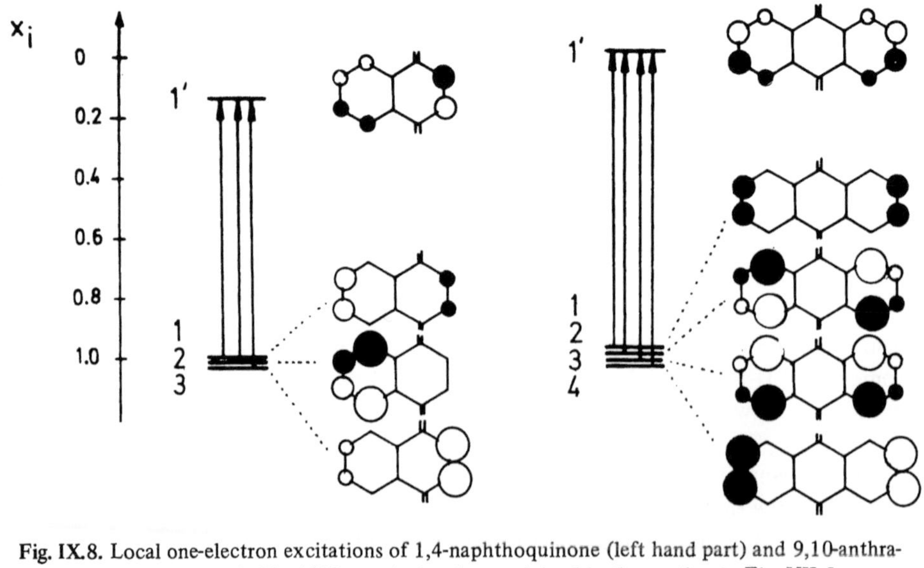

Fig. IX.8. Local one-electron excitations of 1,4-naphthoquinone (left hand part) and 9,10-anthraquinone (right hand part). The MO's are depicted as explained in the caption to Fig. VII.5

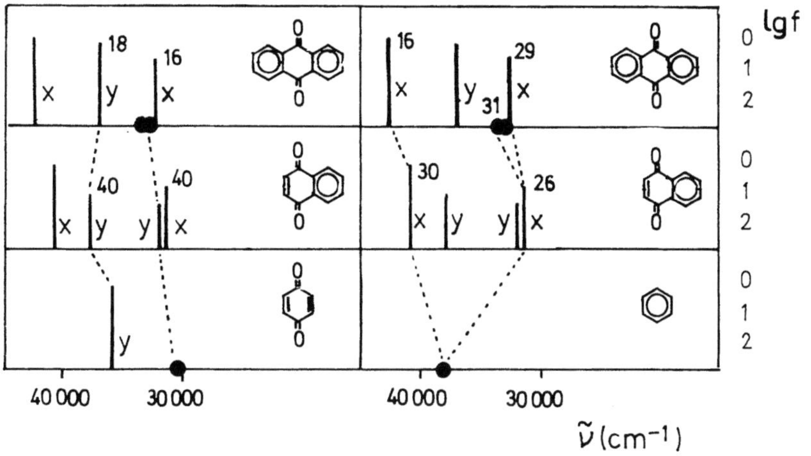

Fig. IX.9. Correlations between electronic transitions according to configuration analyses of the pertinent excited states. The figures indicate the contributions of the lowest excited states of *para*-benzoquinone and benzene, resp., to the lowest excited states of the annelated quinones

caution is, however, required, because the electronic states of *5* are very incompletely described by those of the reference systems.

The more intense lowest energy $\pi \rightarrow \pi^*$ transition of *5* corresponds clearly to the L_b transition of the benzene. The same conclusion can be drawn by modelling according to fragmentation mode *B*, dissecting *5* into two acyl fragments. The assignment of the more intense component of the shorter wavelength absorption, however, is more complicated than formerly [510] assumed.

There is also some theoretical support for the assumption that substitution in the quinoid part is predominantly reflected in a bathochromic shift of quinoid-type transitions [116]. Whereas single substitution in 2(3) position more strongly affects the above mentioned lower energy quinoid transition, twofold substitution in the 2,3 positions is rather discerned in the long wavelength absorption region. In good agreement with this conclusion, a clearly separated weak absorption band observed at 439 nm for 2,3-dihydroxy-*5* (Isonaphthazarin) has been assigned to a second quinoid band [508]. Additional red shift occurs when 2,3-donor substitution is associated with introducing an acceptor substituent in 5-position [509b]. Absorption spectra of 1,4-naphthoquinones containing electron donor groups both in the benzenoid and quinoid rings show a broad complex band system in the 350–360 nm region [509c]. PPP calculations indicate that this results from two overlapping $\pi \rightarrow \pi^*$ transitions.

The assignment of the electronic transitions of 9,10-anthraquinone (*6*) has been the subject of controversy for several years [510]. However, more recent polarization measurements using the photoselection [461, 511, 512] or stretched film technique [513] and theoretical studies [459, 461] have led to consistent conclusions. According to them, the absorption regions at about 320, 270 and 250 nm arise from three allowed $\pi \rightarrow \pi^*$ transitions. The second transition is polarized along the axis passing through the carbonyl group (y-axis in Fig. IX.7) and the first and third transition are polarized perpendicular to this axis (x-axis). According to PPP calculations, the first intense absorption band at about 320 nm results approximately from the 2 → 1′ one-electron transition and the 270 nm band from the 1 → 1′ transition. The lowest-energy one-electron transitions are depicted in the lower part of Fig. IX.8.

Recently found absorption maxima at 205 nm and 180 nm are also excellently described by PPP-type calculations [459]. This outcome makes questionable the n → σ* assignment of the 205 nm absorptions given in Ref. [514].

The lowest excited states of *6* can be similar interpreted by the PPP configuration analyses as discussed for naphthoquinone *5* [442] (cf. Fig. IX.9). In contrast to *5* the lowest-energy quinoid transition of *6* is forbidden, as in the parent compound *4*. On the other hand, due to the presence of two benzene rings two local L_b states exist. Excitation to one of the two combined states is allowed and accounts for the lowest energy absorption band of *6* [511]. This interpretation is in harmony with the doubling of absorption intensity with respect to 1.4-naphthoquinone. In accordance with LHM calculations [515, 516], L_a-type excitations of the benzene fragments only moderately contribute to the description of the shorter wavelength absorption in the near UV. PPP configuration analysis reveals some quinonoid character of the 270 nm absorption band [442] (cf. Fig. IX.9). This outcome supports the former phenomenological classification of this band as "quinoid band" [507, 513, 517, 518], although the correlating states are not the same as formerly assumed.

Since only two bonds are broken, model *B* represents the most appropriate fragmentation mode of anthraquinone. The first and second excited state of the benzoyl fragments combine and mix with CT states. The lowest excited states give rise to the allowed long-wavelength transition (325 nm absorption band) and to one of the forbidden transitions. According to model *B* the "quinoid band" at about 270 nm is attributed to the second π → π* transition of the benzoyl fragments which mix with intramolecular CT transitions.

Table IX.4. Intense longest-wavelength absorption maxima of mono-substituted anthraquinone dyes in methanol [517]

Substituent	λ_{max} in nm (lg ε)	
	1-Substitution	2-Substitution
NO_2	325 (3.63)	323 (3.72)
CN	325 (3.53)	325 (3.72)
Cl	333 (3.70)	325 (3.59)
H	323 (3.65)	323 (3.65)
Me	331 (3.68)	–
OH	402 (3.74)	368 (3.59)
OMe	378 (3.72)	363 (3.60)
O^-	493 (3.70)	465 (3.73)
NHCOMe	400 (3.75)	367 (3.62)
NH_2	475 (3.80)	440 (3.65)
NHMe	503 (3.85)	462 (3.76)
NMe_2	503 (3.69)	472 (3.77)

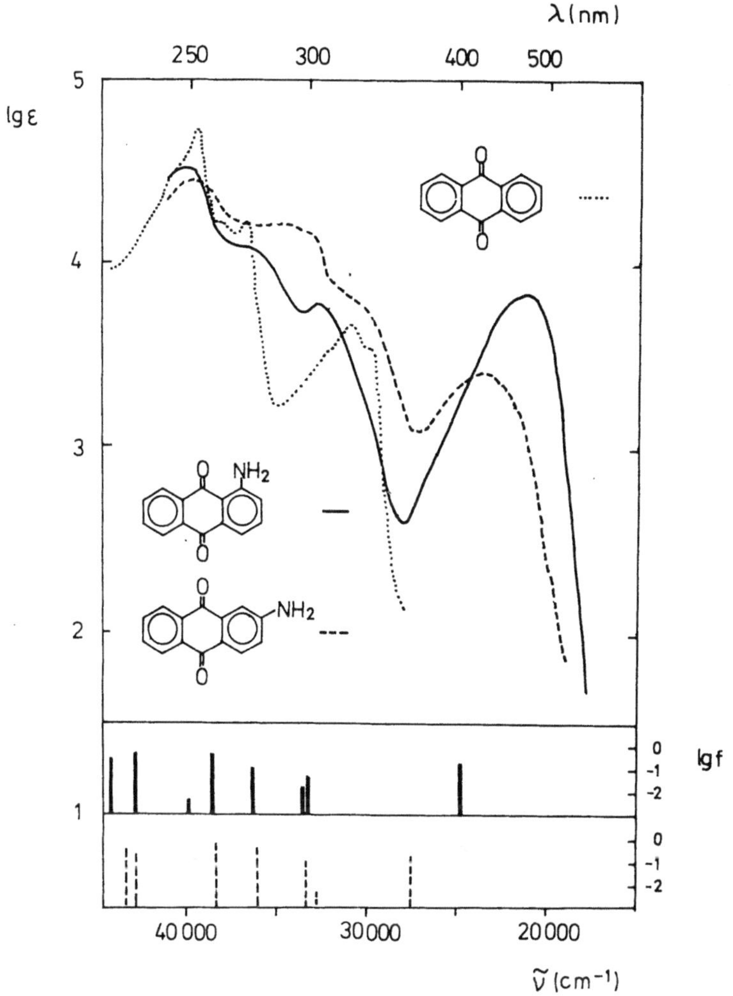

Fig. IX.10. Spectral absorption curves of 1- and 2-amino anthraquinones in dioxane [506] and results of PPP calculations

The importance of substituted 9,10-anthraquinones as disperse dyes induced an extensive investigation of their spectral properties [519]. Whereas acceptor-substituted *6* absorb nearly at the same wavelength as the parent compound, introducing of donor substituents leads to the occurrence of longer-wavelength absorptions (cf. Table IX.4). This fact is illustrated in Fig. IX.10 in which the absorption curves of 1- and 2-amino-9.10-anthraquinone are contrasted with the spectral curve of the parent compound.

The substituent effect of substituted anthraquinones has been extensively studied by PPP-type calculations [520–523]. Some theoretical and experimental results are juxtaposed in Fig. IX.10. According to configuration analyses of the calculated wave functions, only the fourth $\pi \rightarrow \pi^*$ transition of mono-amino anthraquinones at about 270 nm is overwhelmingly localized on the anthraquinone fragment. The 270 nm band corresponds to the "quinoid band" of the parent compound. The remaining $\pi\pi^*$ states of monoamino anthraquinones are mixtures of $\pi\pi^*$ states of anthraquinone. Whereas the allowed longest-wavelength transition of the parent compound is involved in the longest-wavelength transition of the 1-amino anthraquinone, a forbidden transition contributes to the longest-wavelength transition of the 2-amino derivative. Therefore, a lower transition probability is found in the second case. A characteristic feature of both lowest-energy states is the strong participation of CT configurations arising from electronic excitation from the amino group to the anthraquinone unit. The color band of amino anthraquinones can, therefore, be regarded as intramolecular CT band. The strong CT character accounts for its pronounced solvatochromic properties [518,524].

1-Substituted anthraquinones are distinguished from the 2-substituted isomers in the position of the color band. As demonstrated by the data listed in Table IX.4, 1-substituted compounds consistently absorb at longer wavelengths and with higher intensity than the 2-substituted ones.

This finding can be readily explained in terms of the simple one-electron theory. Since the lowest-energy excited state of anthraquinone is mainly described by the $2 \rightarrow 1'$ transition, the strong donor effect in 1-position reflects the efficient overlap of the lone-pair substituent orbital with the occupied MO 2, which bears high expansion coefficients in this position (cf. Fig. IX.8, right-hand part). Substitution in the 2-position of anthraquinone, however, mainly affects the lower-energy occupied MO 4. The $4 \rightarrow 1'$ transition, which is forbidden for the parent compound, is now lowered in energy and gains intensity.

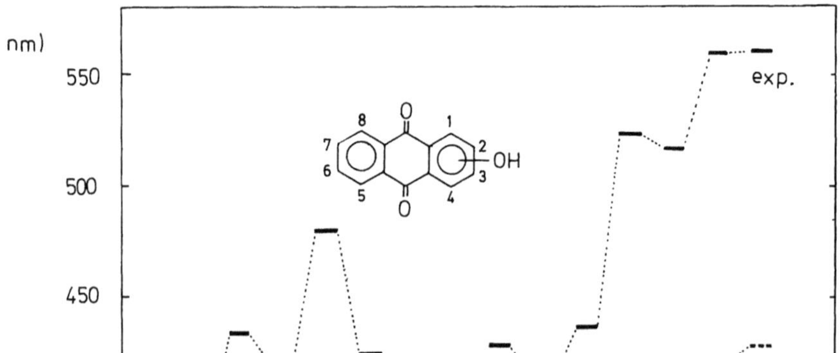

Fig. IX.11. Experimental absorption wavelengths of hydroxy-substituted anthraquinones in ethanol [441] (solid bars) and calculate values [325] (dotted bars)

The degree of bathochromic shift of donor substituents reflects their donor capacity. A linear correlation was found between the λ_{max} values and the ionization potentials of the donor groups X in the molecule CH_3–X. The lower intensity of the color band of 1-dimethylamino-substituted *6* is obviously caused by steric hindrance to co-planarity, which reduces the orbital overlap between the substituent and the π-system of the parent compound [518].

The effectiveness of 1(4,5,8)-position in generation of long-wavelength absorption by donor substitution is also clearly proved by spectral shifts resulting from multiple substitution. Figure IX.11 illustrates the experimental absorption wavelengths of hydroxy-substituted *6* in comparison with values obtained by PPP calculations. The calculated wavelengths parallel those observed experimentally, but they are consistently too low. The difference is partly due to the bathochromic effect of the hydrogen bridge between the OH group in 1-position and the neighbouring carbonyl group, which was not considered in the PPP calculations.

Extensive spectroscopic [456, 520, 521, 525–527] and theoretical studies [520, 521] have been performed on multiply-substituted amino anthraquinones. Some experimental data are collected in Table IX.5. If donor substituents are introduced in 1(4,5,8)-position, the longest-wavelength absorption is approximately polarized in the long molecular axis [513, 520, 528], whereas short axis polarization is observed with substitution in the 2-position [521].

PPP-CA and LHM calculations confirmed the strong CT character of the color band and revealed the correlation between the color-determining electronic state and the electronic states of the parent compound [442]. According to this study, locally excited states of anthraquinone contribute non-negligibly to the lowest-energy state. Amino-substitution causes mixing of the anthraquinone excited states and the most of the excited states of amino anthraquinone do not correlate with a single excited state of the parent compound. As mentioned for mono-substituted *6*, however, correlations are indicated in each case between one of the excited states of multi-substituted anthraquinones and

Table IX.5. Intense longest-wavelength absorption maxima of amino substituted anthraquinone dyes[a]

Substitution	in Dioxane/Hexane	in Ethanol
1-	463 (3.80)	476 (3.72)
2-	412 (3.58)	449 (3.55)
2,6-	383.5 (3.76)	426 (3.00)
2,7-	435 (3.45)	468 (3.63)
2,3-	–	503 (3.38)[b]
1,8-	490 (4.00)	–
1,5-	473 (4.20)	489 (3.91)
1,2-	–	535 (3.92)[b]
1,4-	585.5 (4.03)	592 (4.19)
1,4,5,8-	–	630[c]

[a] Ref. [456] [b] Ref. [564] [c] Ref. [525]

the fourth excited state of anthraquinone, which is responsible for its "quinoid band" at about 270 nm.

Some further regularities in the spectra of amino anthraquinones can be illuminated with the aid of model *B*, considering anthraquinones as composed of two benzoyl fragments. Since the 320 nm absorption of anthraquinone arises from one of the two combinations of the lowest excited states of the benzoyl halves, absorption still occurs in this region when only one benzoyl fragment is substituted. Bands at about 320 nm are not longer observed if both benzoyl fragments are substituted [525]. The absorption is now bathochromically shifted. If two equal monoaminobenzoyl fragments are joined, a similar absorption is observed as for the aminobenzoyl-benzoyl composite molecule, but the intensity is now twice [517]. An extremly strong red shift is brought about by junction of two 2,5-diamino-substituted benzoyl fragments. The resultant 1,4,5,8-tetra-amino-anthraquinone absorbs at more than 600 nm (cf. Table IX.5).

Also these considerations found some theoretical foundation [442]. Popov and co-workers discussed the effect of multiple substitution by perturbation calculations, deriving the term levels of amino anthraquinones by mixing the lowest excited states of the benzoyl fragments with amino-to-benzoyl CT configurations [525].

A successful attempt has been made to estimate the position of the color band by defining incremental values [529]. Starting from substituent increments and interaction terms a sufficient prediction of the wavelengths of di-substituted anthraquinones *6* can be given, but the method fails with application to higher substituted derivatives.

Many of the practically used amino anthraquinones dyes contain aryl substituents at the nitrogen atoms. Introduction of aryl substituents generates red shift of the color band [527] and an additional absorption band emerges in the 400 nm region [513]. Some anthraquinoid dyes contain the anthraquinone chromophore twice such as *25*:

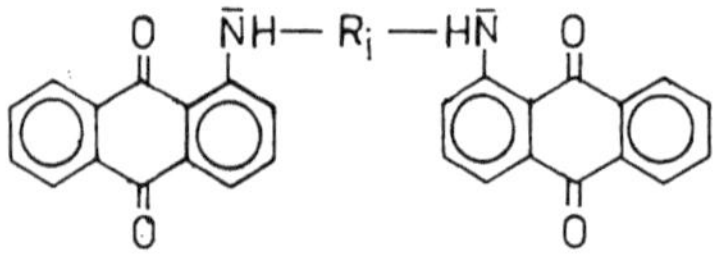

25

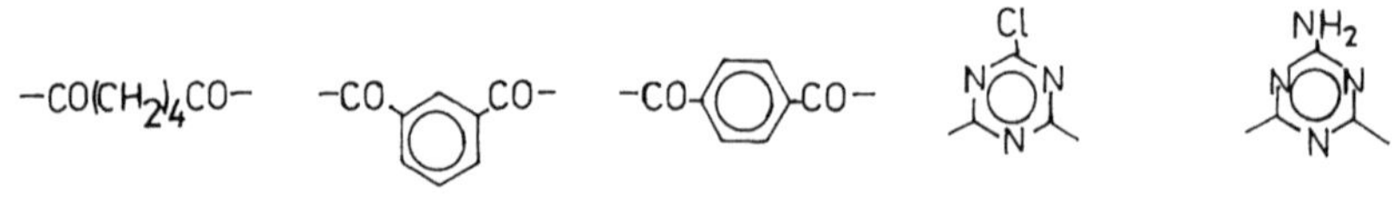

R_1	R_2	R_3	R_4	R_5
411 (4.01)	419 (4.06)	420 (4.21)	422	486
in o-chlorophenol [517]			in chlorobenzene [530]	

The absorption wavelengths are not strongly affected, however. Even insertion of conjugative bridges does not give rise to a marked red shift of the color band. The bridging groups R_2 to R_5 seem to cause a change of the donor properties of the amino group rather than to combine the two anthraquinoid fragments to a delocalized chromophoric system. Similarly the practically used anthraquinone carbazoles *26* and *27* absorb at similar wavelengths to those of 2-amino anthraquinone.

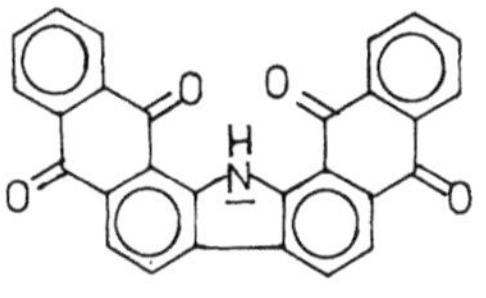

26

456, 411
in chlorobenzene [530]

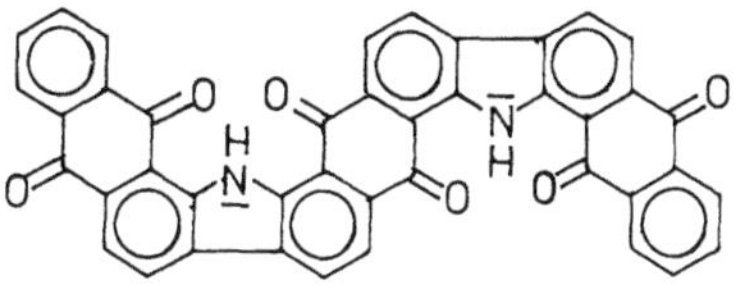

27

460, 412

Indanthrone *28* constitutes an interesting exception.

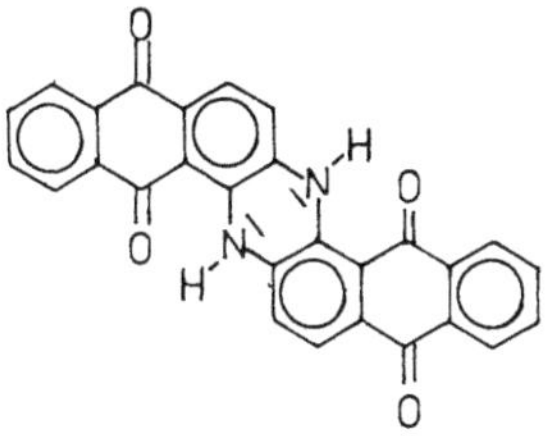

28

645, 690, 740
in chloroform [457]

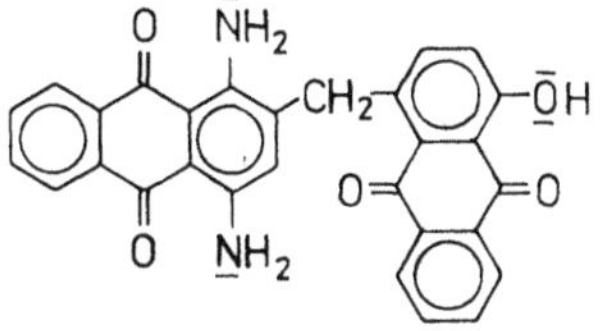

29

412 (3.86), 585 (3.25)
in benzene [531]

Indanthrone displays an intense and structured absorption band in the 700 nm region [457, 532]. The structure of indanthrone has been a subject of controversy for many years. The IR- [533] and X-ray studies [534] as well as the fact that nearly identical uv-vis-spectra are observed for indanthrone and its N-methylated derivative [533] ensured that the structure is represented by the formula *28*. Indanthrone obviously owes its deep color to the strongly bathochromic effect of the amino groups, which combine the anthraquinone fragments in the 1,2-positions. Its longest-wavelength absorption can conceivably be interpreted if indanthrone is regarded as built up from two 2-amino anthraquinone fragments. Because of the nearly linear arrangement of the two sub-chromophores the lowest-energy excited state results from their excitonic combination. This state causes an allowed lowest-energy transition. According to the PPP configuration analysis CT configurations bring about an additional lowering of the excitation energy. The alternative coupled merocyanine model, proposed by Dähne and Leupold [24], is less suited. For this fragmentation mode, PPP configuration analysis indicates a bad description of the excited states and a complex mixture between various LE and CT configurations.

Dyes comprising two chromophorically different quinone fragments may display color properties quite different from those of the constituent parts. Thus, green dyes result from composing blue 1,4-diamino-anthraquinones with yellow 1-hydroxy-anthraquinones. This is exemplified by *29* [531].

The spectral effect of annelation on the color of 1,4-disubstituted benzanthraquinone disperse dyes has been considered recently [534a]. Annelation of anthraquinone dyes in the positions 2,3- or 6,7-produces a small hypsochromic shift, while 5,6-annelation produces a large bathochromic shift.

Since derivatives of isomeric annelated *ortho*-quinones have not yet found application as commercial dyestuffs, we shall dispense with a detailed discussion on their spectral properties. Concerning recent experimental and theoretical studies of the parent structures the references [459, 460, 465, 536] and [537] should be consulted.

The first thio heteroanalogues have been synthesized recently: Monothio-1,4-anthraquinone absorbs weakly at 697 nm (lg ε 1.67) and strongly at 334 nm (lg ε 4.16) in chloroform [537a]. The disperse dye CI Vat Green 7 may be considered as a heterocyclic derivative of dithio-1,4-anthraquinone [537b].

5. Higher Annelated Quinones

Our discussion has so far only included compounds having the two carbonyl groups in one six-membered ring. The carbonyl group may also be situated in different rings. Among a great many known compounds we only mention some examples which, applied as vat dyes [31], gained importance in color chemistry.

Thus anthanthrones are used for dyeing of cellulosic fibres or silk. The parent compound *30* exhibits a well-shaped spectrum, which is excellently reproduced by PPP calculations (cf. Fig. IX.12).

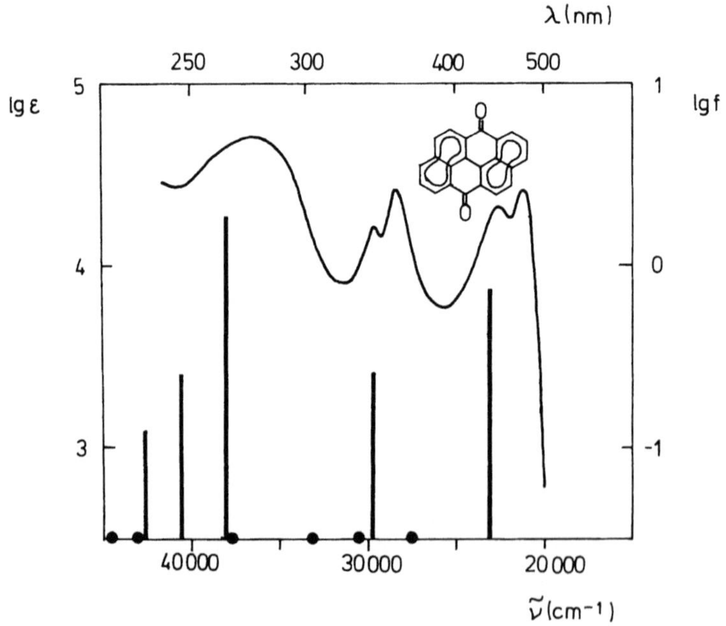

Fig. IX.12. Spectral absorption curve on anthanthrone in dioxane [506] and results of the PPP calculation [325]

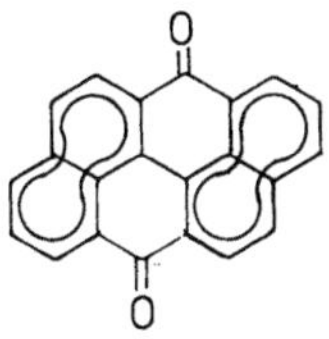

30

In the dibenzopyrenequinones *31* and *32* the naphthalene unit is interposed between two benzoyl groups giving rise to a *trans*- or a *cis*-configurated structure.

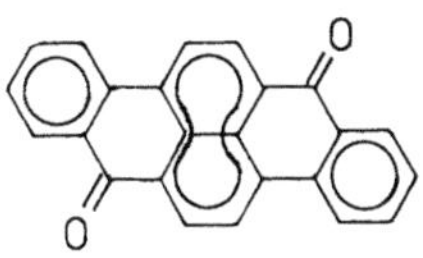

31

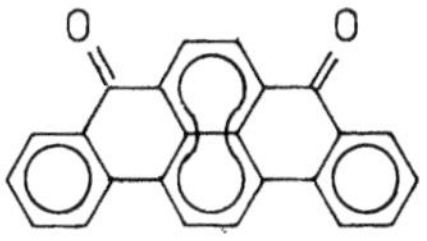

32

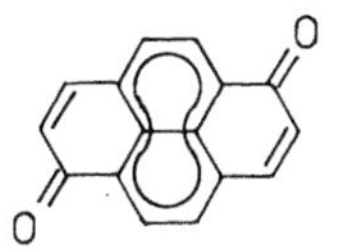

33

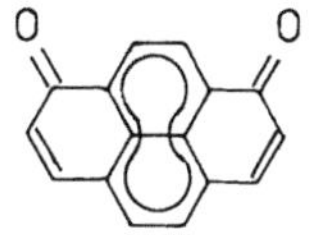

34

The unsubstituted *31* itself represents a widely used vat dye giving reddish-yellow shades on cellulosic fibres, silk and polyamides [31]. The color band moves to longer wavelengths on passing from the *trans* dye *31* to the isomeric dye *32* [506]. The same relationship holds between the isomeric pyrenequinones *33* and *34*, which have been thoroughly studied both experimentally and theoretically [535]. According to PPP calculations the color bands of *31* and *32* arise from a single $\pi \rightarrow \pi^*$ transition polarized in the lengthwise molecular axis. The transition has some CT character. Naphthalene acts as donor and the CO groups as acceptor part of the molecule.

Replacing of the naphthalene units of *31* and *32* by perylene, leading to *35* and *36*, lowers the energy of the CT as well as LE states and a pronounced red shift should occur. The dibenzanthrones, known as isoviolanthrone *35* and violanthrone *36a* absorb, in fact, at about 100 nm longer wavelengths than *31* and *32* [538–540]. Violanthrone gives dark blue shades on cellulosic fibres or silk and has found extensive practical application [31].

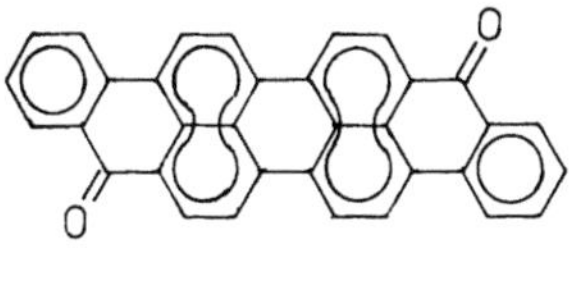

35

588 (4.62)

in dimethylformamide [325]

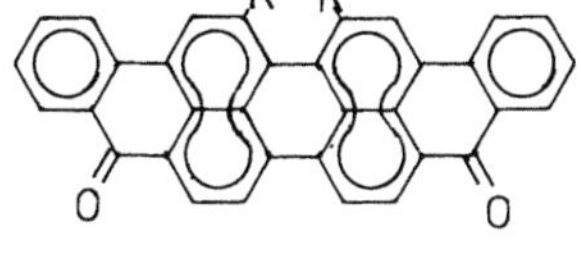

36

36a	R, R' = H:	600 (4.78)
36b	R=H, R' = OCH_3:	598 (4.59)
36c	R, R' = OCH_3:	636 (4.62)
36d	R, R' = OCH_2CH_2O:	618
36e	R=H, R' = NH_2:	620 (4.17)
36f	R, R' = NH_2:	640 (4.00)

in dimethylformamide [540], [325]

The shift of the longest-wavelength absorption of 16,17-dimethoxy-violanthrone (*36c*) with comparison to the parent compound *36a* has been generally attributed to the non-planarity of *36c* due to steric effects [541]. This explanation, however, has been doubted [540] since the assumed weakening of essentially single bonds should cause a hypsochromic shift rather than a bathochromic shift. The doubt is supported by the fact that a substantial blue shift (25 nm) occurs in passing from the parent compound to the 16,17-dimethyl derivative. The bathochromic and hyperchromic shift in going from the mono-methoxy-substituted *36b* to the dimethoxy-substituted *36c* suggest that strong electron repulsion between the adjacent substituents rather than deviation from the co-planarity causes the observed red shift. Introduction of a methylene bridge between the oxygen atoms in *36d* lessens the overlap between the lone-pair orbitals and the p_{π}-orbitals of the ring system. Consequently, the color band moves towards shorter wavelengths in passing from *36c* to *36d*.

Increasing distortion of the amino groups probably similarly occurs in passing from *36e* to *36f*. Although the absorption wavelength slightly increases, the absorption intensity (lg ϵ values) shows the expected decrease.

Another important class of polynuclear quinoid vat dyes are pyranthrones and the bisazaanalogous flavanthrones derived from *37* and *38*, resp. They absorb intensely in the 500 nm region.

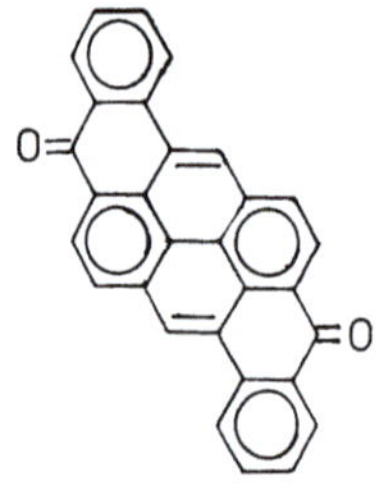

37

476 (4.42)
in chlorobenzene [539]

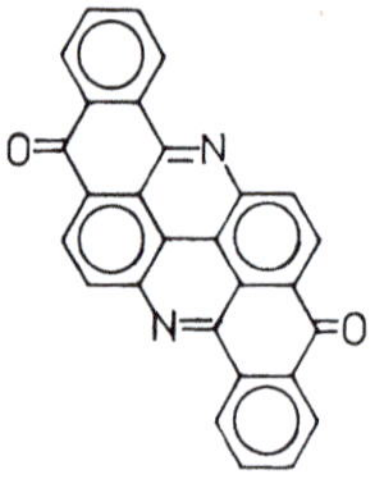

38

469 (4.52)
in chlorobenzene [539]

The color bands exhibit a fine structure [532, 539]. The occurrence of weak $n\pi^*$ absorption band of *37* at about 550 nm, [532] however, still requires further evidence.

6. Medium Effect

The solvent effect is generally small for unsubstituted quinones but remarkable shifts have been observed for donor-substituted quinones such as mono-donor-substituted anthraquinones [518], indophenol [486, 542, 543], phenoxazone [544] and indulin dyes [492]. These dyes display a positive solvatochromism which is generally attributed to the increase of the dipole moment upon excitation. However, positive solvatochromism was also observed for quinones which exhibit no alteration in the dipole moment, such as 2,5-diamino-*para*-benzoquinone [477]. As found for merocyanine and betaine dyes, the spectral solvent shifts correlate with E_T-values [477]. Strong red shift occurs if the donor group OH or NH_2 loose a proton in alkaline medium, as demonstrated for many synthetic or naturally occuring naphthoquinone dyes [441]. The absorption wavelengths of the deprotonated species of anthraquinones have been sufficiently accounted for by PPP-type calculations [545].

Striking change in color generally also takes place when quinoid dyes are dissolved in sulphuric acid. Many quinones, such as pyranthrone and flavanthrone, then absorb at longer wavelengths [532]. The observed colors and spectral data of various other commercial dyes are compiled in Ref. [36].

The red shift observed in strong acids may be due to protonation [546]. In many cases the resulting protonated species are not known with certainty, but systematic investigations of some simple quinones have led to definite conclusions: Whereas 1,4-naphthoquinone and 9,10-anthraquinone undergo only mono-protonation at one of the carbonyl oxygen (even in fuming sulphuric acid), twofold protonation occurs with 1,4-anthraquinone, 5,12-tetracenequinone and 6,13-pentacenequinone [547]. The color turns from yellow to red in the first stage of protonation and to blue in the second stage. The good agreement between experimental and calculated spectral data of mono- and di-protonated quinones and measured absorption maximum suggests that the oxygen atom is the site of protonation, giving rise, for example, to the cations *39* and *40* [548].

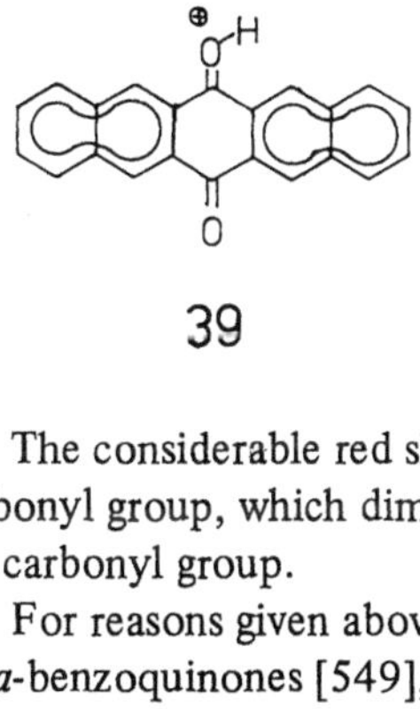

39

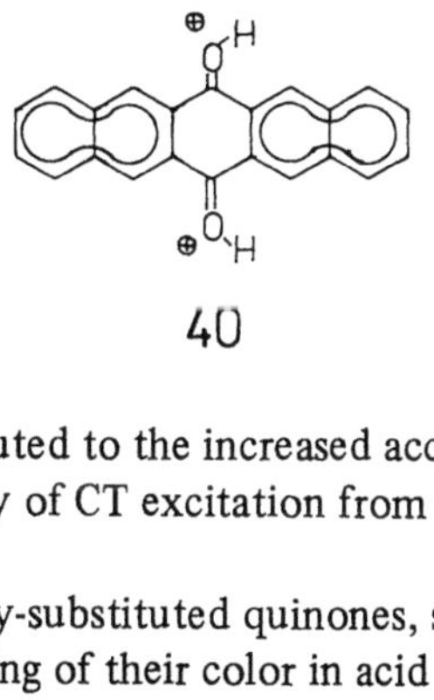

40

The considerable red shift can be attributed to the increased acceptor capacity of the carbonyl group, which diminishes the energy of CT excitation from the hydrocarbons to the carbonyl group.

For reasons given above various hydroxy-substituted quinones, such as 2,5-hydroxy-*para*-benzoquinones [549], undergo deepening of their color in acid as well as alkaline medium:

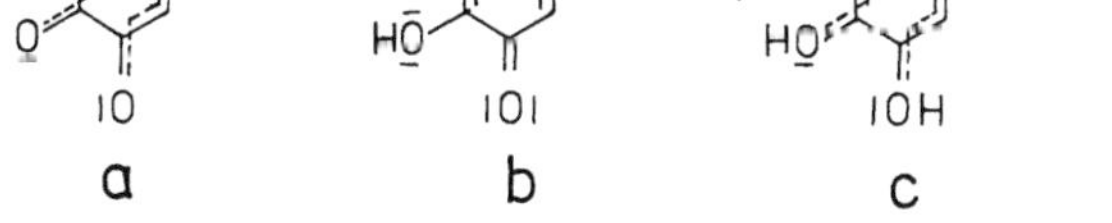

41

red yellow red

Both of the resulting ions can be considered to be built up from two trimethine chains.

In amino substituted quinones the protonation may alternatively occur at the amino group. Protonation at the amino nitrogen, i.e. fixation of the lone-pair electrons responsible for the CT absorptions of amino quinones, is accompanied by decoloration. This has been observed, for example, with 1-amino, 1-methylamino, 1-phenylamino and 2-amino anthraquinones [550]. The absorption spectra of the protonated species are very similar to the spectrum of anthraquinones. This fact has been discussed by means of PPP-type calculations [523].

Due to presence of amino as well as hydroxy groups various blue anthraquinone acid dyes such as *42* show color change both in acids and alkalis [551].

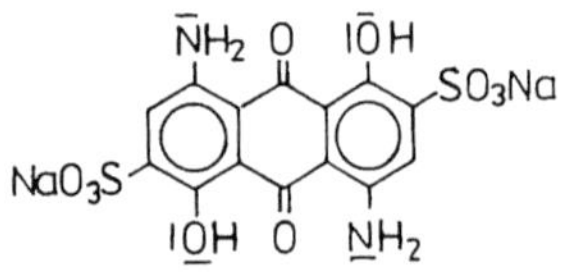

42

Reduction of quinone dyes leads to leuco compounds the spectral feature of which corresponds to that of substituted hydrocarbons. Spectral data on various leuco compounds of commercial dyes are discussed in Ref. [552].

X. Indigoid Dyes

Although one of the dyes of antiquity, the deep blue indigo is still used as vat dye. Structural modifications have made accessible a broad assortment of dyestuffs giving orange, red, blue or bluish-black shades on cellulosic fibres, silk and wool [31, 245, 246]. Indigo dyes are, in general, characterized by high melting points and low solubility in organic solvents. They can be reduced to dihydro derivatives (leuco form). Some indigo dyes display photochromic properties [569].

1. Chromophoric System

Many attempts have been made to explain the deep color of indigo *1* (X = NH), which according to X-ray studies exists in the *trans* (E)-configuration. Remarkable progress in solving this problem has only been achieved since appropriate quantum chemical methods have been available.

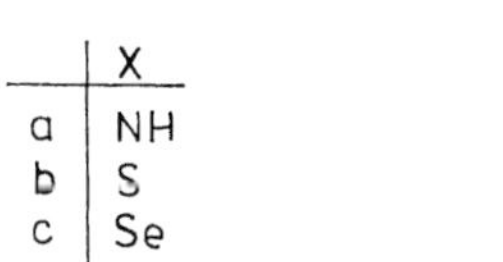

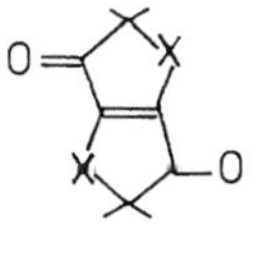

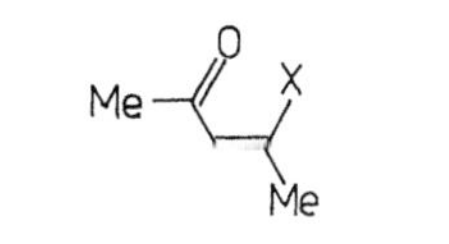

1 2 3

	X
a	NH
b	S
c	Se
d	O

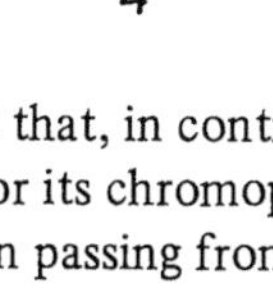

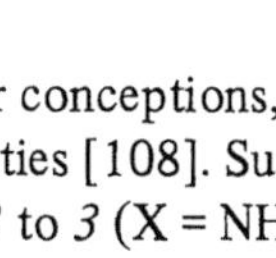

4 5

HMO and PPP calculations revealed that, in contrast to former conceptions, the benzene rings of indigo are not essential for its chromophoric properties [108]. Successive diminution of the conjugated system in passing from *1* through *2* to *3* (X = NH) shows only a slight lowering of the excitation energy. The cross-conjugated bond system of the

Table X.1. Longest-wavelength absorption maxima of indigo, thioindigo and selenoindigo and their partial chromophores (*trans* isomers)

X	λ_{max} in nm (lg ε)			
	1	*2*	*3*	*5*
NH[a]	605 (4.30)	529 (3.83)[c]	487 (4.00)	285[e]
S[b]	546 (4.26)	505 (4.30)	458 (4.34)	297[d]
Se[f]	564 (4.06)		482 (4.04)	

[a] in tetrachlorethane [570]
[b] in chloroform [629]
[c] tetramethyl-pyrrole-indigo
[d] X = SH, in cyclohexane [635]
[e] $X = NH_2$, in cyclohexane [636]
[f] in chloroform [630]

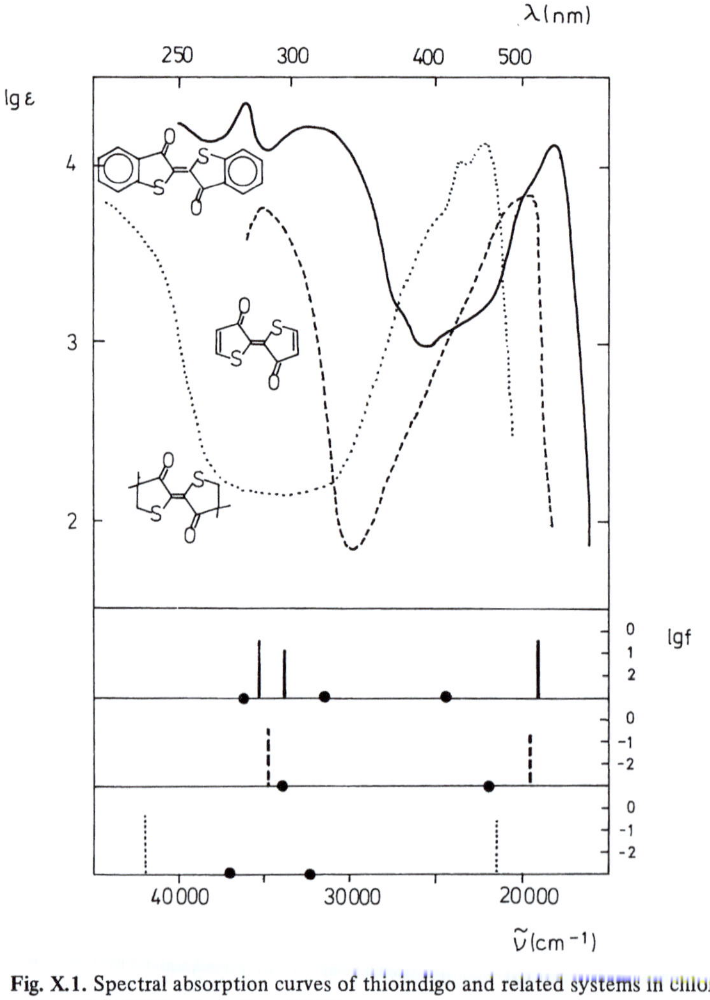

Fig. X.1. Spectral absorption curves of thioindigo and related systems in chloroform [501, 629] and results of PPP calculations [578]

structure *3* (X = NH) has been predicted to be the basic indigo chromophor ("H-indigo chromophore") [108]. The later synthesis of the colored *3a* confirmed this prediction [570]. The trimethine merocyanine *5a* resulting from further diminution of the conjugated system, absorbs only in the near ultraviolet region (cf. Table X.1). Similarly, relative long-wavelength absorptions have been observed for the heteroanalogous sulphur and selenium compounds of the series *3*. The spectra of the sulphur-containing compounds are juxtaposed in Fig. X.1.

The $\pi\pi^*$-chromophore of *3* is characterized by release of π-electrons from the lone-pair heteroatom X upon the lowest-energy excitation, whereas the carbonyl group accepts π-electrons. This particular feature of the basic indigo chromophore follows from the expansion coefficients of the frontier orbitals which determine the color-determining $1 \rightarrow 1'$ transition. They are depicted in Fig. X.2 für the model-type structure *6*. For simplicity equal electronegativity of the heteroatoms X and Y has been adopted in the MO calculation.

The low excitation energy of the parent chromophore *6* may be derived from that of the trimethine chromophore *7*, considering *6* as substituted *7*. The inspection of the expansion coefficients of *7* and those of the substituent orbitals shows that donor substitution in 3-position strongly raises the energy of the HOMO, whereas acceptor substitution in 2-position at the same time, strongly lowers the energy of the LUMO (cf. Fig. X.3). This change in orbital energy causes a strong lowering of the $1 \rightarrow 1'$ transition

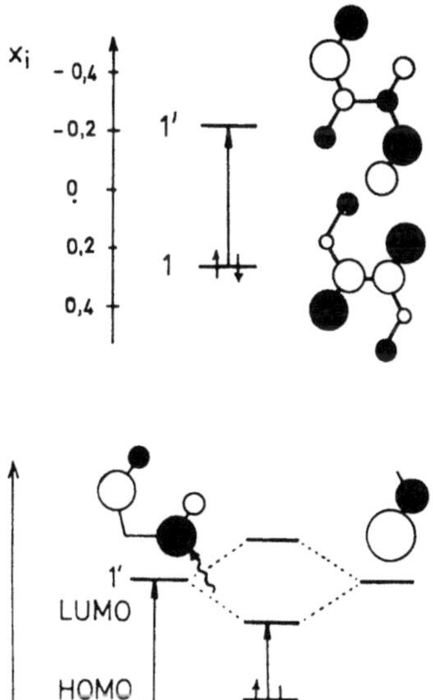

Fig. X.2. Lowest-energy one-electron transition and expansion coefficients of the frontier orbitals of the basic indigo chromophore (H-indigo chromophore). The coefficients are depicted as explained in the caption to Fig. VII.5

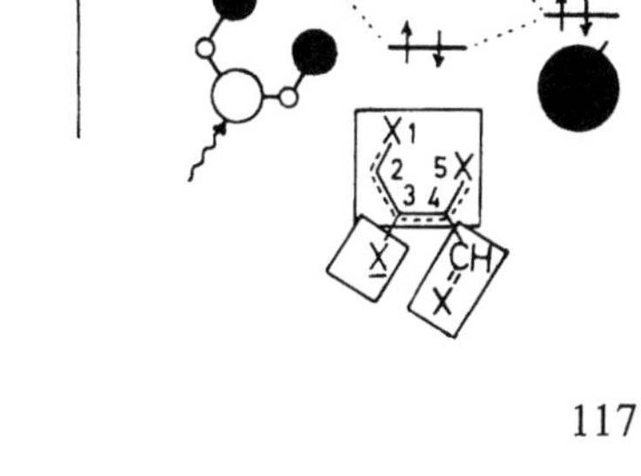

Fig. X.3. Qualitative derivation of the frontier orbitals of the basic indigo chromophore by considering essential interactions between the fragment orbitals (cf. text)

energy in passing from the trimethine *7* to the branched conjugated *6*. Thus, *6* can be considered as a particular chromophore.

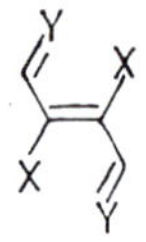 6 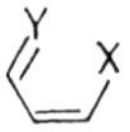 7 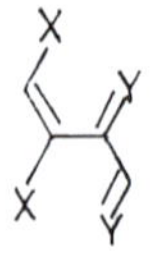 8 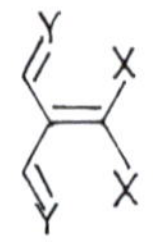 9

Reasoning along these lines leads to the conclusion that the branched π-system *8* should also exhibit a deep color, whereas the isomeric π-system *9* cannot be deeply colored because of the inverse junction of the donor and acceptor fragments to the trimethine fragment.

In fact, contrarily to indigo and the likewise blue unsymmetrical indigo diimine *10* [571], the isomeric compound *11* is yellow [572, 573].

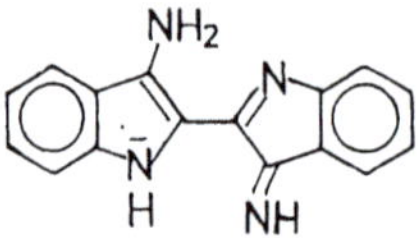 10 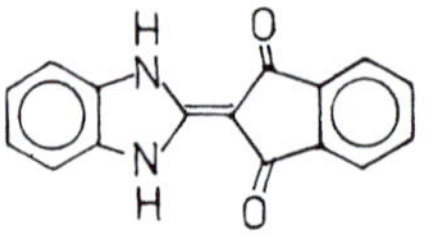 11

617 (3.95)
in chloroform [571]

355
in methanol [573]

Perturbation theoretical arguments also lead to the conclusion that appropriate substitution of longer chain polymethines brings about chromophoric systems of comparably low excitation energy, such as *12* to *14*.

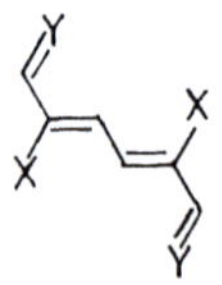 12 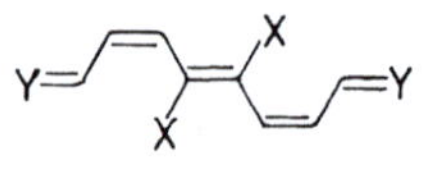 13 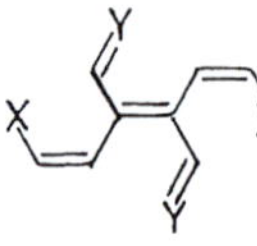 14

Some dyes of this type are considered in Sect. X.3.

The chromophoric system is also affected by double bonds which extend the branched conjugated system beyond the heteroatomic groups, as the acyclic compounds *15* to *17* and the cyclic compound *18*.

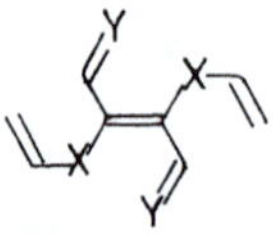 15 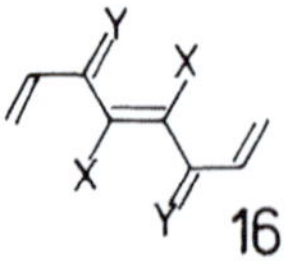 16 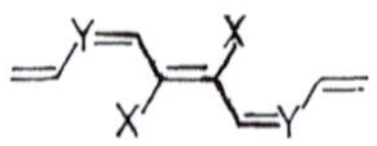 17 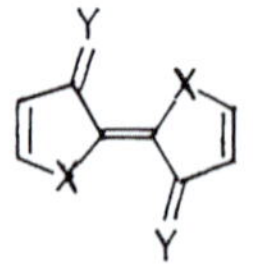 18

The distinct effects of structural modification on the frontier orbital energies and $1 \rightarrow 1'$ transition energies of *6–8* and of *12–18* are depicted in Fig. X. 4.

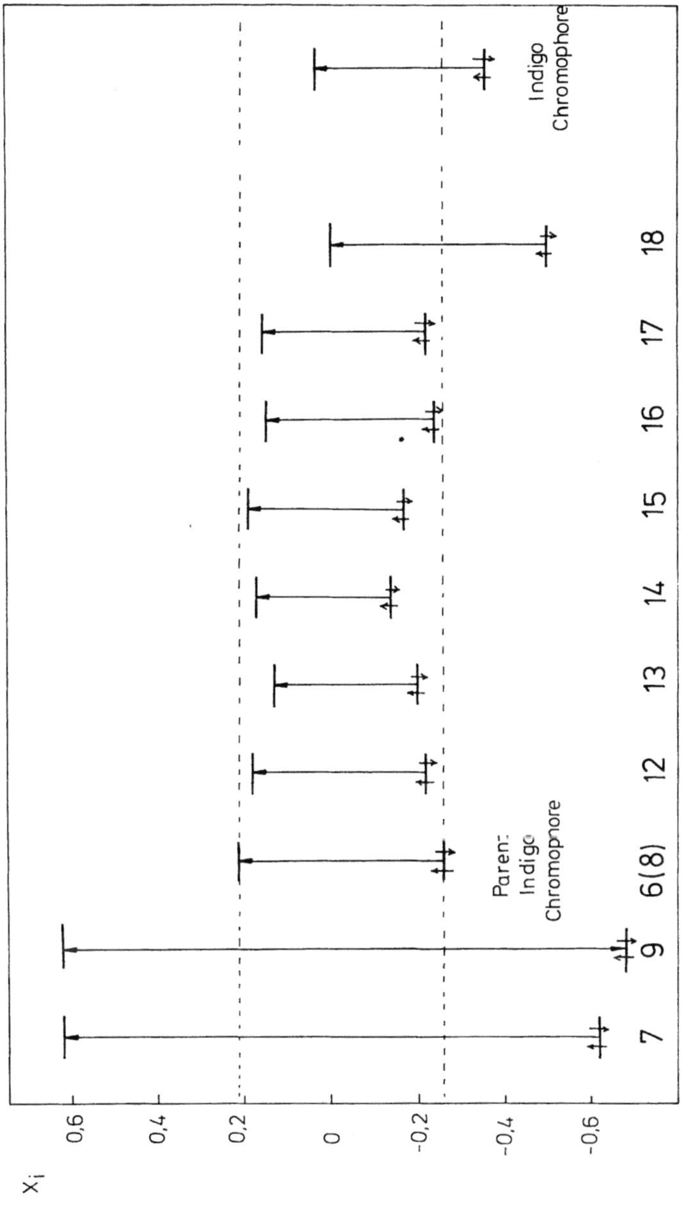

Fig. X.4. HOMO's, LUMO's, and transition energies of models of indigoid structures according to HMO calculations ($\alpha_X = \alpha_Y = \alpha + \beta$; $\beta_{CC} = \beta_{CX} = \beta_{CY}$)

According to Fig. X.4 the HMO transition energy of the cyclic conjugated π-system *18* is considerably higher than those of the corresponding open chain structures *15* and *16*. At the same time, both the HOMO and LUMO energy are markedly lowered. This distinct feature of *18* comes not unexpected in consideration of the frontier orbitals of the parent chromophore. According to the MO expansion coefficients of *6* shown in Fig. X.2, the frontier orbitals of *6* will preferrably interact with ethylene LUMO's giving rise to a downturn in MO energy as mentioned above.

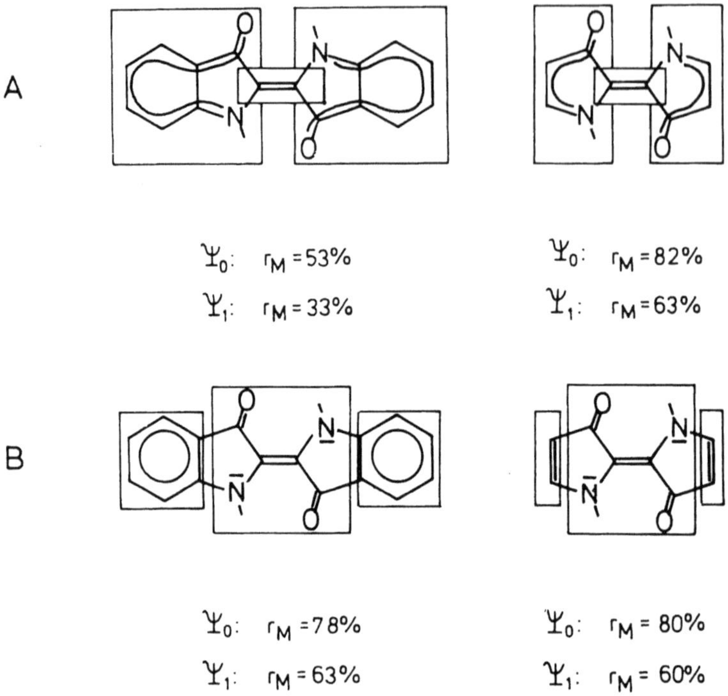

Fig. X.5. Alternative models for interpreting the ground state Ψ_0 and lowest excited state Ψ_1 and their suitability according to the PPP configuration analyses (r_M values)

These conclusions have been established, in the main, on the PPP level [574]. PPP-type calculations allowed, moreover, a critical reexamination of former interpretations of the color of indigoid dyes, such as interpretation in terms of coupled merocyanine chains. This interpretative model was originally presented in the language of the resonance theory (cf. Ref. [575]) but has been recently revived by Leupold and Dähne [478] within their "polymethine conception", based on results of MO calculations. Thus *1a* and *2a* have been considered to contain two heptamethine merocyanine chains which are bridged through the central double bond according to model *A* in Fig. X.5 ("quadrupole merocyanines").

This fragmentation mode was confronted with that of Klessinger and Lüttke [108] which involves the H-indigo chromophore building blocks according to model *B*.

PPP configuration analysis has shown that both models are, at least in part, of comparable suitability [134]. The r_M values of bispyrrole-indigo obtained with respect to model *A* are nearly equal to those obtained with respect to model *B* (cf. Fig. X.5). A more detailed examination of the results of the PPP-CA calculations revealed, however, that only model *B* allows a simple and clear-cut interpretation of the lowest excited state. In this case the configuration contributing predominantly to the color-determining lowest-energy state Ψ_1 is just the locally excited state of the H-indigo chromophore as discussed above. The same holds for the indigo chromophore.

The relatively low r_M values indicate, however, that the ethylene and the benzene fragments of bispyrrole-indigo and indigo, respectively, participate markedly in the excited states. This configurational mixing provides, therefore, also some arguments in favour of Dähne's merocyanine conception according to model *A*, which emphasizes the peripheric conjugation. Thus, this model has also some relevance in examing the electronic structure of some cyclic conjugated indigoid dyes. But it cannot place in question the suitability of the Klessinger-Lüttke model *B*. This model has the striking advantage that it makes possible a systematization of a great variety of chromophoric structures [574].

In order to incorporate the H-indigo chromophore into the "coupled merocyanine concept", Dähne later emphasized the presence of central trimethine merocyanine subchromophores [24]. Although the above discussion of Fig. X.3 could be considered to support of the significance of the central merocyanine chromophore, a superficical application of this approach can lead to wrong predictions. Owing to the relatively strong coupling between the "substituents" and the trimethine merocyanine chain the typical feature of the polymethine chain is partly lost. Whereas polymethines display bond length equalization along the chain, pronounced bond alternation has to be expected for the H-chromophore. This conclusion is in good accordance with pronounced differences of the CC-bond lengths found by first X-ray analyses (cf. Table X.2).

The geometrical arrangement of the heteroatomic groups in *6* is of minor importance for the position of the color band [576, 577]. For X = SR and Y = O, PPP calculations predict the sequence of increasing absorption wavelengths [577]:

6

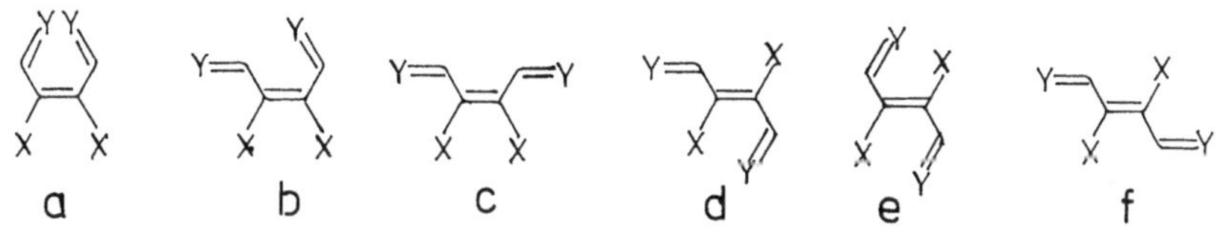

The *trans* isomers *6d* to *6f* should display longer wavelength absorptions than the *cis* isomers *6a* to *6c*. A hypsochromic shift has been in fact observed in passing from *6e* to *6a* when dyes of the configuration *6e* undergo photoisomerization. This is exemplified with the spectra of the photoisomers of *3b* in Fig. X.6. This figure also contains the color

Table X.2. Carbon-carbon bond lengths of indigoid dyes in Å

Bond	*1a*	*1b*	*1c*	*1d*	*3b*	*4b*
X:	NH	S	Se	O	S	S
2-2′ bond	1.37	1.34	1.33	1.23	1.37	1.41
2-3 bond	1.47	1.53	1.47	1.48	1.48	1.46
Ref.	[637]	[637]	[639]	[638]	[640] [640a]	[641]

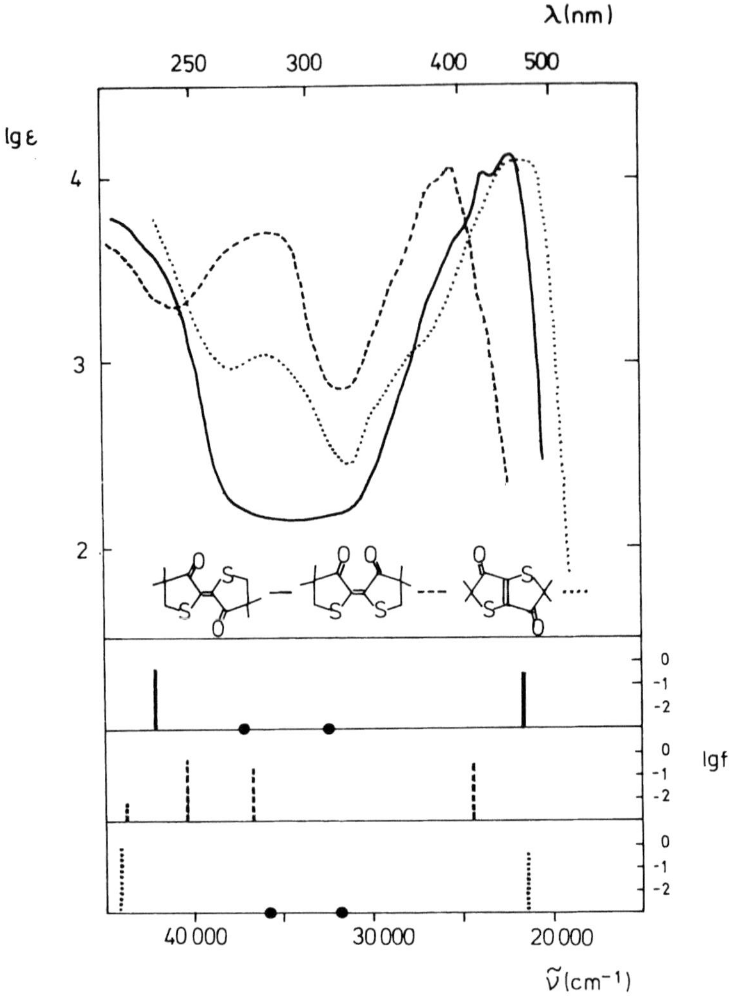

Fig. X.6. Spectral absorption curves of isomeric parent thioindigo chromophores [577, 629] and results of PPP calculations

band of *4* which corresponds to the isomer *6f*. This compound is actually the most bathochromic *trans* compound, its absorption intensity is lower than expected, however.

The blue shift of thioindigo dyes in passing from *trans* to the *cis* isomers is correctly reproduced in PPP calculations [577, 578]. As shown by transition energy partitioning analysis [578], the increase in transition energy mainly arises from the non-bonding S...O interaction, which is removed in passing from the *trans* to the *cis* isomer. The blue shift is quite remarkable since the *cis* isomer probably takes a non-planar structure around the central carbon-carbon bond. The bathochromic shift resulting from torsion around this essentially double bond is overcompensated by the stronger hypsochromic shift due to the removal of non-bonding S...O interaction.

2. Indigo and Derivatives

Indigo owes its color to the intense absorption at about 600 nm, which is followed by a weaker absorption region between 320 nm and 500 nm and an intense absorption region towards shorter wavelengths. Due to the insolubility of indigo in cyclohexane, absorption spectra in this solvent are only known for alkyl-substituted derivatives.

As shown in Fig. X.7, the results of the PPP calculation are not in good agreement with the absorption maxima found experimentally. The deviation amounts to about 100 nm for the color band. From the fact that the vapour of indigo is red in color, it has been inferred that the disagreement between experimental and theoretical results partly arises from solvent and aggregation effects (cf. Sect. X.4). The deviation is considerably lower if nitrogen is replaced by sulphur (cf. Fig. X.1).

The results of PPP calculations of indigo and thioindigo suggest that only one $\pi \rightarrow \pi^*$ transition accounts for the long-wavelength absorption of each compound. The weak absorptions at about 420 nm evidently result from $\pi \rightarrow \pi^*$ transitions which are forbidden for symmetry reasons. Apart from differences between theoretical and experimental results the general features of the absorption spectrum are sufficiently reproduced.

The close similarity of the absorption curves of heteroanalogous indigo dyes is demonstrated in Fig. X.8.

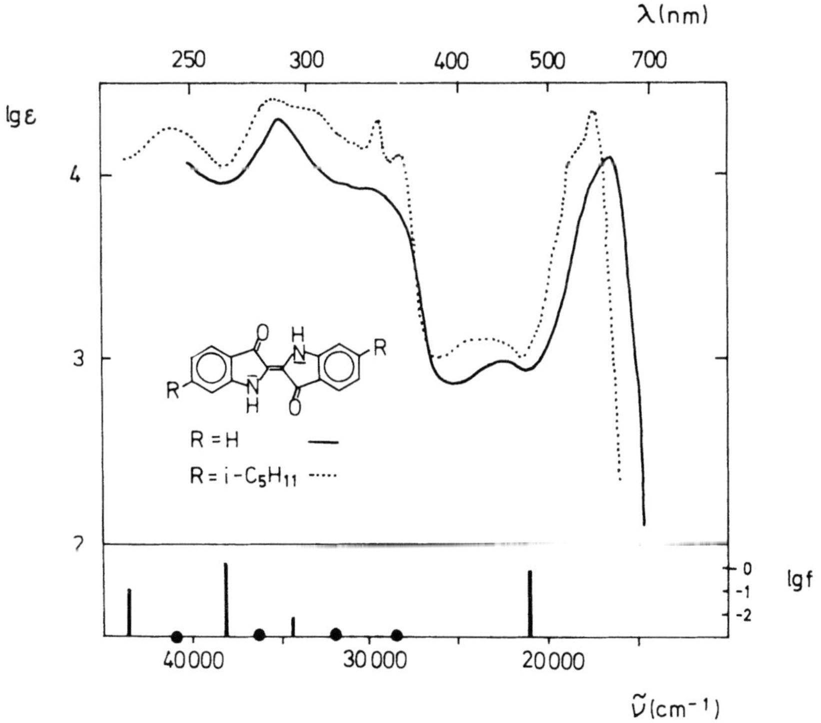

Fig. X.7. Spectral absorption curve of indigo in chloroform [581] and of 6,6′-di-tert-pentyl-indigo in cyclohexane [588] and results of the PPP calculation of indigo [325]

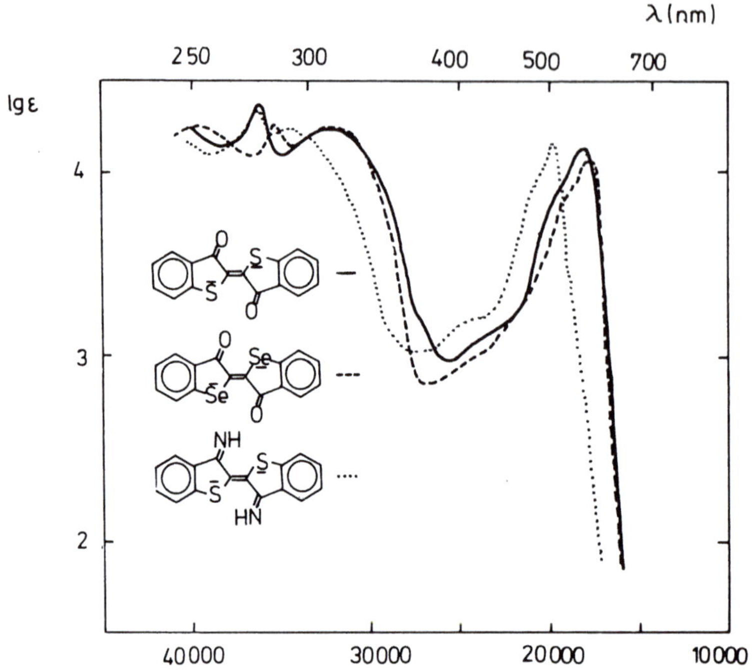

Fig. X.8. Absorption curves of thioindigo [581], selenoindigo [630] and thioindigo diimine [571] in chloroform

Table X.3. Longest-wavelength absorption maxima of indigo and heteroanalogous compounds (in tetrachloromethane) [108]

Compound	*1a*	*1b*	*1c*	*1d*
λ_{max} in nm (lg ϵ)	588 (3.91)	541 (4.12)	558 (4.04)	410 (4.20)

Table X.4. Longest-wavelength absorption maxima of vinylogous indigo dyes (λ_{max} in nm and lg ϵ)

Compound	X	n = 0	n = 1	n = 2
19[a]	NH	615 (4.32)[b]	573 (4.35)[c]	
	S	554 (4.17)[d]	534 (4.36)[d]	532.5 (4.56)[d]
	Se	573 (4.10)[e]	545.5 (4.25)[e]	
20	S[f]	450 (4.13)	446 (4.50)	454 (4.67)
	Se[g]	476 (4.17)	463	

[a] in benzene [588]
[b] 5-i-$C_{14}H_{29}$-5′-n-$C_{16}H_{33}$-derivative
[c] 5,5′-di-i-$C_{14}H_{29}$-derivative
[d] 5,5′-di-tert-C_4H_9-derivative
[e] 5,5′-di-C_2H_5-derivative
[f] in cyclohexane [629]
[g] in cyclohexane [630]

The comparison of their absorption data assembled in Table X.3 shows that the absorption wavelengths decrease in the sequence: N > Se > S > O, in good accordance with PPP-type calculations [579]. The decrease in wavelengths is linearly related to the increase of the C=O stretching vibration frequencies [108].

Also in the series of the vinylogous compounds *19* and *20* the bathochromic effect of the NH group is larger than that of Se and S (Table X.4). Remarkably, a hypsochromic effect occurs upon introducing one vinylene group.

The bathochromic effect of the nitrogen atoms can be strengthened by methyl substitution, as shown for *21*. However, the observed shift may be partly due to a distortion around the central doubly bond. The absorption wavelength of di-Me indigo is shifted from 650 nm to 670 nm in chloroform when one Me group is replaced by the chelating BPh_2 group [579a]. Blue shift occurs, however, when the donor effect of nitrogen is reduced by N,N'-acylation (cf. Table X.5).

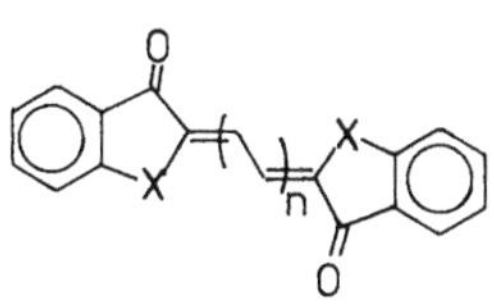

19

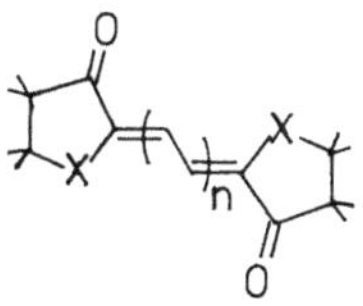

20

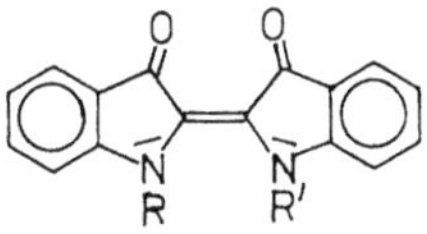

21

An interesting cyclic derivative of indigo is *22*. It absorbs at shorter wavelengths than indigo (indigo in chloroform: 610 nm, lg ε 4.24 [580]).

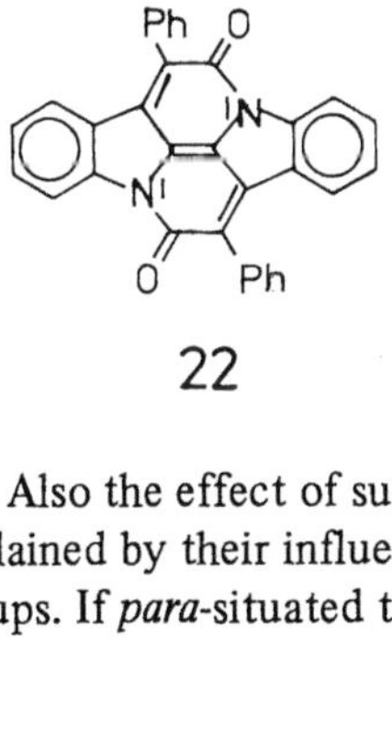

22

537 nm (4.27)
in chloroform [581]

Also the effect of substituents in the 5,5'- or 6,6'-positions of indigo can easily be explained by their influence on the donor or acceptor properties of the NH and C=O groups. If *para*-situated to nitrogen, acceptor substituents in the 5,5'-positions lower the

Table X.5. Longest-wavelength absorption data of N,N'-disubstituted indigo dyes *21* in nm (lg ε)

R,R'	Me,Me	$-CH_2-CH_2-$	–CH=CH–	MeCO,MeCO	–CO–CO–
λ_{max} (lg ε)	650[a] (4.25)	570[a] (4.04)	556[a] (3.9)	559[b]	429[c]

[a] in chloroform [580]
[b] in tetrachloromethane [642]
[c] in chloroform [642]

Table X.6. Longest-wavelength absorption maxima of substituted indigo dyes (*1a*) and substituted thioindigo dyes (*1b*)

Substituent	λ_{max} in nm			
	1a[a]		*1b*[b]	
Positions:	5,5′	6,6′	5,5′	6,6′
NO_2	580	635	513	567
H	*605*	*605*	*543*	*543*
Me	620	595		
F	615	570		
Cl	620	590	556	539
Br	620	590	553[c]	
J	610	590		
SEt	–	–	573	531
OEt (OMe)	645	570	584	473
NH_2	–		638	452

[a] in tetrachloroethane [583]
[b] in dimethylformamide [582]
[c] in dimethylformamide [616]

donor properties of the nitrogen atoms, but donor substituents, conversely, reinforce them. Consequently, a blue shift has been observed in the first case and a red shift in the second case. The same holds for thioindigo dyes (cf. Table X.6). Wizinger [582] discussed the red shift in connection with Kaufmann's rule of the "repartition of auxochromes", which predicts long-wavelength absorptions for 1-acceptor-2,5-bisdonor-substituted benzenes.

The substituent effect in the 6,6′-positions is just opposite to that in the 5,5′-positions (cf. Table X.6). In this case acceptor substituents exert red shift and donor substituents blue shift. Since the 6(6′)-position is situated *para* to the carbonyl group, the shift is evidently related to the strengthening or weekening of the acceptor property of this group.

This interpretation of the substituent effect is in line with conclusions reached by inspection of the frontier orbitals of indigo. As depicted in Fig. X.9, the HOMO possesses high expansion coefficients in the 5,5′-positions, the LUMO, however, in the 6,6′-positions. Due to the favoured HOMO-HOMO interaction with donor substituents in the 5,5′-positions and favoured LUMO-LUMO interaction with acceptor substituents in the 6,6′-positions the above mentioned bathochromic effects are conceivable.

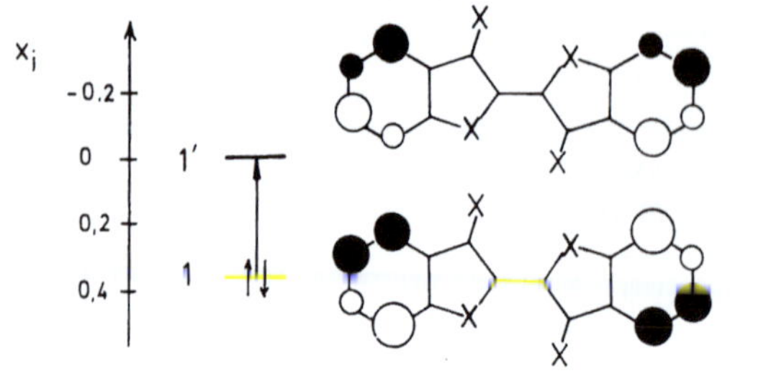

Fig. X.9. Lowest-energy one-electron transition and expansion coefficients of the frontier orbitals of indigo (cf. caption to Fig. VII.5)

Predictions concerning the substituent effects in the 4,4'- and 7,7'-positions are more complicated because bulky substituents cause steric hindrance and an alteration of solvent and aggregation effects [575, 583–585]. Comprehensive experimental data are only available for chloro-substituted derivatives [586]. Whereas red shift has been observed by chloro substitution in the 4,4'- and 5,5'-positions (5,5' > 4,4'), blue shift occurs in the 6,6'- and 7,7'-positions (6,6' > 7,7'). Based on incremental shifts the absorption wavelength data of tetra-, hexa- and octa-substituted chloroindigo dyes are sufficiently reproduced. 6,6'-Dibromo indigo, a main component of Tyrian Purple, displas anisotropy of the light absorption. The absorption at 640 nm is polarized parallel to the central C=C bond, while the absorption at 540 nm is perpendicularly polarized [586a]. The absorption of the various isomeric dibromo indigo dyes are discussed in Ref. [586b].

Also the effect of heteroatom replacement within the benzene rings of indigo has been studied [587]. Aza groups in the 5,5'- or 6,6'-positions cause similar spectral effects to those of acceptor groups at the same positions. In accordance with the lowering of electron density upon excitation in the 5,5'-positions and the increase of electron density in the 6,6'-positions (cf. Fig. X.9), blue shift occurs by replacing =CH by =N in the first case and red shift in the second case. Even the weak blue shift in the 4,4'-positions and the pronounced blue shift in the 7,7'-positions agree with conclusions drawn by virtue of first-order perturbation theory.

Due to the mainly localized nature of the indigo chromophore annelation brings about only small spectral shifts. This is exemplified for thioindigo dyes in Table X.7. However, perinaphtho-thioindigo *23* absorbs at considerably longer wavelength than thioindigo. Since its photoisomeric *cis*-form absorbs at more than 100 nm lower wavelengths, this compound exhibits pronounced photochromic properties [569]. The difference in absorption wavelength between the *trans* and *cis* form is even larger when naphthalene is replaced by pyrene [569a].

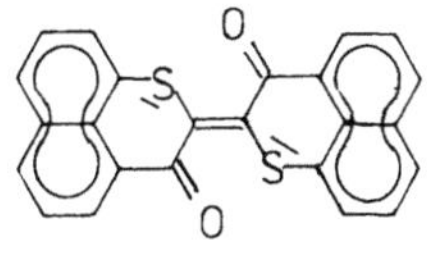

23

Table X.7. Longest-wavelength absorption maxima of thioindigo dyes and related compounds

Dye	Colour[a]	λ_{max} in nm
2b	red	505[b]
1b	red-violet[c]	547[c]
6,7,6',7'-dibenzo-*1b*	brownish-red	540[d]
4,5,4',5'-dibenzo-*1b*	bluish-red	565[a] 558[d]
5,6,5',6'-dibenzo-*1b*	bluish-violet	605[a]
23	deep-blue[c]	644[a]

[a] in tetrachloroethane [643]
[b] in chloroform [629]
[c] in chloroform [577]
[d] in chlorobenzene [644]

One of the pecularities of indigoid systems is the small change of color connected with prolongation of the central chain (cf. Table X.4). Insertion of one double bond even causes, in general, blue shift, which was attributed by HMO calculations to the particular electronic structure of the H-indigo chromophore [108, 588], or by PPP calculations, alternatively, to the spectral effect of the non-bonding S...O interaction [578].

Thioindigophanes have been synthesized recently [578a]. The color band does not show any excitonic splitting and appears at shorter wavelength than that of the corresponding alkyl-substituted thioindigo.

3. Indigo-like Compounds

The indigo parent chromophore can be mainly modified in two distinct ways, either by modifying of the conjugative external bridges or by cyclization of vinylogous derivatives.

The effect of conjugation can be estimated by perturbational considerations, similar as demonstrated in the previous sections. According to the FMO expansion coefficients of the parent chromophore and of the bridging substituent groups ($\overline{X}$, C=C, C=X) visualized in Fig. X.10 either stronger ground state or excited state stabilization may

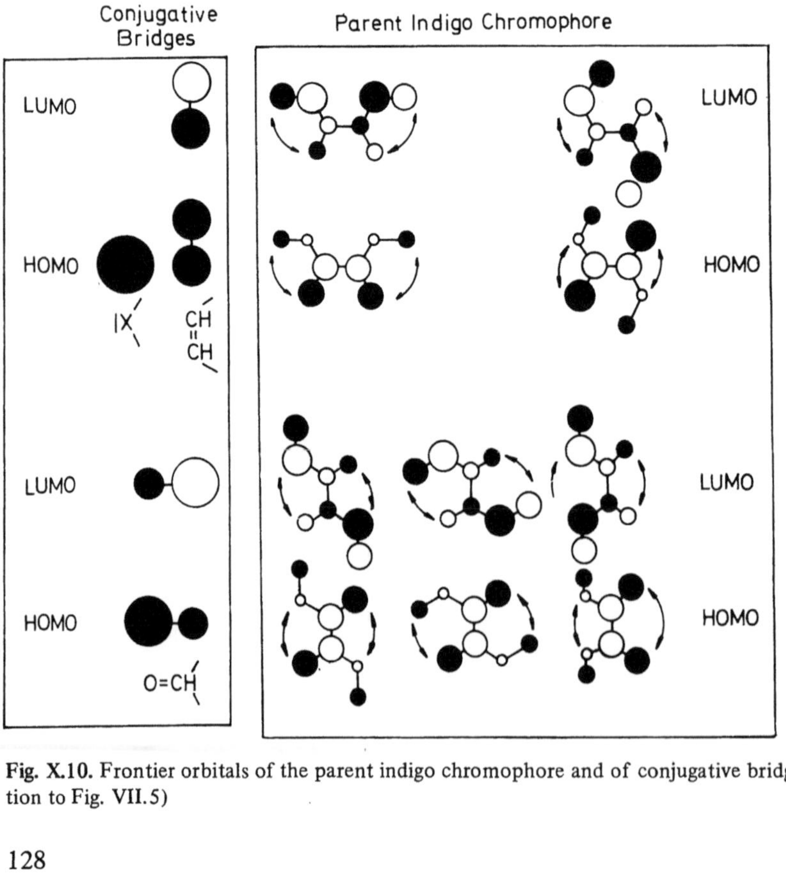

Fig. X.10. Frontier orbitals of the parent indigo chromophore and of conjugative bridges (cf. caption to Fig. VII.5)

occur depending on the nature of the conjugative bridge and the position of its linking. This should be reflected in hypsochromic and bathochromic shift, resp., with respect to the absorption of the parent chromophore.

Unfortunately, the known structures of this series are rather heterogenous and a comparison between prediction and experimental findings is possible only in few cases.

The compounds *24–27* are derived from the *trans*-configuration parent indigo chromophore.

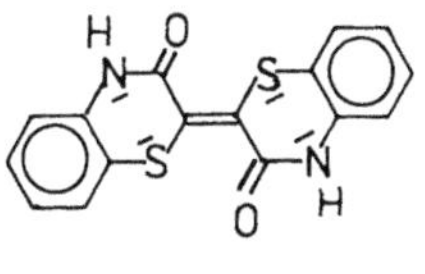

24

450
in dimethylsulfoxide [589]

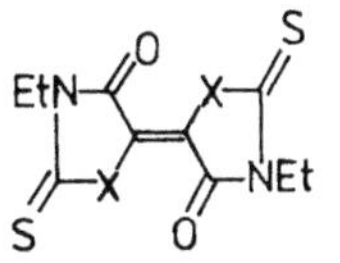

25

a: X = S: 449 (4.66) [590]
b: X = NHPh: 472 (4.46) [591]
in ethanol

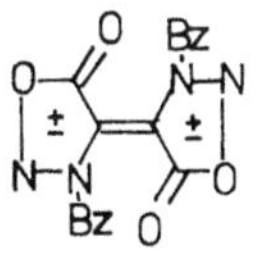

26

350 (4.17)
in chloroform [592]

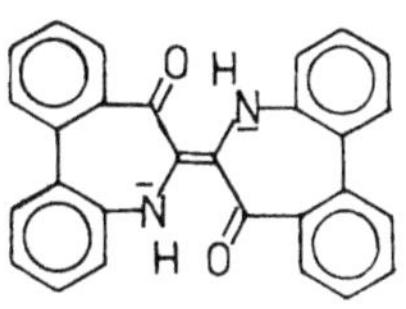

27

495 (3.67)
in chloroform [593]

The dye *24* absorbs at shorter wavelengths than thioindigo, but substituents exert similar effects to those found for indigo dyes [589].

In *28* and *29* the acceptor group C=N instead of C=O is incorporated in the conjugative bridges. Structures of the type *28* are natural pigments of human hair [594]. The absorption wavelength increases from 454 nm to 507 nm in passing from *28* to its vinylogous derivative [595, 596]. The chromophore of the orange pigment *29* is obviously related to that of *28*.

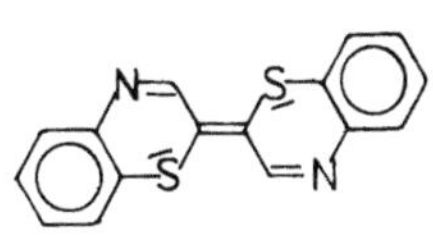

28

454 (4.34)
in dioxane [595]

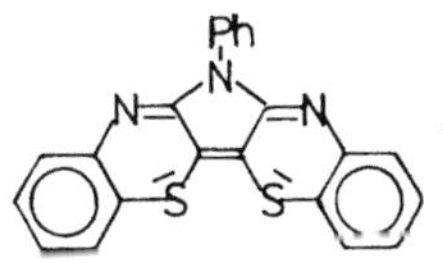

29

orange [597]

Fully nitrogen-containing indigoids of the same structural type are encountered in the dyes *30* and *31*. The endiamine *32* [600] differs structurally from the last mentioned compounds in that the C=N groups are involved in aromatic rings.

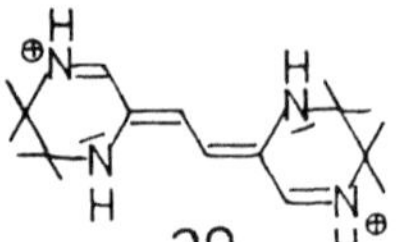

30

550
in puffer, pH 1 [598]

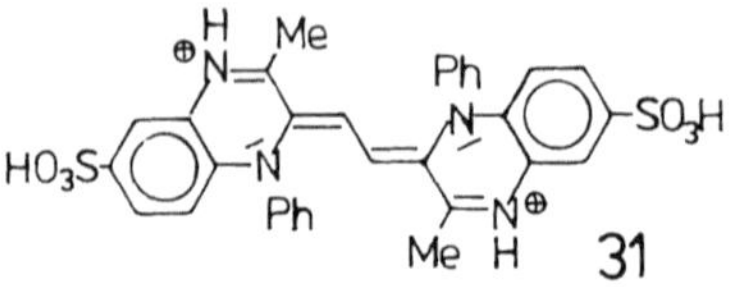

31

641
in 0.1 N HCl [599]

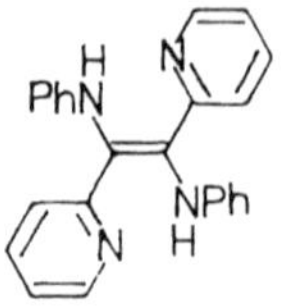

32

410
in Ref. [600]

The well studied yellow to blue aza heterocyclic endiols (OH instead of PhNH in *32*) are related to them. Based on the approximate coincidence of the calculated lowest excitation energies with that of the H-chromophore, they were also recently classed as indigo dyes [601]. A closer inspection of the wave functions shows, however, that the energies of the ground as well as the excited states are affected by external bridging [116]. In good accordance, marked spectral shifts occur with substitution or annelation of the pyridine rings [601].

The yellow color of *33* reflects the high ground state stabilization, which results from an effective HOMO-LUMO interaction between the fragment MO's of the benzene ring and the H-chromophore. The heteroanalogous sulfur compound absorbs at 448 nm [601a]. The spectral benzoannelation effect is similar to that found in the quinacridone series (cf. p. 87). On the other hand, compounds such as *34* and *35* belong to those structures which are predicted to be bathochromically shifted in comparison with the H-chromophore with corresponding configuration. Introducing a carbonyl group between the carbonyl groups of the H-chromophore or a direct linkage between the carbonyl carbon atoms should cause a hypsochromic shift. This has been observed for the squarate *37* in comparison with the sulphur containing H-chromophore (λ_{max}: 458 nm, cf. Table X.1), but not for the croconate *36*.

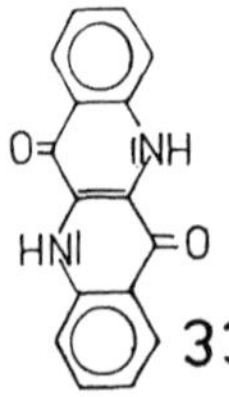

33

444 (4.37)
in dimethyl-
formamide [602]

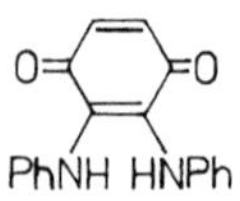

34

deep green [603]

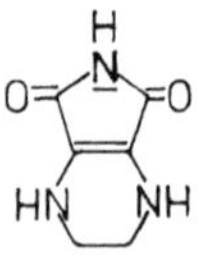

35

red [604]

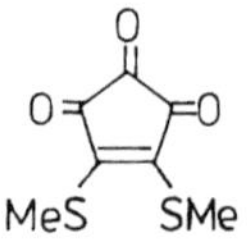

36

480
in methylene
chloride [605]

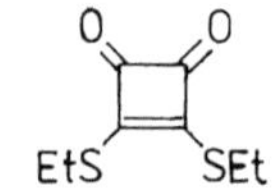

37

311 (4.36)
in cyclo-
hexane [606]

The spectral absorptions of isomeric indigo dyes of the general formula *38* have been studied recently in detail [574]. Formally, they are derived from the vinylogous indigo-H chromophores. They absorb between 400 and 550 nm depending on the kind of linkage (Table X.8).

$R_i = R_j$

38

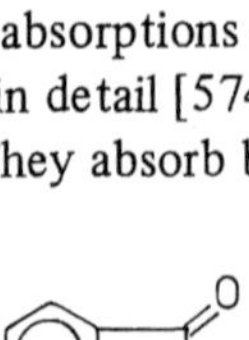

R_1

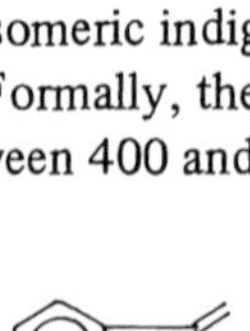

R_2

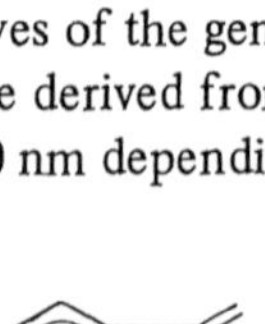

R_3

Table X.8. Longest-wavelength spectral data of isomeric indigoid dyes *38* (in dimethylsulfoxide) [574]

38	λ_{max} in nm (lg ϵ)		
	R_1	R_2	R_3
R_2	516 (4.16)	413 (4.47)	455 (3.85)
R_3	551 (4.04)	455 (3.85)	491 (3.66)

The symmetrical dyes *38* ($R_i = R_j = R_2, R_3$) are derivatives of the parent structures *13* and *14*. Two double bonds of *13* and *14* are replaced by benzene rings and, moreover, the terminal groups are joined. This cyclization is connected with a hypsochromic shift. The shift results from the fact that the donor and acceptor parts of the molecule are directly linked. The introduction of these bonds stabilizes the ground state more strongly than the lowest excited state. This conclusion can easily be drawn by perturbation theoretical arguments, examining the energy change of the frontier orbitals of *13* or *14*.

The *trans* structure of isoindigo (*38*: $R_i = R_j = R_3$) and of indirubin (*38*: $R_i = R_1$, $R_j = R_3$) has been confirmed by X-ray studies [607]. Spectral data of substituted *38* have been reported in Ref. [583].

The effect of heteroatom substitution has been demonstrated with *39* [608]:

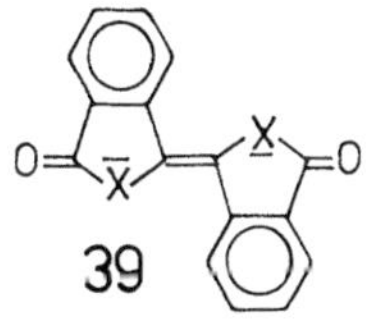

39

a	X = O:	362 (4.46), 375 (4.38)
b	X = S:	392 (4.46), 406 (4.47)
c	X = NH:	393 (4.54), 413 (4.47)
d	X = NMe:	432 (4.48)

in dimethylsulfoxide [608]

The wavelength shift reflects the increasing donor effect of the heteroatoms in the sequence O < S < NH < NMe. Additionally, bathochromic shift occurs by replacement of the exocyclic oxygen of *39b* and *39c* by sulphur. Absorption maxima at 468 nm (in methylene chloride) [609, 610] and 518 nm (lg ϵ 4.68) (in dimethylsulfoxide) [608] have been recorded for the thio analogues of *39b* and *39c*, resp. Deprotonation of dithio-*39c* results in a red shift (573 nm, lg ϵ 4.57) (in dimethylsulfoxide) [608]. The diimino-*39c* described in Ref. [611] exists in the tautomeric enamine form *10* [608]. The longest-wavelength absorption maxima of derivatives of diimino-*39c* range from about 450 to 520 nm [608, 612]. The results of PPP calculations suggest that, in contrast to indigoid systems, the benzene fragments contribute markedly to chromophoric properties of *39c*. [613]

The isoindigoid systems *40* result from bridging a polyenic chain by two sulphur atoms. In this context the likewise sulphur bridged indophenine *41* should be mentioned. The molecular structure of *41* has been established recently [614]. The related Mesoxophenine absorbs at 536 nm (lg ε 4.87) in chloroform [616].

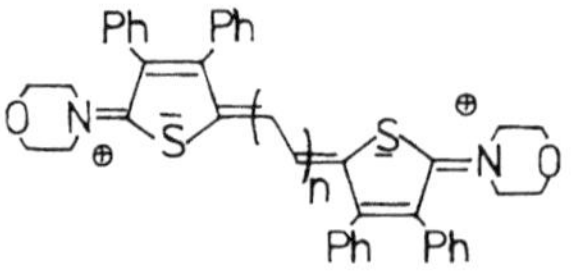

40

n = 0 470 (4.60)
n = 1 513 (4.79)
in acetonitrile [610]

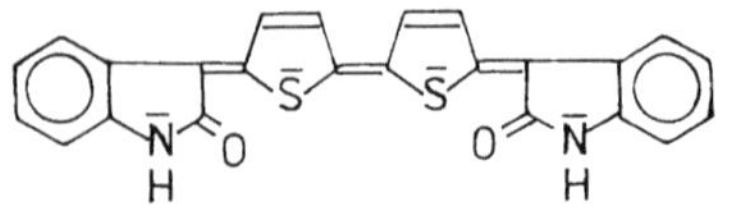

41

blue in acetic acid [615]

Related to isoindigo are also Pechmann's dyes such as *42* and *43*, and Pyrazole Blue *44*.

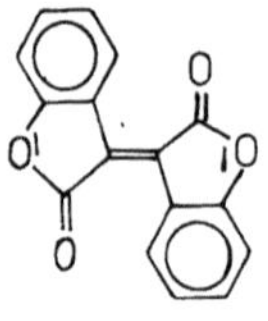

42

orange [618]

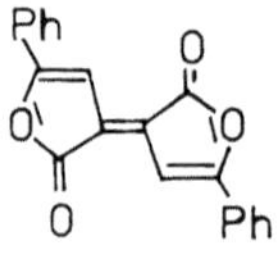

43

red [617]
(500, 535)
in toluene

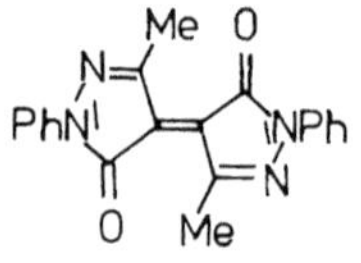

44

blue
(600 in benzene [619]

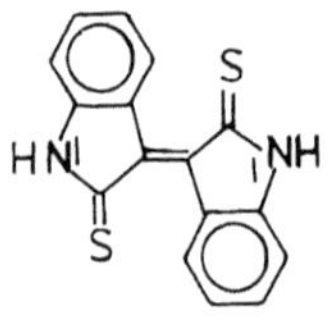

45

green [620]

Red shift should occur on passing from isoindigo to the thiocarbonyl derivative. A green colored compound was assigned the formula *45* [620], but the structure does not yet seem to be settled.

In accordance with the more general conception of the indigo chromophore, *ortho*-dipheno-naphthoquinone *46* (known as "Russig's indigo" [621], should be mentioned.

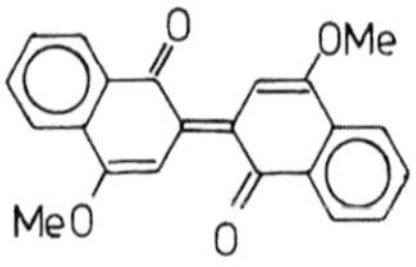

46

633
in benzene [622]

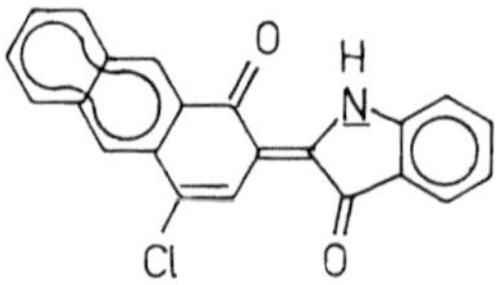

47

652
in chloroform [616]

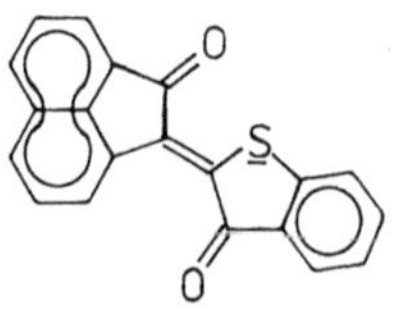

48

512 (4.05)
in benzene [619]

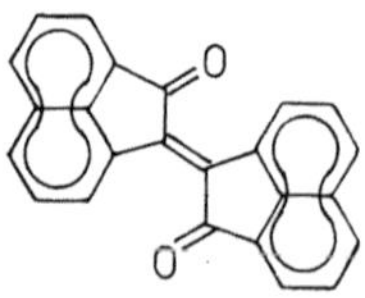

49

476 (3.68)
in benzene [623]

The paramagnetic susceptibility, however, led most recently to the conclusion that the deep blue color arises rather from a stable biradical than from the non-radical *46* [624].

Further compounds on the borderline between indigoid and quinoid dyes are *47* and *48*. The longest-wavelength absorption maximum of *48* is approximately the mean between the wavelength of symmetrical thioindigo *1b* and the quinone dye *49*. Long-wavelength absorptions have been found for the so-called pyridine dyes, such as *50a*. According to Ref. [622] the hydroxy-compounds *50b* and *51* are chromophorically related.

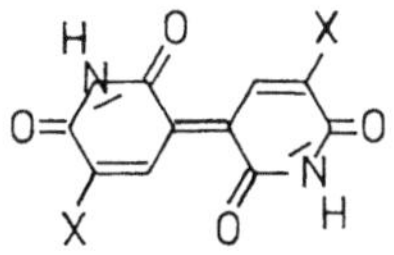

50

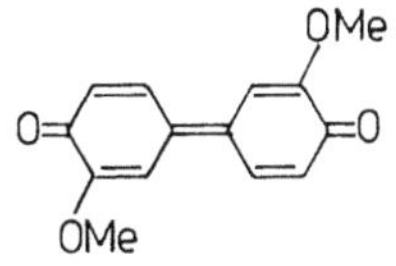

51

a X = NH_2	612 in dimethylsulfoxide [622]	462 in methanol [622]
b X = OH	478 in tetrahydrofuran [622]	

Aurone (*52*, X = O) and heteroanalogues, however, may be considered as merocyanines rather than as indigoid dyes.

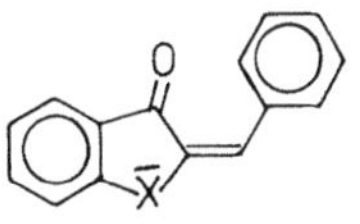

52

X = O 376 (4.20) [625]
X = S 434 (4.07)
X = Se 440 (3.96)

The spectra of *52* and derivatives have been studied experimentally as well as theoretically [625, 626].

4. Medium Effects

It is well known that the color of indigo changes from red to blue depending on the experimental conditions. Indigo exhibits red color in chloroform, and blue color in more polar solvents or in the solid state [627]. The change in color has been substantiated by spectral data [628]. Similar solvent effects have also been found for heteroanalogous indigo dyes [628] and smaller indigoid systems derived from them [570, 629, 630]. The absorption curve of the indigo does not only differ between solution and the solid state but also depends on the nature of the solid phase. Crystalline indigo absorbs at remarkably longer wavelength than amorphous indigo [631] (Fig. X.11).

This extremly large shift observed for indigo itself raised the question as to the importance of the intra- and intermolecular hydrogen bonds according to formulae *53* and *54*, resp.

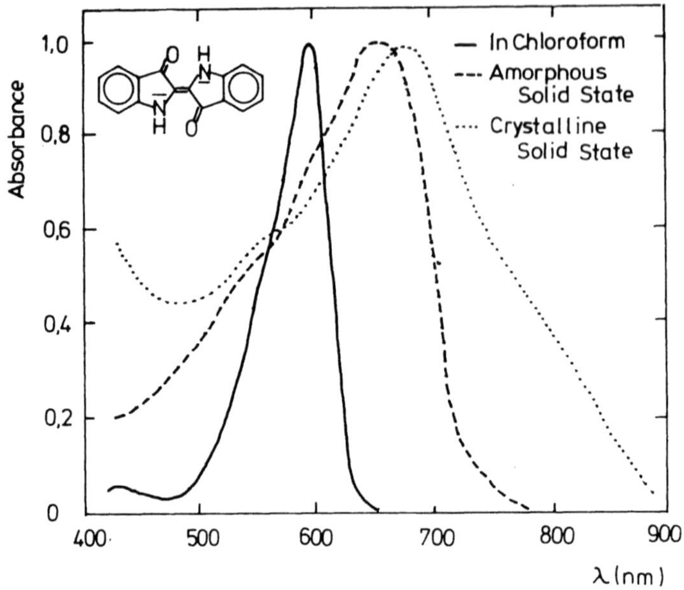

Fig. X.11. Spectral absorption curves of indigo in chloroform and in the amorphous and crystalline solid state [631]

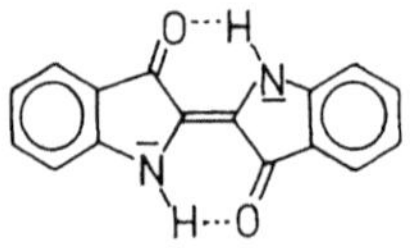

53

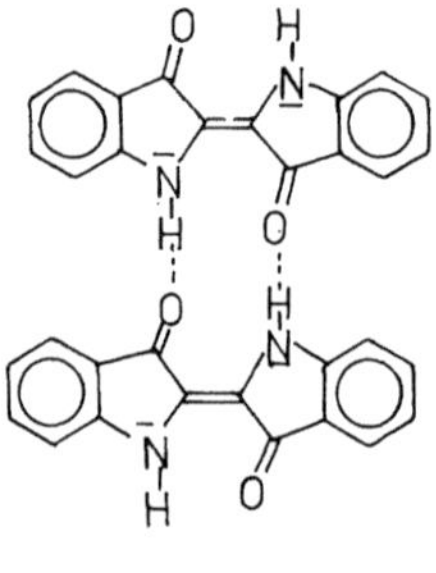

54

HMO calculations suggest that the dimer *54* should absorb at longer wavelengths than the monomer [631]. IR studies clearly support the existence of intermolecular bridges in indigo dyes except those which contain bulky 7,7′-substituents [575, 628].

The extreme long-wavelength absorption of indigo in the crystalline state has been attributed to the fact that the molecules are completely H-bonded [631]. Later experimental results, however, cast some doubt on the importance of intermolecular hydrogen bonding for explaining the strong red shift observed in transition from the vapour phase to the condensed phases: Firstly, even in the vapour state, where the colour is considerably brightened, the association is not completely removed [632]. Secondly, C-alkyl-substituted indigo dyes which have a sufficient solubility in non-polar solvents seem not to undergo dimerization in cyclohexane but absorb nevertheless in the 600 nm region [588].

The slight positive solvatochromism observed for alkyl-substituted indigo dyes in passing from non-polar solvents to polar ones has been attributed to dispersion forces between the solute and the solvent molecules [588]. The shift is stronger for indigo than for thioindigo dyes and increase in passing to the vinylogous systems.

The influence of the intramolecular hydrogen bonds on the color is not proved with certainty [588, 631]. The comparison with the N-methylated derivatives and with vinylogous systems, which cannot have such bonds, did not allow a convincing conclusion about these bonds in the indigo molecule since the structural modifications are also accompanied by spectral shifts. Whereas a hypsochromic shift has been anticipated in Ref. [631] for theoretical reasons, a bathochromic shift of 400 cm^{-1} was estimated in Ref. [588].

Indigo dyes show a marked change in colour when dissolved in sulphuric acid [36]. Obviously they undergo protonation. The monoprotonated indigo shows a deeper color than the unprotonated compound [580]. Less is known about the nature of the protonated species. Thioindigo dyes are assumed to undergo a double protonation in concentrated sulphuric acid, giving rise to cations auch as *55* [633] and *56* [629].

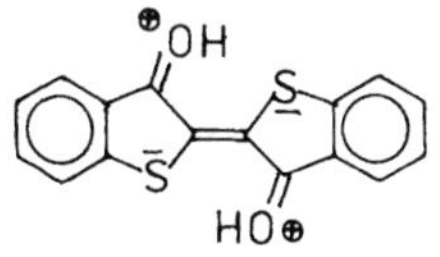

55

641
in sulphuric acid [633]

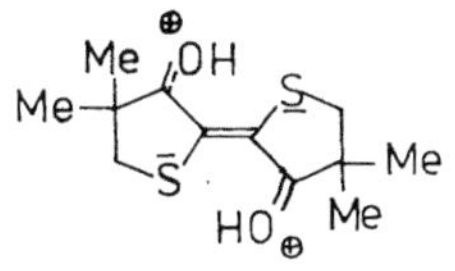

56

618
in sulphuric acid [629]

The strong red shift may be explained by the increased acceptor capacity of the carbonyl group. In contrast to diprotonated thioindigo the dye *56* isomerizes to the *cis*-form, which absorbs at 529 nm [629]. This process is accompanied by a striking change of color from deep-blue to red. Also the protonation of the selenium analogues of *56* has been studied [630].

Pronounced red shift also occurs if indigo dyes are treated with strong alkali. The mono-deprotonated indigo *57* is of green color (773 nm in tert-butanol/Na tert-butoxide) [622].

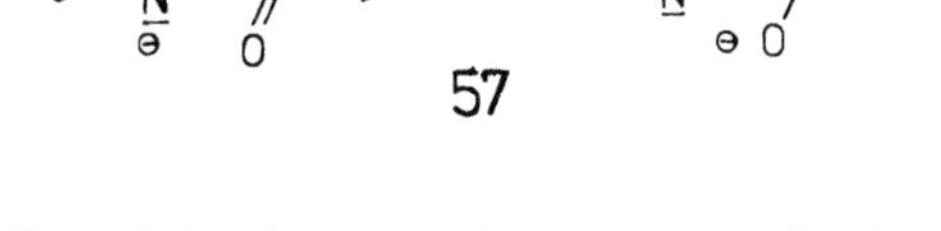

57

One of the characteristic properties of indigo *1a* and of some related systems is their easy reduction or oxidation to leuco- (*58*) and dehydro-indigo (*59*), resp.

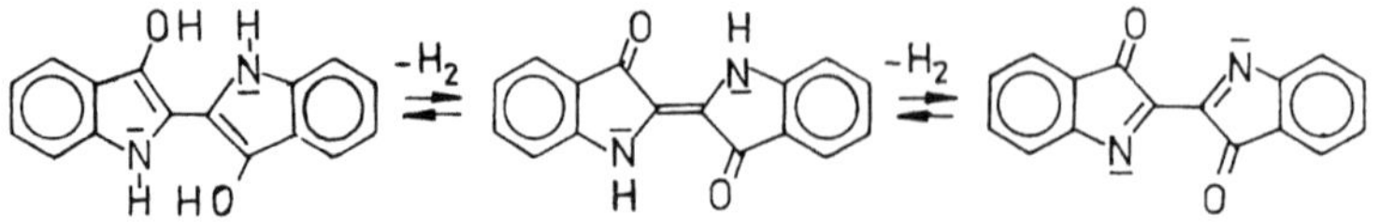

58 1a 59

304.5 in ethanol (N,N'O,O'-tetra-methyl-derivative) [108]

442.5 in ethanol [108]

The blue color of indigo disappears in both processes. The reduction process leads to the leuco compounds which are applied for dyeing or printing. Reoxidation yields the original dye on the respective substrate.

Attempts have been made to describe theoretically the short-wavelength shift on passing from indigo to *58* or *59* as well as the preferential tendency of indigo to undergo reduction [108, 634].

XI. Diphenylmethane, Triphenylmethane and Related Dyes

Various compounds considered in this section belong to the earliest synthetic dyes which were discovered in the nineteenth or early twentieth century. The yellow, red, violet, blue or green dyes are characterized by extreme brilliance and high color depth, but due to low fastness to light they partly lost their importance in dyeing of wool, silk or cotton. Some of them, however, are still manufactured for coloring paper and leather, printing inks, indicators in analytical chemistry and various other applications [31, 245, 246]. More recently they have found application in textile technology in dyeing polyacrylonitrile and its copolymers.

1. Chromophoric System

Diphenyl and triphenylmethane dyes are derived from bis-phenylmethyl cation *1* or bis-phenylmethyl anion *2*, and tris-phenylmethyl cation *3* or tris-phenylmethyl anion *4*, resp.

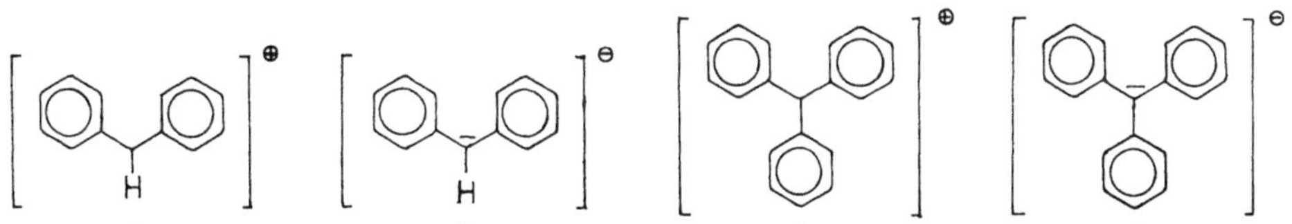

1 2 3 4

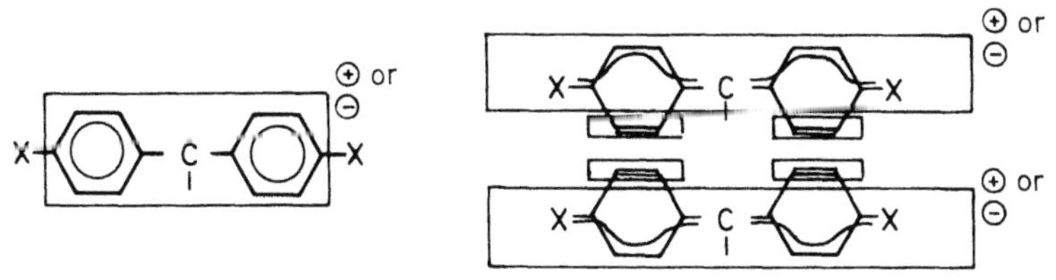

Model A Model B

Fig. XI.1. Alternative fragmentation modes for describing the electronic structure of diphenylmethane dyes

The topologically equivalent ions *1* and *2* as well as *3* and *4* differ only in the number of π electrons.

These ions can be considered as the basic chromophores of various more complex dye structures. The interpretation refers to model *A* in Fig. XI.1.

An alternative fragmentation mode *B* is favourable for donor-substituted diphenylmethane and triphenylmethane dyes and has been revived more recently [23]. According to model *B* dyes are thought to contain straight-chain or branched-chain polymethine subchromophores, which are modified by ring closure [23, 60, 645], as schematically illustrated in Fig. XI. 1.

This kind of modelling enables revelation of the close similarity between diphenyl and triphenylmethane dyes and polymethine dyes. This relationship is especially striking in the vinylogous diarylmethane dye series, in which the polymethine structure predominates.

2. Unsubstituted Monoaryl, Diaryl and Triarylmethane Dyes and Vinylogues

The spectra of the yellow cationic parent compounds Ph_2CH^+ (*1*) and Ph_3C^+ (*3*) are both characterized by a sharp and intense absorption band in the 400 nm region (Fig. XI.2).

Theoretical studies revealed that the color band of *1* comprises three $\pi \rightarrow \pi^*$ transitions, whereas five $\pi \rightarrow \pi^*$ transitions are responsible for the longest-wavelength absorption of *3* [646, 647] (cf. Fig. XI. 2).

π Electron calculations assign the lowest-energy transitions in *1* to electron jumps from the highest occupied orbitals into the unoccupied ("non-bonding") orbitals (LUMO). The simple Hückel model predicts the HOMO of *1* to be triply degenerate. In the PPP model, this is still approximately true (cf. Fig. XI.2). One of the three approximately degenerate transitions is weak and short-axis polarized (y), whereas two are long-axis (x) polarized. A fivefold degeneracy is predicted for the lowest excited states of *3* on the Hückel level. Since *3* has a propeller-shaped geometry [648], in good agreement with theoretical predictions [649], PPP calculations provide two forbidden transitions, two transitions polarized along the threefold axis, and a degenerate transition polarized in a plane perpendicular to this axis [647, 650].

In spite of the larger number of π electrons in Ph_2CH^- (*2*) and Ph_3C^- (*4*) in comparison with Ph_2CH^+ (*1*) and Ph_3C^+ (*3*), the differently charged ions absorb at similar wavelengths (cf. Fig. XI.2). This result is in good agreement with theoretical conclusions. Since *1* and *2* (and correspondingly *3* and *4*) belong to odd-alternant hydrocarbons, characterized by a non-bonding MO (NBMO) and the remaining MOs in pairs (pairing theorem), the transitions between occupied MOs and the NBMO (cation) and between the NBMO and unoccupied MOs are equivalent (cf. Fig. XI.3). The experimentally found slightly higher excitation energies of cations in comparison to anions were explained by charge effects [650].

Absorption data of some representative bis-arylmethyl and tris-arylmethyl cations are compiled in Table XI.1. Further data are available in the literature [651, 652]. Ob-

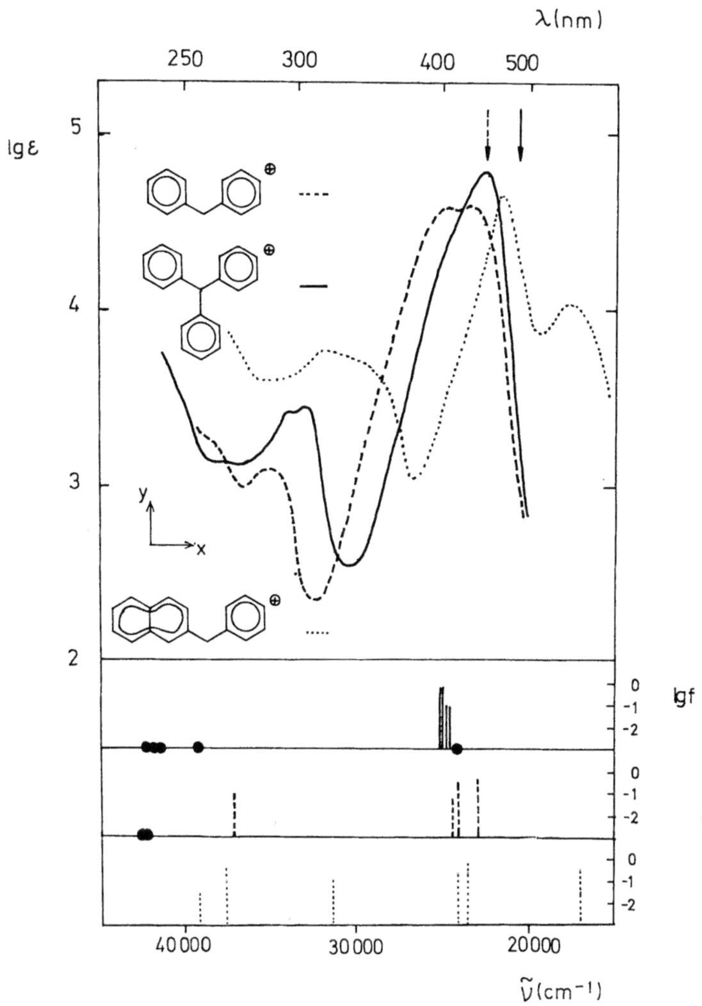

Fig. XI.2. Absorption spectra of the bis-phenylmethyl [39], phenyl-2-naphthyl-methyl [653] and tris-phenylmethyl [39] cations in sulphuric acid and results of PPP calculations. The arrows indicate the position of the color bands of the bis-phenylmethyl and the tris-phenyl-methyl anions [660]

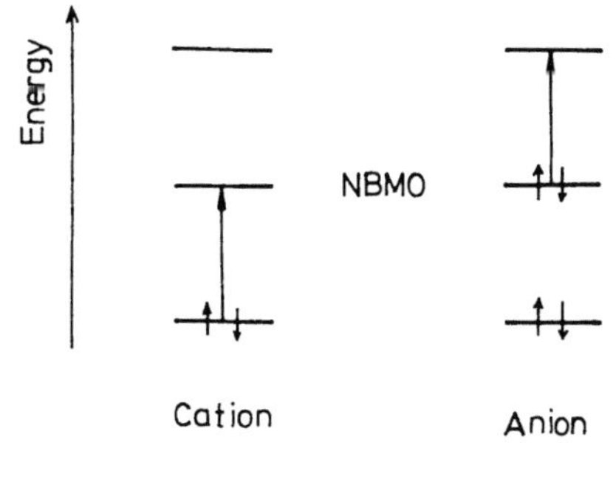

Fig. XI.3. Frontier orbital levels and one-electron transitions of odd alternant hydrocarbons

Table XI.1. Longest-wavelength absorption maxima of mono-, di-, and tri-arylmethyl cations $R_1R_2R_3C^+$

R_1	R_2	R_3	λ_{max} in nm (lg ϵ)	Solvent	Ref.
phenyl	methyl	methyl	390 (3.15)	FSO_3/SbF_5	[657]
phenyl	phenyl	H	440 (4.58)	FSO_3/SbF_5	[657]
phenyl	phenyl	phenyl	429 (4.59)	FSO_3/SbF_5	[657]
1-naphthyl	phenyl	H	572 (4.37)	H_2SO_4	[653]
2-naphthyl	phenyl	H	568 (4.00)	H_2SO_4	[653]
9-anthryl	phenyl	H	666 (4.17)	H_2SO_4	[653]
3-phenanthryl	phenyl	H	588 (4.51)	H_2SO_4	[653]
1-pyrenyl	phenyl	H	641 (4.43)	H_2SO_4	[653]
p-biphenylyl	phenyl	phenyl	507 (4.74)	$HClO_4$	[650]
p-biphenylyl	p-biphenylyl	phenyl	538 (4.86)	$HClO_4$	[650]
p-biphenylyl	p-biphenylyl	p-biphenylyl	541 (4.93)	$HClO_4$	[650]

viously, progressive red shift occurs with increasing extensions of the aromatic system. The descent in excitation energy is quite well reproduced by HMO transition energies, and linear correlations have been obtained both for cations [646, 653, 654] and anions [655].

Interestingly, the color band splits in passing from the symmetrical to unsymmetrical bis-arylmethyl cations. This is exemplified in Fig. XI.2 by juxtaposing the spectra of the bis-phenylmethyl and phenyl-2-naphthyl-methyl cation. The results of the PPP calculation show that the three lowest-energy transitions are no longer degenerate in the latter case. Exploration of the nature of the three electronic transitions by analyzing their wave functions (OKK analysis) or squared transition density matrix (LSU analysis) has revealed that the first and second transition are mainly localized at the naphthylcarbenium fragment, whereas the third transitions predominantly involves the phenylcarbenium fragment [116].

Based on the HMO approximation, Mason and Grinter derived a simple relationship between the position of the color band and the number and nature of participating aryl residues [650]. It follows from the frontier orbitals, which are depicted in Fig. XI.4. Since all the cations considered are related to odd-alternant hydrocarbons, which have an unoccupied NBMO, the color band results from electronic excitation from the HOMO to the NBMO. Because the HOMOs are nodal at the central carbon atom for the structures Ar_3C^+, Ar_2C^+H and Ar_2C^+Ph, their HOMO energies are equal to those of the hydrocarbon residue ArH. Thus, within the model employed, the transition energies of these cations should be equal. Moreover, they are expected to be half that of the hydrocarbon ArH corresponding to the respective aryl fragment Ar. The cation ArC^+H_2, ArC^+HPh, ArC^+Ph_2 should, however, absorb at higher energies as those bearing two or three aryl residues.

As exemplified in Table XI.1 (R = biphenyl) this is actually observed. The mono-biphenylmethane dye is less bathochromic than the bi- and tri-biphenylmethane dyes. The two last mentioned dyes hardly differ in the position of the colour band. On the other hand, whereas biphenyl absorbs at about 250 nm ($\tilde{\nu}_0$ = 40,000 cm^{-1}), the respec-

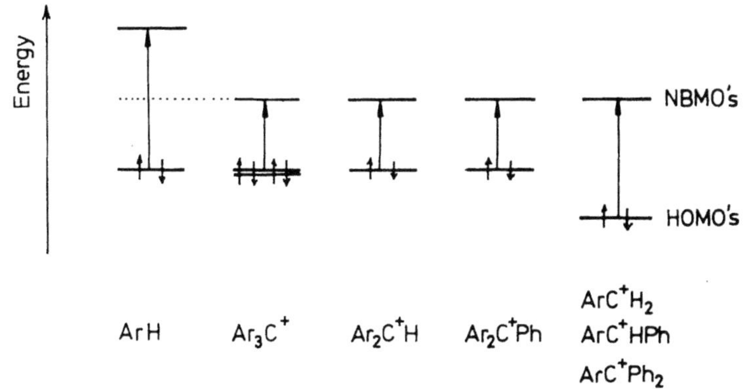

Fig. XI.4. Frontier orbital levels and one-electron transitions of some bis-arylmethyl and tris-arylmethyl cations

tive cations with two or tri biphenyl residues absorb in the 500 nm region ($\tilde{\nu} = 20{,}000\ \mathrm{cm}^{-1} = \tilde{\nu}_0/2$).

Rigidized bis-phenylmethyl cations, such as *5* to *8*, result when acenes or phenes are dissolved in strong mineral acids.

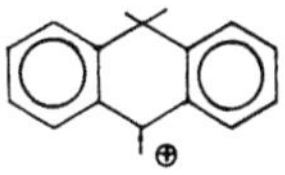

5

408 (4.58)

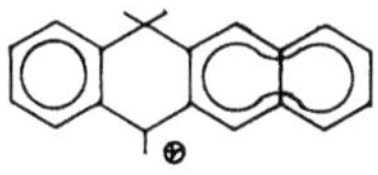

6

448 (4.70)

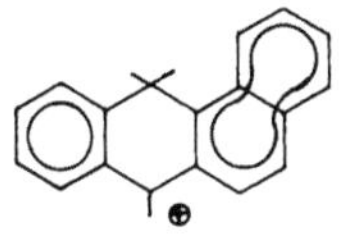

7

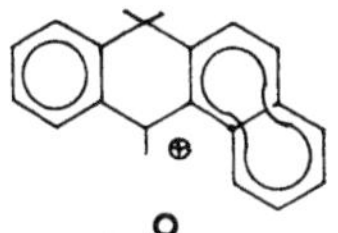

8

527 (4.25)
in hydrofluoric acid [656]

Whereas the nature of the strong acid is of secondary importance for the spectra of arylmethyl cations, the absorption wavelengths of the corresponding anions vary greatly with the change of the solvent or with the change of the gegenion. Thus, the influence of ion pairing and solvation forces upon the transition energies are more marked in the arylmethyl anions than in the corresponding cations [650, 657–659]. If the particular medium effect has not been taken into account theoretically, semiempirical methods reproduce spectral data of cations more closely than those of anions [650, 653, 660].

The solvent effect may also account for the different vinylene shift of the color band in the series *9* and *10* (Table XI.2).

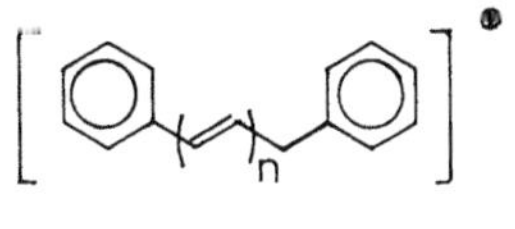

9

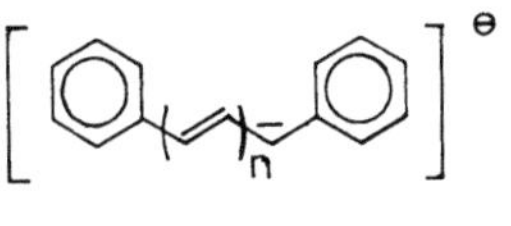

10

A more recent theoretical study [661] has shown that the observed spectral differences between *9* and *10* are not due to the fact that the pairing theorem is not valid

Table XI.2. Longest-wavelength absorption maxima of bis-phenylmethyl ions and vinylogues

Compound	λ_{max} in nm (lg ε) n: 0	1	2	3	4	5	6	Ref.
9	441 (4.63)	498 (4.88)	559 (5.10)	610 (5.11)	660 (5.11)	714 (5.11)	758 (> 5.11)	a
10	501	535	568	600	635			b
16 X = OMe	512 (5.10)	581 (5.29)	655 (5.33)	725 (5.40)	792 (5.46)			c
16 X = NMe_2	610	693	790	883				d

[a] in sulphuric acid [747]
[b] in ether [748]
[c] in methylene chloride [749, 658]
[d] in acetic acid [691]

within more elaborate PPP approximations [662]. A better understanding of the solvent effects is probably required.

The cations *11* to *13* can be considered as containing the cationic diphenylmethane or triphenylmethane chromophore twice or three times.

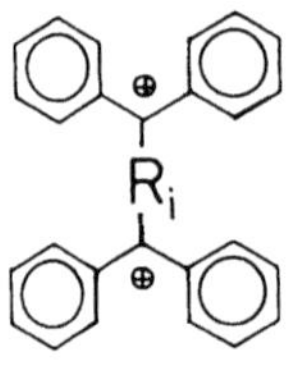

11

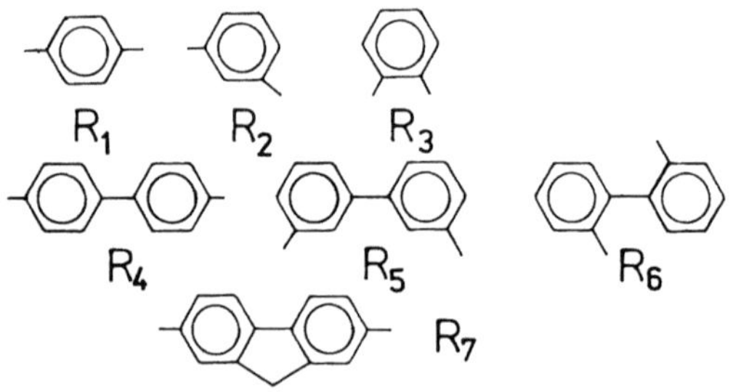

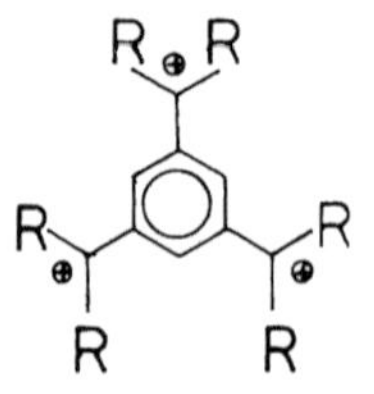

12

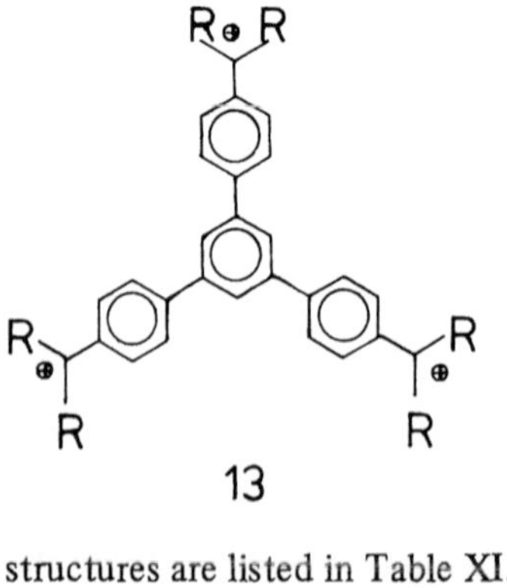

13

Some representative data of such structures are listed in Table XI.3. Insertion of benzene between the two charged carbon atoms in *11* ($R_i = C_6H_4$) and phenyl substitu-

Table XI.3. Spectral data of di- and trications *11–13* obtained in sulphuric acid

Compound	R_i or R	λ_{max} in nm (lg ε)			Ref.
11a	R_1	455 (4.77)			[750]
11b	R_2	440 (4.85)	419 (4.86)		[750]
11c	R_3	455 (4.64)	373 (4.49)		[750]
11d	R_4	530 (4.95)	442 (4.73)	410 (4.54)	[654]
11e	R_5	430 (4.83)			[654]
11f	R_6	428 (4.55)	390 (4.38)		[654]
11g	R_7	568 (5.18)	525 (4.66)	440 (4.54)	[654]
12a	C_6H_5	446 (4.9)	394 (4.8)		[751]
12b	p-C_6H_5-C_6H_4	592 (5.2)	487 (4.9)		[751]
13	C_6H_5	553 (5.3)	500 (5.4)		[751]

tion in *12* (R = C_6H_5) bring about only slight red shift in comparison with the monochromophoric dyes. For steric reasons, the same holds upon insertion of o,o'- or m,m'-disubstituted biphenyl into *11* (R_i = R_5, R_6). The p,p'-substituted biphenyl derivative (R_i = R_4) absorbs at considerably longer wavelengths, however. Additional red shift occurs when the biphenyl moiety is constrained to a planar structure, as encountered in the fluorenyl-bridged *11* (R_i = R_7).

3. Substituted Arylmethane Dyes and Heterocyclic Analogues

Replacement of hydrogen atoms in the benzene rings of bis-phenylmethyl cations and vinylogues, or in the tris-phenylmethyl cation, by heteroatomic groups X leads to the substituted arylmethane dyes *14* to *16*.

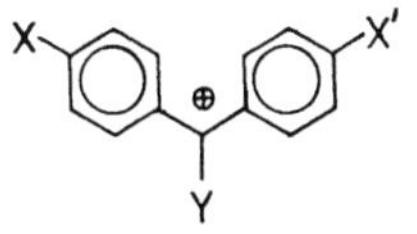

14

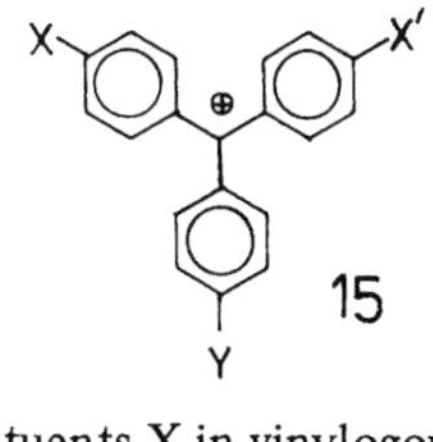

15

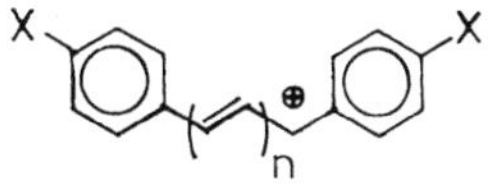

16

Introduction of donor substituents X in vinylogous diphenylmethane cation (*16*) gives rise to a marked red shift, increasing in the sequence H < OMe < NMe_2 and amounting to more than 150 nm with NMe_2 substitution (Table XI.2). An approximate 100 nm vinylene shift has been observed as in the series of vinylogous polymethine dyes. Recently, the experimental data have been reproduced and discussed by PPP calculations [662a].

Increasing bathochromic shift has also been observed with introduction of one donor substituent (X) or two donor substituents (X, X') in the tris-phenylmethyl cation (*15*, Y = H). If, however, a third donor substituent Y is introduced into *15*, hypsochromic shift occurs. Some examples are given in Table XI.4.

Table XI.4. Longest-wavelength absorption maxima of *para*-substituted tris-phenylmethyl cations *15* [39, 752]

X	X'	Y	λ_{max} in nm
H	H	H	431
NH_2	H	H	485
NH_2	NH_2	H	578
NH_2	NH_2	NH_2	538
NMe_2	H	H	490
NMe_2	NMe_2	H	620
NMe_2	NMe_2	NMe_2	585
NPh	H	H	502
NPh	NPh	H	633
OMe	H	H	475
OMe	OMe	H	501
OMe	OMe	OMe	486
SMe	H	H	547
SMe	SMe	H	594
SMe	SMe	SMe	586

Table XI.5. Longest-wavelength absorption maxima of bis- and trisphenylmethyl cations depending on the substituents Y (X,X' = NMe_2)

Y	λ_{max} in nm (lg ϵ)		
	14[a]	*15*[c]	
	x Band	x Band	y Band
H	607.5 (5.17)	621 (5.02)	427.5 (4.30)
Me	606 (4.04)	616.5 (5.03)	437.5 (4.40)
F	588 (4.53)	620 (5.02)	430.5 (4.32)
Cl	626 (4.80)	627.5 (5.02)	433 (4.34)
Br	630 (3.78)	628 (5.02)	435 (4.34)
J	–	628 (5.02)	437.5 (4.40)
OH	–	602.5 (5.03)	470 (4.56)
OMe	–	608 (5.03)	465 (4.83)
SMe	592 (4.68)	618 (5.01)	492 (4.56)
NH_2	433 (4.76)[b]	–	–
NMe_2	–	590	–
CF_3	693 (4.34)	645 (4.92)	425 (4.07)
CN	716 (4.73)	–	–
COOEt	646 (4.65)	–	–

[a] in different solvents [753] [b] in sulphuric acid [754] [c] in acetic acid [672]

Similar substituent effects have been observed for the Y-substituted dyes *14* and 15 (Table XI.5). Whereas the longest-wavelength absorption is increasingly blue-shifted with increasing donor capacity of the substituent, red shift results from acceptor substitution.

The spectral data of *15* reveal a further characteristic feature: If the substituents X and Y are different, two absorption bands are, in general, observed in the visible region. Whereas the longest-wavelength absorption of *15* appears nearly in the same wavelength region as those of the corresponding solely X-substituted diphenylmethane and triphenylmethane dyes ("x band"), the introduction of Y gives rise to a new absorption band at shorter wavelengths ("y band") [2].

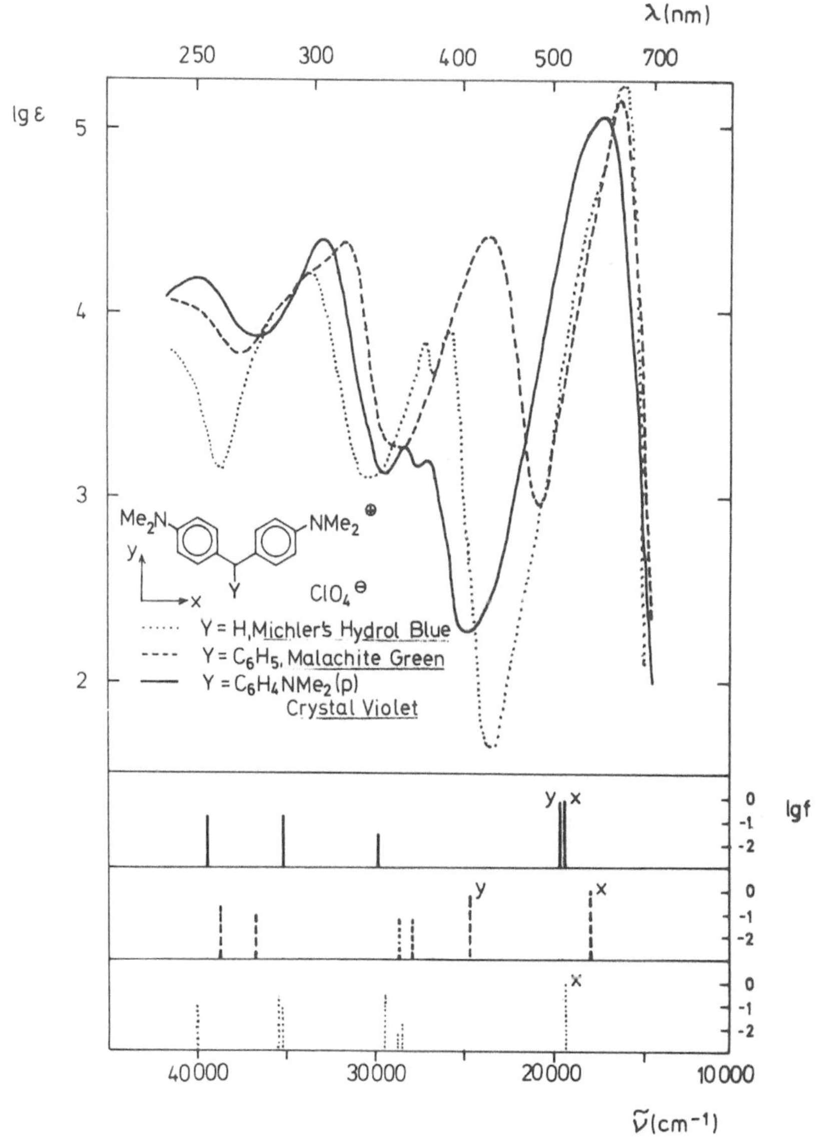

Fig. XI.5. Absorption spectra of Michler's Hydrol Blue [616], Crystal Violet [39] and Malachite Green [39] in chloroform and results of PPP calculations [116]

This pecularity is illustrated in Fig. XI.5 where the spectral curves of Michler's Hydrol Blue *14* (X = NMe_2, Y = H), Crystal Violet *15* (X = X' = Y = NMe_2) and Malachite Green *15* (X = X' = NMe_2, Y = H) are juxtaposed.

The appearance of color bands at nearly the same position for all of the three dyes is in line with the Mason Grinter rule discussed in Sect. XI.2. Also the distinct substituent effect of Y on the x and y band of *15* can be readily explained in the simple MO picture. Along these lines the x band is approximately described by the $1 \rightarrow 1'$ transition and the y band by the $2 \rightarrow 1'$ transition (Fig. XI.6). Due to the nodal properties of the HOMO no interaction occurs between HOMO and the donor orbital. The x band is consequently not affected. The subjacent occupied MO, however, is capable of overlapping, thus giving rise to an pronounced substituent effect on the y band.

Inspection of MO's involved into the two electron transitions shows, moreover, that the x band of Malachite Green and related dyes is polarized perpendicular to the twofold axis (x axis in Fig. XI.5), whereas the shorter wavelength y band is polarized along this axis (y axis). This conclusion is in line with polarization measurements [497, 663–667]. A more recent polarization study on Crystal Violet and Parafuchsin *15* (X = X' = Y = NH_2) showed, however, that the symmetrically substituted molecules do not have the regular propeller-shaped geometry [667]. One of the aminophenyl rings is probably more twisted out of the molecular plane than the others.

For sake of brevity only few spectral data are selected in the following from data of a great number of substituted diphenylmethane and triphenylmethane dyes described in the literature.

Introduction of additional donor groups into the phenyl rings in passing from bis- to tetrakis-methoxy-substituted diphenylmethane dyes *17* gives rise to red shifts as expected by Kaufmann's distribution rule [668], i.e. increasing red shift in the order: 4 < 2,4 < 2,3 < 2,5 (cf. Table XI.6).

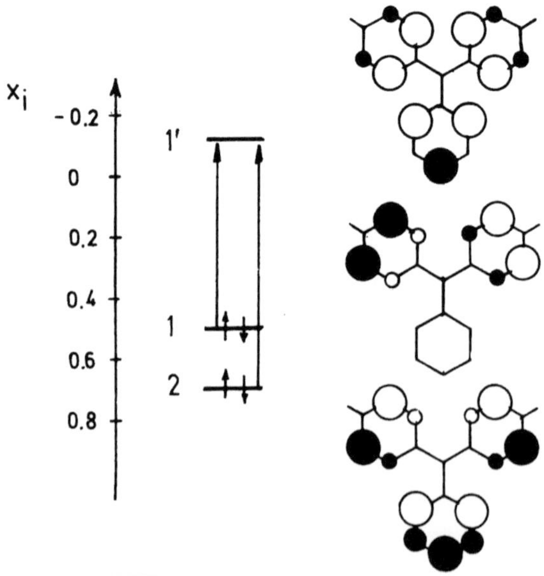

Fig. XI.6. Lowest-energy electronic transitions of Malachite Green and expansion coefficients of the frontier orbitals at the *meso*-phenyl group (cf. caption to Fig. VII.5)

Table XI.6. Spectral effect of singly and multiply methoxy-substitution of the *meso*-methyl-diphenylmethyl cation *17* [668] and of Malachite Green *18* [669]

Substitution Pattern	λ_{max} in nm *17* x Band	Substitution Pattern	λ_{max} in nm (lg ε) *18*[a] x Band	y Band
4,4′	487	–	621 (5.02)	427.5 (4.30)
2,2′,3,3′	530	2	625 (5.06)	442 (4.11)
2,2′,4,4′	500	3	622.5 (5.03)	435 (4.26)
2,2′,5,5′	585	4	608 (5.03)	465 (4.53)
		2,3	626.5 (5.06)	422 (4.11)
		2,4	615 (5.06)	484 (4.34)
		2,5	626.5 (5.06)	404 (4.00)
		2,6	625.5 (5.08)	429 (4.00)
		3,4	609.5 (5.02)	485 (4.42)
		3,5	624 (5.02)	433 (4.20)
		2,4,5	618 (5.06)	400 (4.00)
		2,4,6	621.5 (5.08)	493 (4.18)
		3,4,5	619 (5.02)	467 (4.30)

[a] in acetic acid

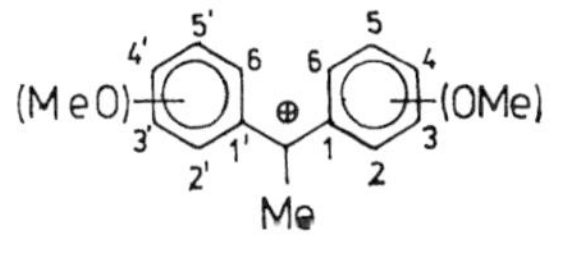

17

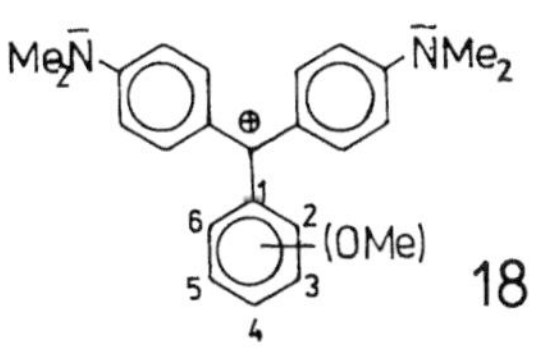

18

Interestingly, a completely different sequence is observed for s the y band of multiply methoxy-substituted Malachite Green dyes *18* [669] (cf. Table XI.6). The 2,4-substituted dye now absorbs at longer wavelengths than the 2,5-substituted one. Examination of the HOMO levels in both cases makes the difference understandable. The degenerate HOMO's of bis-phenylmethyl cations confer the phenyl group spectral properties such as observed in other substituted benzenes. The donor effect on Malachite Green, however, is limited to a single occupied MO (MO 2 in Fig. XI.6), which exhibits high expansion coefficients only in the *ortho*- and *para*-positions.

Results of PPP calculations [116] are depicted in Fig. XI.5. Two degenerate lowest-energy excited states account for the color band of Crystal Violet, which represents a superposition of the x and y band [670]. As expected, the lowest-energy transitions differ in energy when the third substituent is removed on passing from Crystal Violet to Malachite Green. Examination of the electronic transitions by OKK and LSU analysis supports the proposition that the phenyl group of Malachite Green is only weakly involved in the lowest energy transition polarized in x direction, but more strongly in the higher-energy y-polarized electronic transition [116].

The empirical examination of spectral data of symmetrical and non-symmetrical triphenylmethane dyes revealed a relationship between the position of the longest-wavelength absorption and that at shorter wavelengths [671]. According to this study the equation $\lambda_{max} = \lambda_0 \times 1.1^N$ approximately holds, where λ_0 denotes the wavelength of the UV maxima at about 300 nm and N an empirical integer which varies with the absorption band considered (N = 7, 0 and −2 for symmetrical dyes).

Numerous substituted diphenylmethane and triphenylmethane dyes have been studied by Barker, Hallas and coworkers [669, 672–689]. According to them, bulky substituents adjacent to the central carbon atom (*ortho* position) cause steric hindrance and, therefore, departure from the molecular coplanarity [672]. First-order perturbation theory predicts red shift of the first absorption band in that case [690]. In good agreement, 2-methyl and 2,2′-methyl derivatives of Michler's Hydrol Blue and 2-methyl, 2,2′-dimethyl and 2,2′2″-trimethyl derivatives of Crystal Violet show progressive bathochromic shifts and reduction of maximum intensities of the color band [672]. The red shift is nearly additive. The uniformity of the shift suggests that axial rotation is shared uniformly between the benzene rings in all of the dyes. More complex situations arise, however, by introducing crowding substituents in the less symmetrical dyes of the Malachite Green type, where one of the rings might undergo most of the rotation.

Methyl substitution in the *meta* position, adjacent to the terminal dimethylamino groups, leads to complex effects: Red shifts are observed for Crystal Violet, whereas symmetrically substituted derivatives of Michler's Hydrol Blue and Malachite Green show moderate and small blue shifts, resp. [673]. Attempts failed to explain these shifts by perturbational arguments based on the iso-π-electronic odd-alternant hydrocarbons [690] and by HMO calculations [689] as long as departure from the electronic symmetry has not been taken into account. Simultaneous central and terminal steric distortion resulting from twofold methyl substitution has been recently discussed [689].

The spectral effect of substitution at the phenyl residue of Malachite Green has been extensively studied. Whereas substituents in *meta* and *para* position of the *meso*-phenyl group exert nearly additive effects, modifying the absorption wavelengths of the color band according to Hammett-type substituent constants, substituents in *ortho* position force the phenyl group to rotate about the contral CC bond [672, 687]. As demonstrated with the substitutents CH_3, CF_3, NO_2, CN, COOMe, SO_2Me and Ph in the *ortho* position of the *meso*-phenyl group, the distortion, in general, leads to an increase of the intensity of the first band (x band) and decrease of the intensity of the second band (y band) [683]. The same spectral effect results from replacing the unsubstituted phenyl ring of *15* by the more extensive naphthyl and phenanthryl rings [688].

Since the x band and y band of *19a* show higher and lower intensity, resp., than the corresponding *15*, the phenyl ring of *19a* is evidently predominantly deconjugated. Consequently, *19a* absorbs like the bridged parent *19b* of Michler's Hydrol Blue [676, 677, 684].

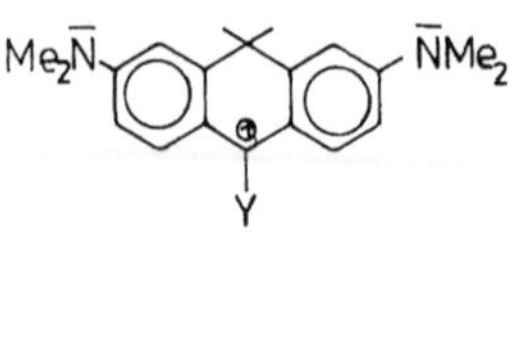

19

a Y = Ph: x 613.5 (5.17)
y 407.5 (3.89)
in acetic acid [677]

b Y = H: x 607.5 (5.17)
in acetic acid [676]

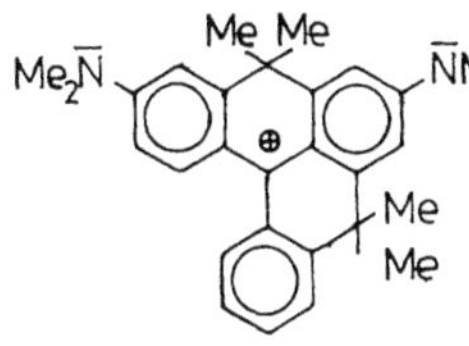

20

x 669.5 (4.89)
y 463.5 (4.45)
in acetic acid [677]

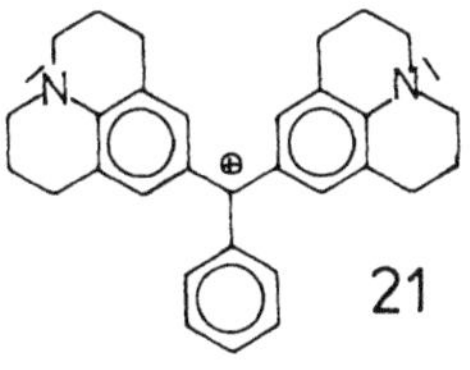

21

x 657 (5.04)
y 440 (4.24)
in acetic acid [686]

The absorption bands undergo, however, red shift when the phenyl group of *19a* is forced to coplanarity. Similar, red shift results from incorporation of the terminal amino group into the julolidine ring. Examples afford the dyes *20* and *21*.

Naphthalene analogues of Crystal Violet such as *22* and *23* absorb at similar wavelengths as Crystal Violet. However, the absorption maximum of *24* is remarkable red-shifted due to the more extended conjugation in the x direction.

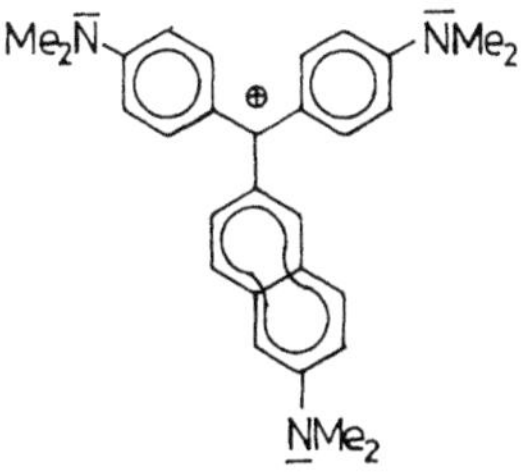

22

613 (5.01)
in 98% glacial acetic acid [680]

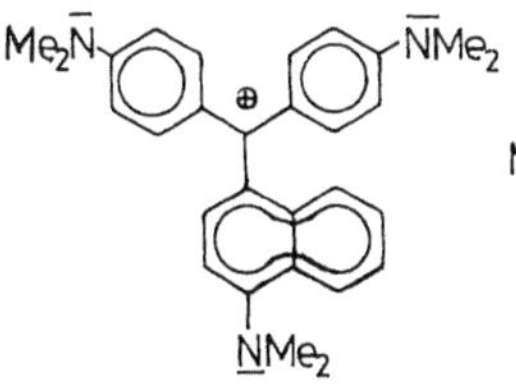

23

623 (4.99)
in ethanol/HCl [685]

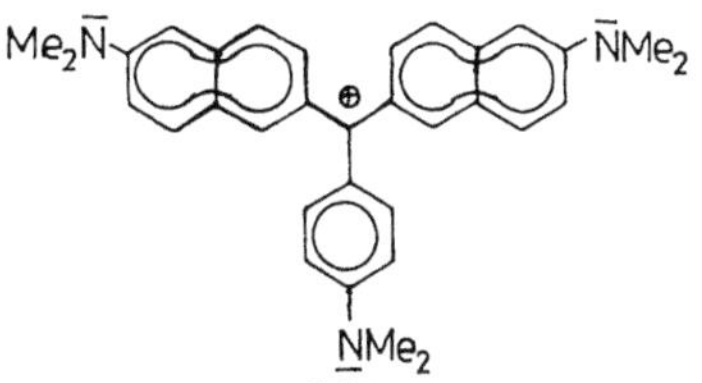

24

666 (4.05)
in 100% acetic acid [680]

In Therephthal Green *25* both x and y band are red-shifted in comparison with Malachite Green *15* ($X = NMe_2$, $Y = H$).

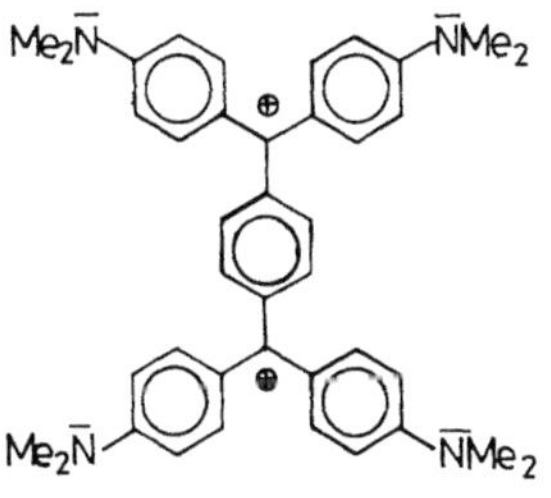

25

x 646 (5.05)
y 483 (4.62)

in 98% acetic acid [678]

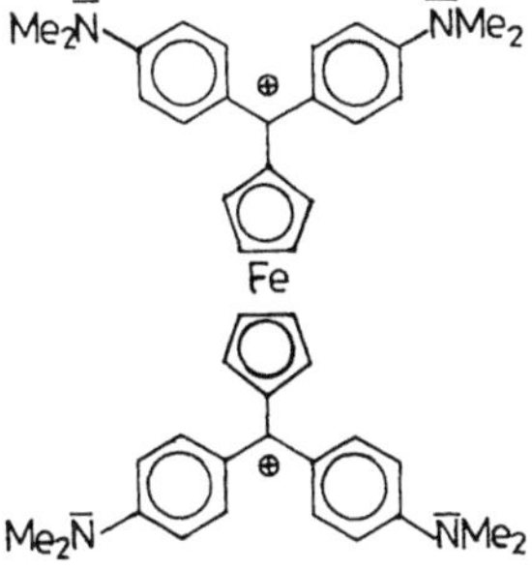

26

x 654 (4.72)
y 527 (4.53)

The y band undergoes an additional remarkable displacement to longer wavelengths on passing from *25* to the ferrocene analogue *26.*

Further interesting compounds are the vinylogues of the Crystal Violet and Malachite Green [121, 691, 692], such as *27* and *28,* which both show vinylene shifts of about 100 nm like vinylogues of Michler's Hydrol Blue (*16,* X = NMe_2, cf. Table XI.2).

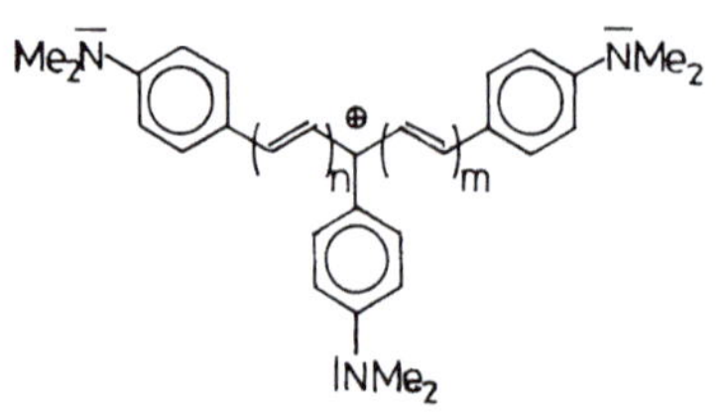

27

m	n	λ_{max} (nm)
0	0	591
1	0	690
1	1	770

in acetic acid [691]

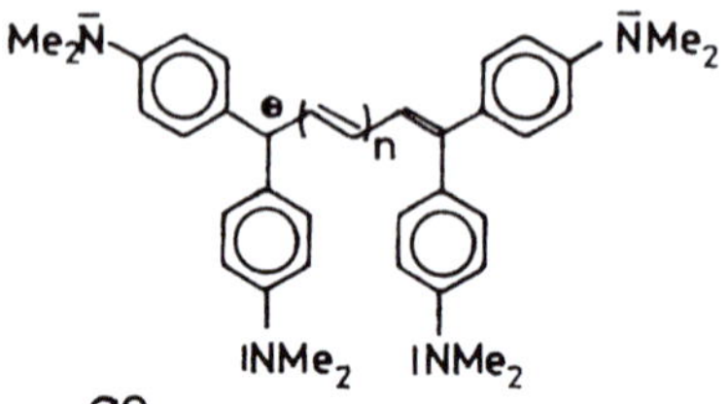

28

n	λ_{max} (nm)
0	740
1	810
2	911

in chloroform [692]

At first sight it seems to be quite inconsistent to discuss spectral data of the anionic dyes Benzaurin *29* or Aurin *30* along with those of cationic dyes derived from bis-phenylmethyl and tris-phenylmethyl cations, because they seem to be more related to those of the corresponding anions.

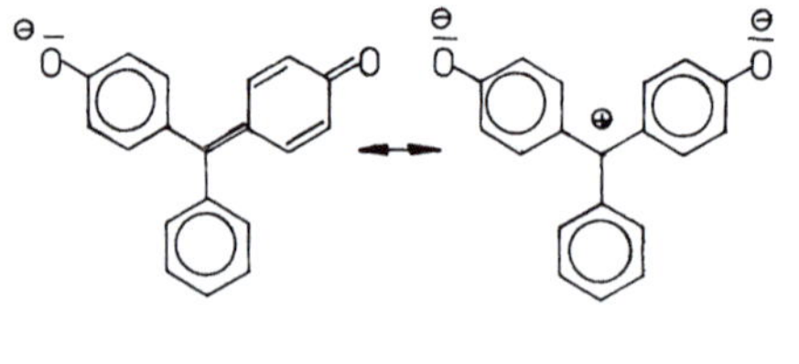

29a 29b

585 nm
in pyridine/water [693]

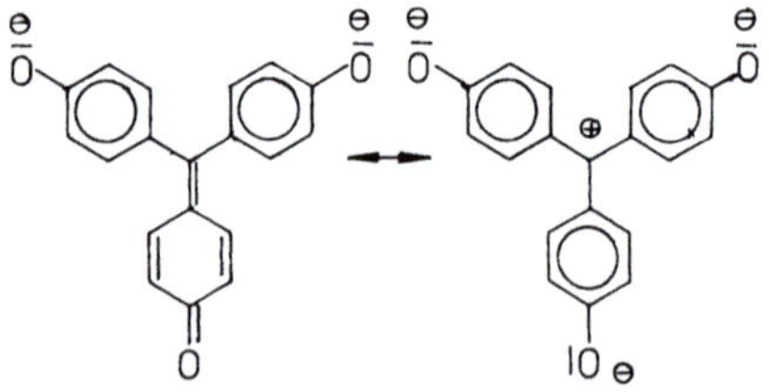

30a 30b

531 nm
in ethanol/KOH [36]

However, as regards the number of π-electrons, *29* and *30* are iso-π-electronic to Malachite Green *15* (X = X' = NMe_2, Y = H) and Crystal Violet *15* (X = X' = Y = NMe_2), resp. This close connection becomes evident in the mesomeric formulae *29b* and *30b*. Comparison of the spectral data of *29* and *30* with those of the a forementioned iso-π-electronic cation *15* indicates that the donor capacity of the negatively charged oxygen is slightly lower compared to that of the NMe_2 group.

The widely used indicator phenolphthalein *31* is less deeply coloured than *29.*

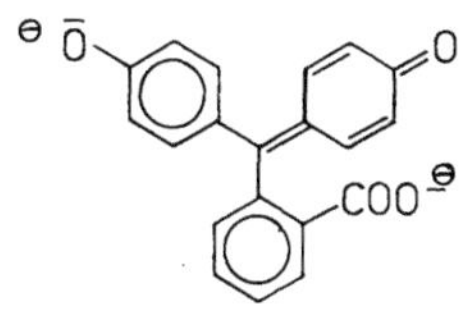

31

552 (4.44)
in water pH 10 [694]

Data on anionic dyes within the diphenylmethane series are scanty. The anions *32–34* have been synthesized recently [695]:

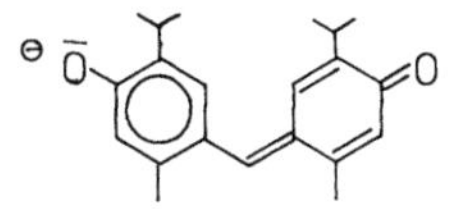

32

577 (4.84)
in ethanol/KOH

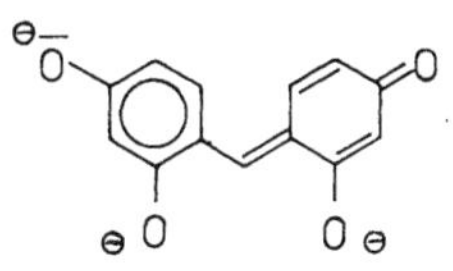

33

545 nm
in ethanol/KOH

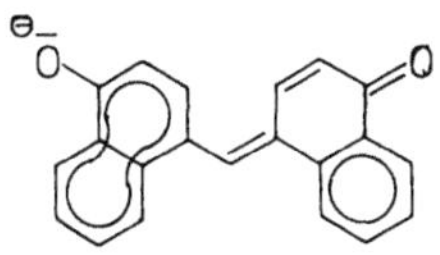

34

634 nm
in water/NaOH

Intermediate cases between the symmetrical cationic or anionic dyes represent unsymmetrical neutral compounds such as 4,4′-bis(dimethylamino)-fuchsone *35*.

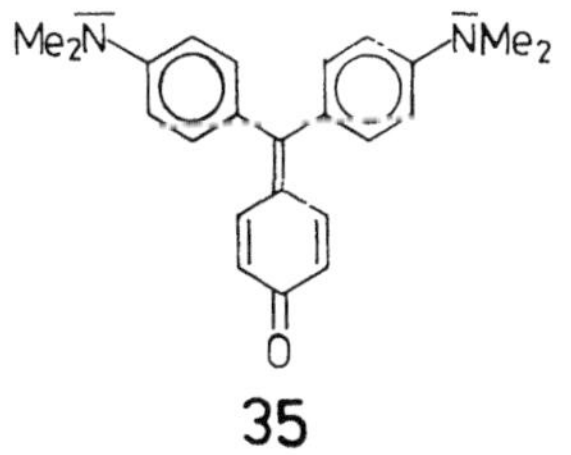

35

465 nm
in isooctane [497]

The color band of *35* becomes narrow and more intense in methanol and moves to a considerably longer wavelength (559 nm). In dilute acid the spectrum of *35* resembles that of methoxy Malachite Green (610 nm). The Fuchsone *35* owes its longest-wavelength absorption to a y-polarized transition, whereas x-polarization is found for the triphenylmethane dye resulting from protonation [497].

The great variety of dyes is enlarged by replacing one or several of the aryl residues of the bis-phenylmethyl or tris-phenylmethyl cation by heterocyclic groups, as exemplified with some sulphur-containing compounds in Table XI.7.

The resulting cations absorb consistently at longer wavelengths than the iso-π-electronic arylmethyl cations. PPP-type calculations on the bis-(2-thienyl)methyl cation *36*

Table XI.7. Longest-wavelength absorption maxima of heterocyclic analogues of diphenylmethane and triphenylmethane dyes $R_1R_2R_3C^+$ (solvent: sulphuric acid)

R_1	R_2	R_3	λ_{max} in nm (lg ϵ)	Ref.
2-thienyl	phenyl	H	470 (4.29)	[697]
2-thienyl	phenyl	phenyl	455 (4.52)	[697]
2-thienyl	2-thienyl	H	487 (4.33)	[697]
2-thienyl	2-thienyl	phenyl	475 (4.31)	[697]
2-thienyl	2-thienyl	2-thienyl	472 (4.55)	[697]
2-thionaphthyl	phenyl	H	483 (5.02)	[653]
3-thionaphthyl	phenyl	H	529 (4.66)	[653]
2-dibenzothienyl	phenyl	H	538 (4.48)	[653]
4-dibenzothienyl	phenyl	H	600 (4.13)	[653]

have revealed that the three nearly degenerate low energy $\pi \rightarrow \pi^*$ transitions of the bis-phenylmethyl cation *1* are split into a more intense lower energy transition and two nearly degenerate higher-energy transitions [696]. This is in good agreement with two long-wavelength absorption maxima of *36* found experimentally at about 370 and 490 nm [697].

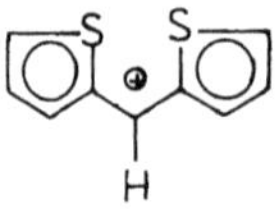

36

Introduction of the NMe_2 in heteroanalogous bis-arylmethyl cations gives rise to pronounced red shift similarly to that observed for the carbocyclic cations. Comparison of the spectral data of the iso-π-electronic ions *37a–d* shows a remarkable constancy of the wavelengths, whereas vinylene shifts of about 100 nm have been observed for *38*.

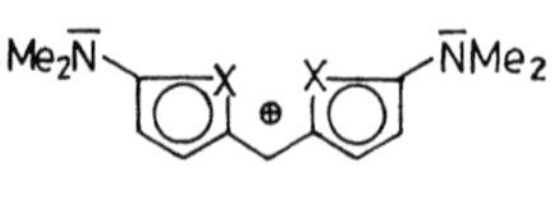

37

	X	
a	S	587 (5.15)
b	O	583 (5.10)
c	Se	600 (5.10)
d	CH=CH	606

in nitromethane [698]

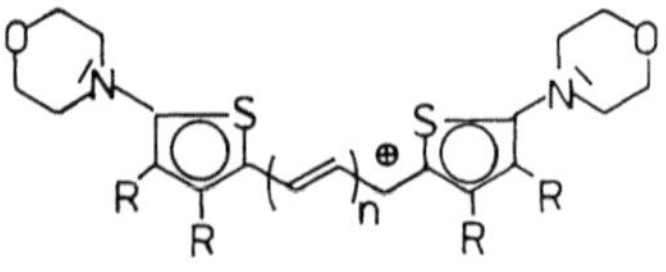

38

	a R = H	*b* $R = C_6H_5$
n = 0	599 (5.16)	630 (4.96)
n = 1	692 (5.27)	728 (5.25)
n = 2		840 (5.15)

in methylene chloride [609, 699]

These results might suggest that the basic nonamethine-cyanine (λ_{max}: 625 nm [701]), undecamethine-cyanine (λ_{max}: 734.5 [701] and tridecamethine-cyanine chromophore (λ_{max}: 848 nm [701]), resp., determines the absorption features of *37* and *38*, with regard to their π-systems according to model *B* of Fig. XI.1. Although the PPP configuration analysis of *37a* confirms the predominance of a polymethine-type excited state at long wavelength, the relatively incomplete expansion of the excited state wave function in terms of the wave function of the molecular fragments clearly indicates a low appropriateness of this fragmentation mode [116]. Thus, the above-mentioned coincidence of wavelengths is rather fortuitous and does not reflect strongly localized polymethinic subsystems in *37* and *38*.

Inspection of the spectral data of *39* to *42* leads to a similar conclusion. Considering their π-systems according to model *B*, about 100 nm red shifts should be expected in passing from *39* (pentamethine substructure) through *40* and *41* (heptamethine substructure) to *42* (nonamethine substructure). This has not been found experimentally, however:

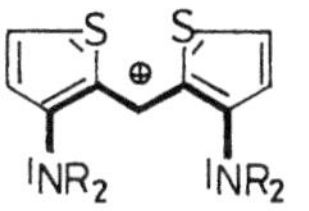

39

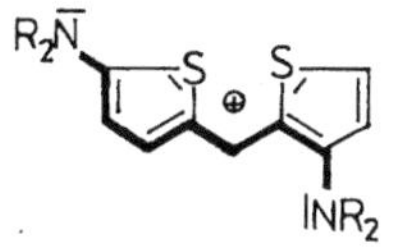

40

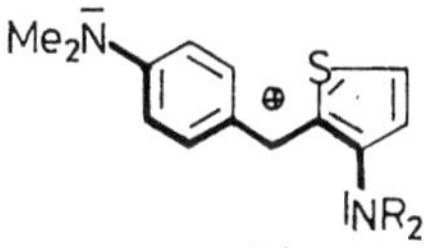

41

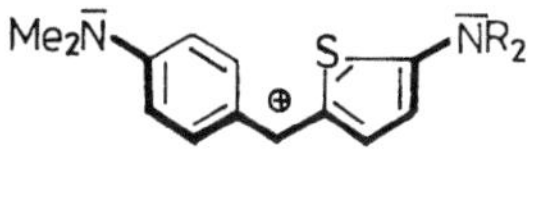

42

586 (4.45) 580 (4.67) 591 (4.92) 588 (4.91)

in methylene chloride [609]

NR_2 = 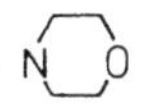

All the compounds absorb in the same spectral region.

The position of the color band of phenyl-heteroarylmethyl cations depends on the donor properties of the aryl fragments. The wavelengths of *43*, for example, increases with replacement of the substituent X in the sequence H < Me < Cl < OMe < SMe < NMe_2. The red shifts correlate with the ionization potential of the corresponding monosubstituted benzenes [700].

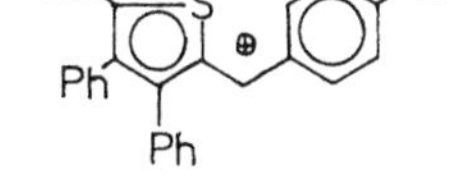

43

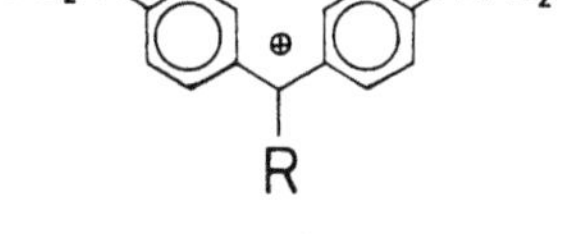

44

Another series of heterocyclic dyes result from replacing of the unsubstituted phenyl ring of Malachite Green (*44*, R = C_6H_5) by heteroaromatic fragments. Some spectral data tabulated in Table XI.8 show that heteroaromatic groups only slightly affect the position of the x band in the 600 nm region. As for the parent compound, the y-type band appears at shorter wavelengths. Dyes which result from replacing each phenyl group of triphenylmethane dyes by heterocyclic groups have been described recently [702].

The bisferrocenyl-methyl cation *45* represents an interesting chromophore of remarkably deep color:

Table XI.8. Longest-wavelength absorption maxima of heterocyclic analoga of Malachite Green (*44*)

R	λ_{max} in nm (lg ϵ)		Solvent	Ref.
	x Band	y Band		
thienyl(2)	629 (4.93)	469 (4.45)	CH_3COOH	[755]
thienyl(3)	617 (4.99)	444 (4.36)	CH_3COOH	[756]
thionaphthenyl(2)	645 (4.88)	492 (4.49)	CH_3COOH	[756]
thionaphthenyl(3)	627 (4.95)	476 (4.27)	CH_3COOH	[756]
benzthiazolyl(2)	666	450	EtOH	[757]
2-methyl-benzthiazolyl(5)	618	572	EtOH	[757]
2-methyl-benzthiazolyl(6)	624	450	EtOH	[757]
5-phenyl-1,3,4-triazolyl(2)	643 (4.19)	438 (4.36)	HCl/EtOH	[758]

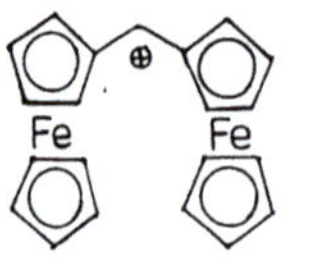

45

605 (3.89)
in chloroform [703]

The vinylogue of *45* absorbs even at 785 nm [703].

The compounds mentioned above belong to the bis-arylmethyl and tris-arylmethyl cations, resp. Far less experimental material is accessible for compounds derived from the corresponding anionic species. Examples are nitro-substituted bis-phenylmethyl anions *46*, which exhibit a blue or green color [704]:

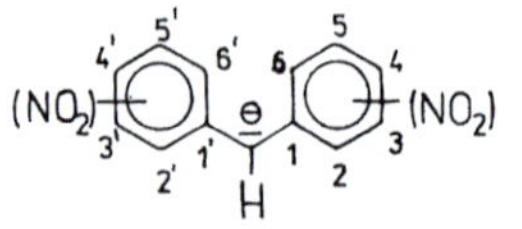

46

a 2,2′,4-trinitro 645
b 2,4,4′-trinitro 580
c 2,2′,4,4′-tetranitro- 700

in KOH/ethanol [704]

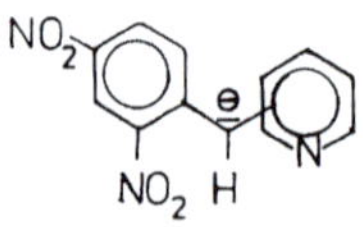

47

a *ortho-* 650
b *para-* 640

The nitro-substituted benzene residue can be replaced by the pyridine residue without altering appreciably the longest-wavelength absorption maxima, as demonstrated with *47* [704].

The less deeply colored bis-chinolylmethyl anions such as *48* possess *cis*-configuration because of forming a chelate structure with the gegenion [705, 706].

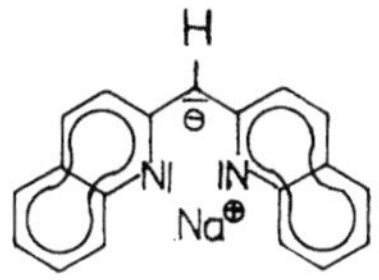

48

543 in dioxane [706]

It should only be mentioned that the most heterocyclic monomethine dyes discussed in Sect. XII might also be considered as heteroaryl-methyl anions. This relationship is visualized by means of the appropriate mesomeric formulae of the polymethine dyes *49*.

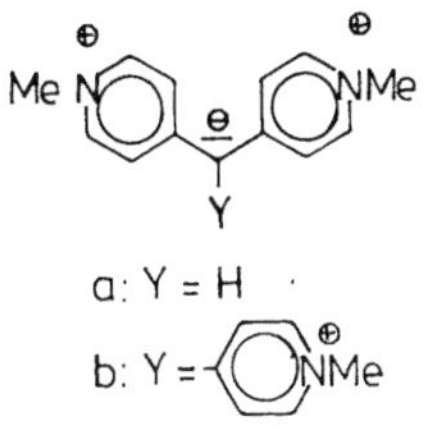

49

a:	483 (5.30)	in ethanol [707]
b:	513 (4.80)	in ethanol [707]

Although this classification may also be of help for understanding spectral properties accompanied by internal and external substitution, the wide knowledge about vinylogous derivatives favours a discussion as polymethine dyes.

4. Heteroanalogous Arylmethane Dyes and Heteroatom Bridged Derivatives

Internal substitution of the methine group of diarylmethane and triarylmethane dyes by heteroatoms of appropriate valence properties leads to iso-π-electronic structures which also may be colored. They can be derived from the cations or the anions as follows:

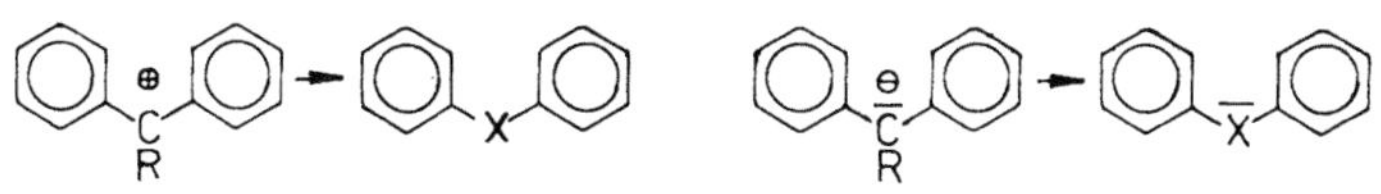

1 50 2 51

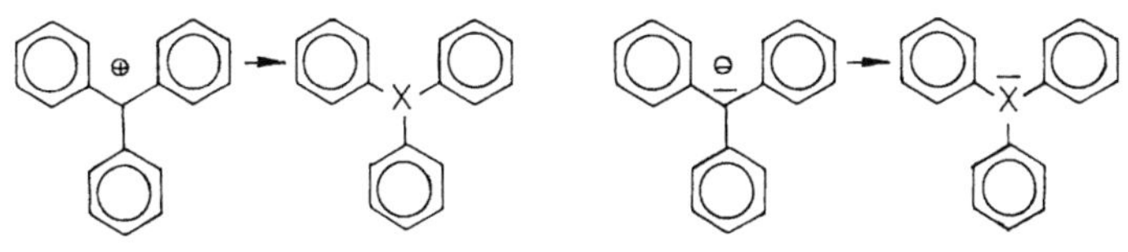

3 52 4 53

Compounds of the formulae *51* and *53* with X = N, P, As and Bi show absorption features which imply a non-planar and consequently weakly-conjugated π system [708], but arylamines bearing acceptor groups possess a yellow color:

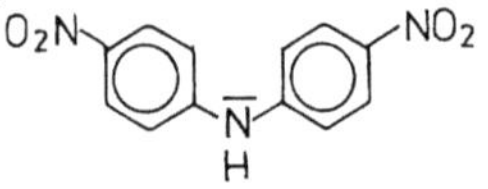

54

402 (4.58)

in ethanol [439]

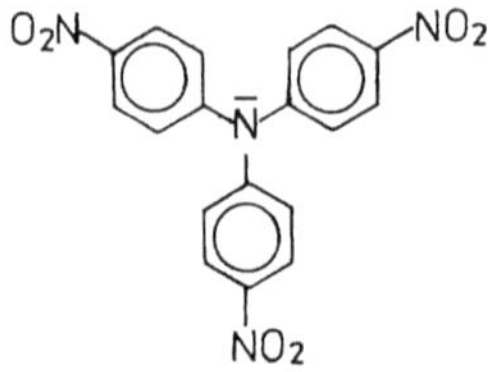

55

405 (4.38)

in ethanol [709]

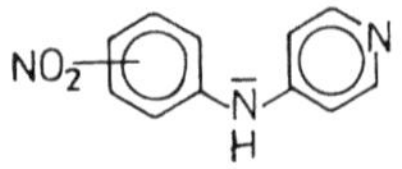

56

ortho: 402 (3.66)
meta: 340 (4.06)
para: 364 (4.32)

in ethanol [710]

The compounds *57* and *58* exemplify structures of the type *50* or *52*, resp. They appear as derivatives of the corresponding phenylmethyl cations.

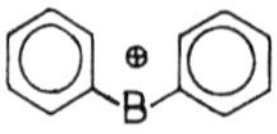

57

337
in methylene chloride [711]

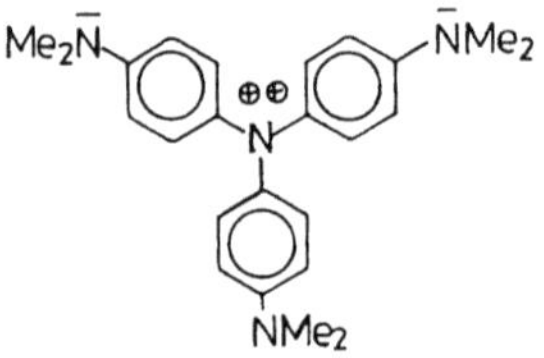

58

920 (4.28)
in methanol [712]

The compounds *59, 60* and *61* are dyes of the indamine, indophenol and indaniline type, resp. They can be considered alternatively as *para*-benzoquinone derivatives (see Sect. IX.3).

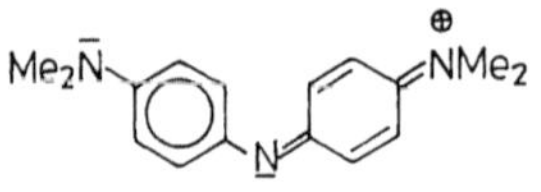

59

725
in methanol [713]

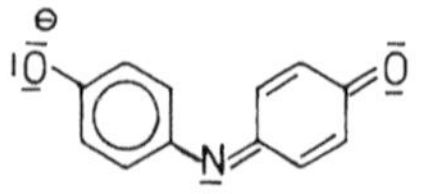

60

640
in methanol [714]

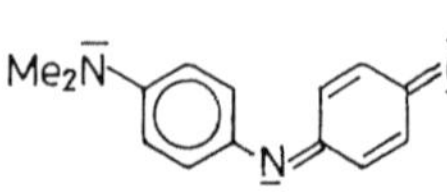

61

608
in methanol [715]

Replacement of the central methine groups of cationic diphenylmethane dyes by the less electronegative boron is accompanied by a blue shift, whereas substitution by the more electronegative nitrogen causes red shift. The shift amounts to 100 nm on going from Michler's Hydrol Blue to Bindschedler's Green (*59*) and increases up to 300 nm on going from Crystal Violet to the dication *58*.

Table XI.9. Longest-wavelength absorption maxima of o,o'-heteroatom bridged Phenol Blue base, Bindschedler's Green and Michler' Hydrol Blue

Substituent	λ_{max} in nm (lg ϵ)		
X	*62*	*63*[d] [713]	*64*[d] [713]
H,H	637 (4.50)[a]	725	603
S	590 (4.80)[b]	668	~550
O	573 (4.90)[c]	665.5	549.5
NMe		564.5	491

[a] in water, pH 11 [557] [b] in 0.01 N NaOH [759] [c] in ethanol [561]
[d] R = Me

On the other hand, insertion of an heteroatomic bridge in the *ortho,ortho'*-position of Bindschedler's Green (*59*) and the deprotonated Phenol Blue (*60*) produces blue shifts (cf. Table XI.9). The hypsochromicity increases in passing from Methylene Blue (*63*, X = S) and Capri Blue GN (*63*, X = O) to Neutral Red (*63*, X = NMe) (R = Me). Accordingly, appropriate substitution of Michler's Hydrol Blue brings about increasing blue shift in passing from Thiopyronine *64* (X = S) and Pyronine (G *64*, X = O) to Acriflavine (*64*, X = NMe, R = Me) (Table XI.9).

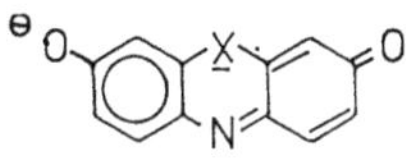

62

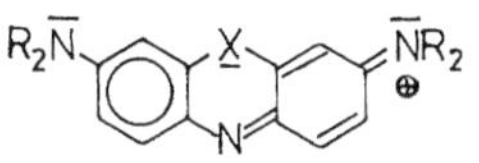

63

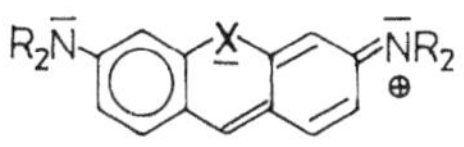

64

Progressive blue shift likewise occurs in the sequence Selenine (*63*, X = Se), Thionine (*63*, X = S), and Oxonine (*63*, X = O), (R = H) [716].

The above mentioned dyes belong to the selenazine, thiazine, oxazine, diazine and thioxanthene, xanthene and acridine dyes, resp. Various of these have attracted much interest owing to their particular photophysical and photochemical properties. Polarization measurements have been performed to attain a better understanding of the nature of their excited states [717, 717a, 744].

The spectral shifts of the above mentioned dyes with respect to the diphenylmethane parent dyes can easily be explained by second-order perturbation theory [70, 122]: Whereas azasubstitution lowers the energy of the LUMO, the presence of an additional occupied p_π orbital in cases of *ortho,ortho'*-bridging raises the LUMO and increases so the $1 \rightarrow 1'$ transition energy (cf. Fig. XI.7).

As the bond order of bis-phenylmethyl and tris-phenylmethyl cations increases between the (non-bonded) *ortho*- and *ortho'*-positions upon excitation, ring closure to the fluorenyl cation should bring about red shift. This has indeed been found. The dye *65* absorbs at 200 nm longer wavelengths than Malachite Green (*15*, *X* = *X'* = NMe_2, Y = H) [718].

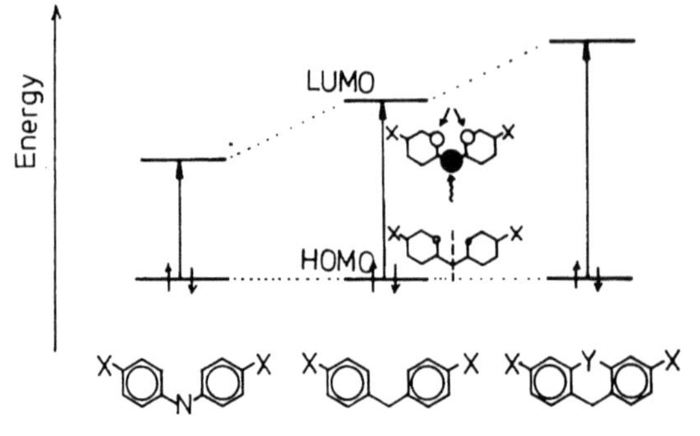

Fig. XI.7. Frontier orbitals of Michler's Hydrol Blue and orbital shift arising from substitution of the central methine group (left-hand side), and introduction of an heteroatomic bridge in the o,o′-positions (right-hand side)

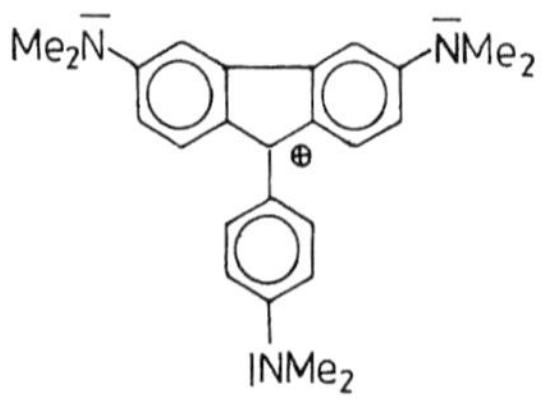

65

470, 647, 850
in acetic acid [718]

The absorption wavelengths calculated for various of the aforementioned bridged dyes with the aid of the simple HMO method are in surprisingly good agreement with the experimental data [70, 719, 720]. The spectra of acridine dyes have been discussed by PPP calculations [720a, 720b].

Anionic dyes which contain oxygen substituents absorb at shorter wavelengths than the corresponding cationic dyes with amino groups. This can be inferred from Table XI.9 by comparing the longest-wavelength absorptions of *62* and *63*. This relationship is also reflected in the absorption wavelength of Fluorescein (*66a*) and Rhodamine B (*67*). The non-symmetric intermediate between *66a* and *67* with X = NEt_2 and X = O absorbs at lower wavelengths (517 nm) than the symmetric dyes, in good harmony with the Brooker deviation rule [720c]. A very strong red shift, however, has been observed in passing from Fluorescein (*66a*) to Dithiofluorescein (*66*b) [721].

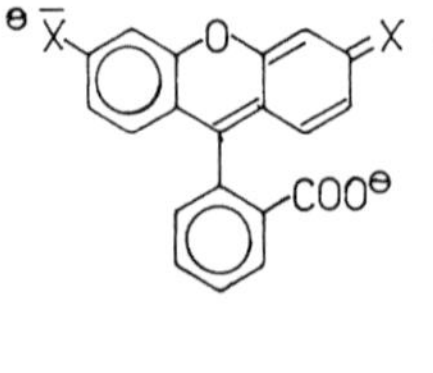

66

a X = O 500
b X = S 635

in basic ethanol [721]

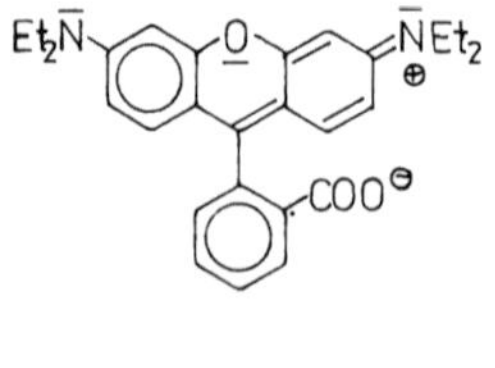

67

543

in basic ethanol [721]

Spectral comparison led to the conclusion that the carboxyphenyl substituent is held in a position nearly perpendicular to the xanthene chromophore because of the bulky carbonyl group.

If we pass from derivatives of the bis-phenylmethyl cations to those of the bis-phenylmethyl anions the effect of internal aza substitution and external heteroatomic bridging becomes, as expected, just reversed. Thus, bathochromic displacement of the colour band has been observed when sulphur or oxygen bridges are inserted between the nitrobenzene moieties of *68a*:

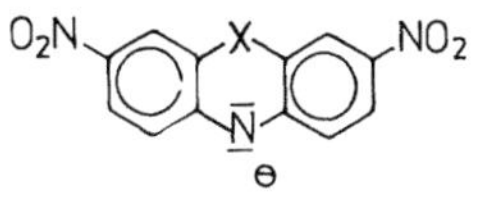

68

a X = H, H 598
b X = O 704
c X = S 692

in ethanol, dimethylformamide [122]

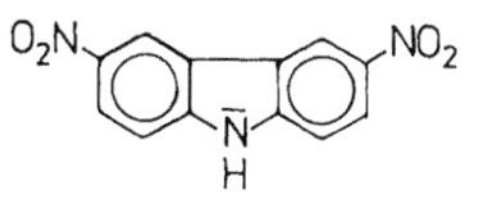

69

360 (4.26)
in ethanol [439]

On the other hand, in contrast to *68*, hypsochromic shift has been observed with ring closure in passing from *para,para'*-dinitro-diphenylamine (*54*) to 2,7-dinitro-carbazole (*69*) [439].

5. Medium Effect

The spectral data of many dyes discussed in this section depend strongly on the experimental conditions. Addition of nucleophilic components may lessen the absorption intensity [722–724], and formation of ion pairs [725, 725a], aggregates [716, 726–734, 734a, b] and heterocomplexes [734c] displace the position of the color band of triphenylmethane, xanthene, acridine and thiazine dyes. The longest-wavelength absorptions of the meropolymethinic fuchsone *35* [735, 736] and indophenol dyes *60* [542, 714, 736a] undergo pronounced spectral solvent effects.

The spectra of diphenylmethane and triphenylmethane dyes greatly depend on the pH value. Spectrophotometric studies on Malachite Green *15* (X = X' = NMe_2, Y = H), for instance, revealed the existence of several protic equilibria, involving the formation of the carbinol *70* and dication *71* [737].

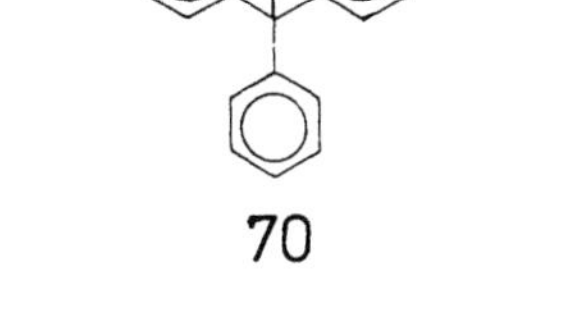

70

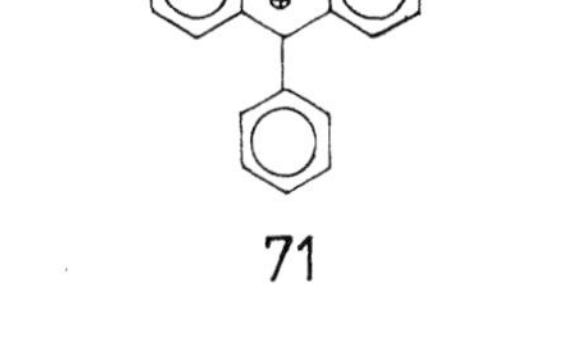

71

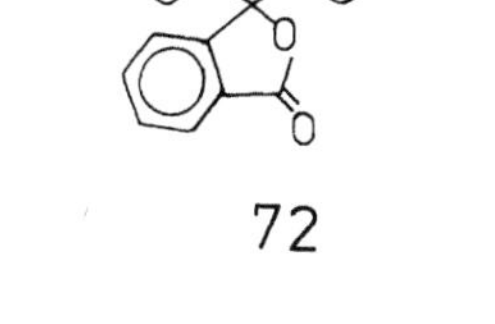

72

Mono-protonation of Crystal Violet type dyes at one of the amino groups is indicated by the appearance of the y band. Since the spectra of these dyes are generally measured in acidic media, partial protonation often occurs [675, 680, 685, 688]. Diminished conjugation of terminal dimethylamino groups because of steric distortion leads to a higher contamination of protonated species [673, 689].

Xanthene dyes that carry a free COOH group (such as the acids of *66a* and *67*) equilibrate with several other compounds [721, 738]. In polar solvents like ethanol the carboxyl group undergoes a typical acid-base equilibrium. The zwitterionic Rhodamine B (*67*), for example, absorbs at an about 10 nm longer wavelength than the corresponding acid [721]. Protic solvents, high concentration and low temperature displace the equilibrium towards the zwitterion [738a]. In non-polar solvents the zwitterionic form is not stable. Instead, the colorless inner lactone *72* is formed. Lactones can, in turn, photochemically revert to the colored species (photochromic dyes) [739].

Protonation as well as deprotonation has been observed for acridine dyes, such as *73* [731–733, 740]. The colour is brightened by both reactions.

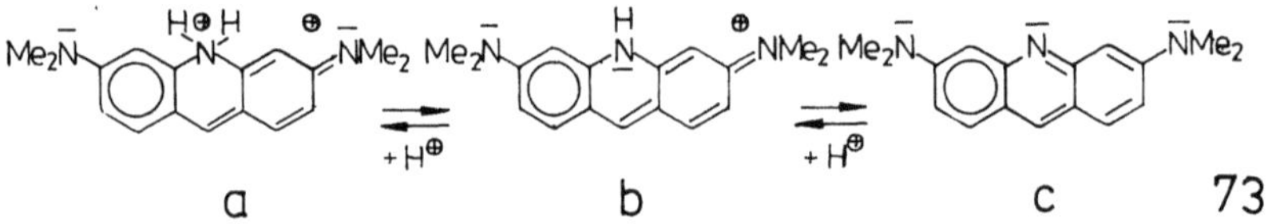

73

Similarly oxazine, thiazine and diazine dyes undergo protonation when acids are added. Whereas Capri Blue *63* (X = O, R = Me) assumes a red colour in acidic solution, a shift to longer wavelengths has been observed for Methylene Blue (*63*, X = S, R = Me) [741] and Thionine (*63*, X = S, R = H) [742, 742a]. This was explained by the assumption that Capri Blue adds the first proton at one of the amino nitrogen atoms, whereas protonation occurs at the aza nitrogen in Methylene Blue and Thionine. Similar spectral effects were found with addition of Lewis acids, such as BF_3 and $SnCl_4$ [742]. More recent studies have revealed that even optically excited states of thiazine dyes are protonated at the intracyclic nitrogen atom [742, 743].

The diazine dye Safranin T (*74*) undergoes protonation at the aza nitrogen [744]. Addition of acids turns the color of the dye solution from red to blue, and isobestic points clearly indicate the presence of a well-defined protic equilibrium. Further addition of acid displaces the color band to shorter wavelengths. The brightening of the color has been attributed to a second protonation at one of the amino nitrogen atoms giving rise to *74c*.

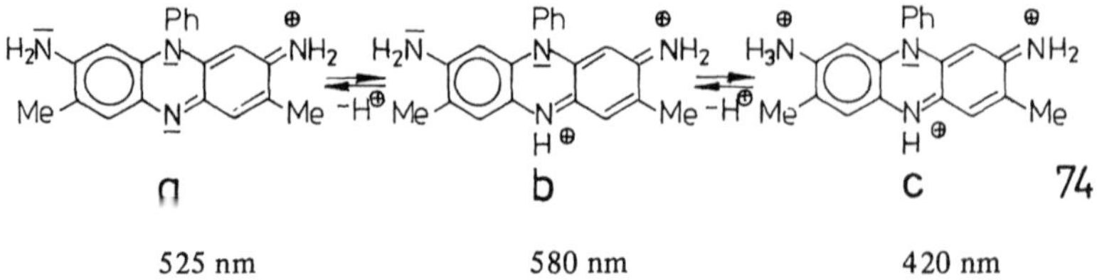

74

In Bindschedler's Green *59* the first step of protonation at the aza nitrogen to the dication *59b* does not cause appreciable deepening of the color, but a yellow colored trication *59c* is assumed to result from the second protonation [549, 745].

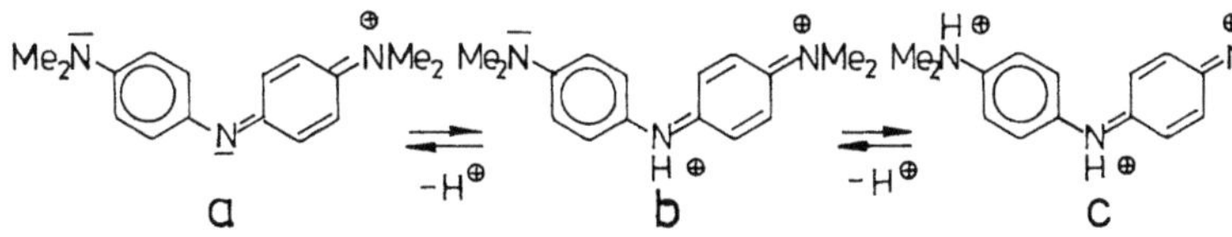

59

725 nm 400 nm

The anionic Phenol Blue (*60*) shows an interesting behaviour towards acids. It undergoes the following acid-base equilibria [746]:

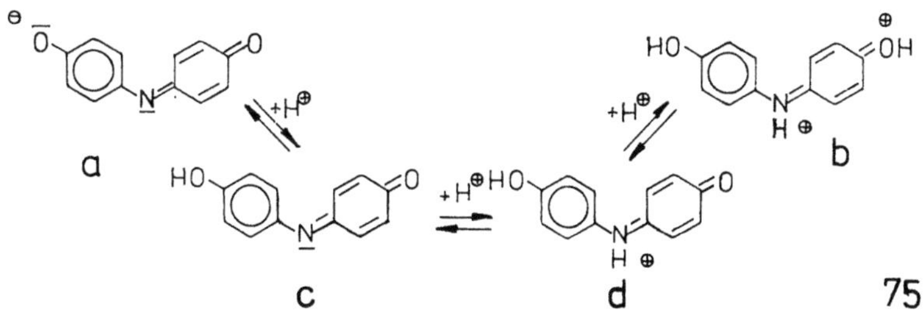

75

blue red red blue

According to Schwarzbach the symmetrical ions *60a* and *60c* display the deepest colour [745]. This fact seems to reflect the formation of polymethinic π systems.

XII. Polymethine Dyes

The polymethine dyes comprise a great variety of dyes. Since the discovery of their property of photographic sensitization (Vogel 1873) they represent a subject of extensive experimental research [760, 760a, 761]. W. König, F. M. Hamer, L. S. Brooker and A. I. Kiprianov made an important contribution to their exploration and paid much attention to their spectral behaviour. Scheibe [762, 763] and Dähne [23–25] contributed essentially to the basic understanding of the electronic structure of polymethines. The progress in the field of cyanine and related dyes has been reviewed by Sturmer [764].

A more recent field of application concerns the dyeing of polyacrylonitrile fibers [765, 766]. Much interest also attached to the photochemical formation of polymethine dyes from colorless precursors. Thus, some spiro compounds are used as light recording materials because of their capability to photoisomerize to meropolymethines [287, 767, 768]. On the other hand, diaza polymethines known as formazans are formed by hydrogenation of tetrazolium salts in biological applications [769–771].

1. Chromophoric System

The parent chromophores of polymethine dyes are the streptopolymethines, such as the cationic streptopolymethine-cyanines *1*, the anionic streptopolymethine-oxonols *2* and the neutral streptopolymethine-merocyanines *3*.

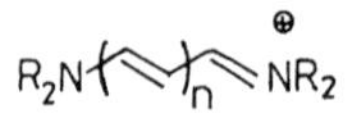

1

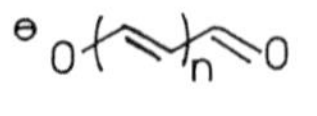

2

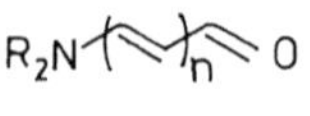

3

The idealized polymethine chain has a unique electronic structure. In contrast to polyenes, it displays charge alternation and bond length equalization. In the case of two electronegative end groups and N atoms along the chain, e.g. (N – 2) methine groups, they contain (N + 1) π electrons (Dähne 1966) [25]. The above mentioned compounds *1* to *3* belong to the (N + 1) π polymethines. Obviously they are either symmetrical or unsymmetrical, as indicated by the general formulae in Fig. XII.1.

Fig. XII.1. General formulae of symmetrical and unsymmetrical (N + 1) π streptopolymethines

$$\overbrace{X{-}(CH)_{N-2}{-}X}^{(N+1)\pi} \quad \cdot \quad \overbrace{X{-}(CH)_{N-2}{-}Y}^{(N+1)\pi}$$

Dähne's polymethine conception has been established experimentally as well as theoretically [27]. As for polyenes, the Hückel wave functions and orbital energies of idealized streptopolymethines are known in closed form [68, 69, 772].

Streptopolymethines display strongly allowed long-wavelength absorptions, which mainly arise from electronic excitation from the HOMO to the LUMO. This $1 \rightarrow 1'$-transition is polarized along the longitudinal molecular axis. The following two higher-energy electronic transitions are degenerate in the simple HMO picture, but split into two separate transitions when electron interaction is taken into account. In Platt's notation [135], these lowest-energy transitions are designed as A → B and A → C transitions, as illustrated for simple polyenes in Sect. VI.1 (Fig. VI.1). The two A → C transitions are allowed, but they have a considerably lower transition probability than the A → B transition at longer wavelength.

The A → B type absorption band of streptopolymethines is narrow, thus indicating a low change in bond length upon excitation. Only one vibrational progression has been discerned for simple chain molecules [773]. The maxima of the colour bands are bathochromically displaced with increasing chain length, as exemplified in Table XII.1.

In contrast to polyenes the absorption wavelengths λ increase linearly with the number of carbon-carbon double bonds n:

$$\lambda = A \cdot n + B \qquad \text{(XII.1)}$$

For symmetrical polymethines the shift amounts to about 100 nm when the conju-

Table XII.1. Longest-wavelength absorptions of some simple streptopolymethines in nm (lg ε) [701] and correlation data according to Eq. (XII.2) [775]

n	*1*[a]	*2*[b]	*3*[a]
0	224 (4.16)		
1	312.5 (4.81)	267.5 (4.43)	283 (4.57)
2	416 (5.08)	362.5 (4.75)	361.5 (4.71)
3	519 (5.32)	455 (4.88)	421.5 (4.75)
4	625 (5.47)	547.5 (4.80)	462.5 (4.81)
5	734.5 (5.55)	(644)	491.5 (4.83)
6	848 (5.34)		512.5 (4.86)
k	104.5	93.8	196
q	0.03	−0.15	−1
a	1	1	0.54

[a] in methylene chloride [b] in dimethylformamide

gated chain is lengthened by one double bond (100 nm "vinylene shift"). The validity of Eq. (XII.1) has been discussed by deriving optimized polynomials [774].

For unsymmetrical polymethines Eq. (XII.1) is no longer valid. The vinylene shift now decreases with increasing chain length. Thus, the unsymmetric merocyanines *3* behave intermediately between cyanines and polyenes.

Dähne and Radeglia demonstrated the polyene-like absorption behaviour of *3* in non-polar solvents by means of Eq. (XII.2) [775].

$$\lambda = k\,(n_\pi + q)^a \qquad \text{(XII.2)}$$

The constant k describes the basic absorption wavelength of an effective electron pair (identical with the vinylene shift in the case of polymethines), n_π the number of pairs of π-electrons and q the contribution of additional terminal groups. The exponent a then reveals the polymethine-like (a ~1) or polyene-like (a ~0.37) nature of the conjugated chain molecules within the series considered. According to the data listed in Table XII.1, the series *1* and *2* represent polymethines, whereas *3* represents a borderline case. Evidences for the polyenic nature of *3* have been actually found in NMR-studies and X-ray studies. They revealed a pronounced bond-length alternation of *3*, in sharp contrast to *1* [27].

The experimental lowest-energy excitation energies of *1* to *3* are proportional to HMO transition energies [71]. Furthermore, the general absorption feature is well reproduced by PPP calculations. Too short absorption wavelengths were calculated for long chain polymethines, however [81]. The energy gap at infinite chain length is commonly assumed to be zero, but arguments against it have been advanced on the basis of theoretical considerations concerning bond alternation [776] and about electron-electron interaction [777]. Dähne concluded from polarographic studies of cationic polymethines that such compounds are no longer stable if the energy gap between the ground and lowest energy excited state is lower than about 0.9 eV, e.g. compounds absorbing at more than 1,380 nm should not exist [778].

Streptopolymethines generally have all-*trans* configuration. Mono-*cis*-isomers absorb at higher wavelengths than all-*trans* compounds [779, 780]. This red shift is reproduced by simple MO calculations, in which resonance integrals between non-adjacent atoms are additionally taken into account [781]. The introduction of these interaction terms is not necessary for interpreting the shifts in the framework of the PPP method, however.

Both internal and external substitution at the polymethine chain give rise to pronounced spectral shifts of the color band. The shift can be qualitatively predicted by considering the substituent effect as perturbation of the lowest-energy one-electron transition, which proceeds between the HOMO and LUMO [122–124, 877]. Starting from the explicitly known polymethine wave functions [68, 69], only the expansion coefficients, or the change of charge or bond orders upon excitation in the positions of substitution have to be considered. Whereas the bond orders of idealized polymethines equally decrease for all bonds upon excitation, the change of charge orders alternately differs in sign along the polymethine chain. As illustrated with the HOMO and LUMO expansion coefficients of streptopentamethine in Fig. XII.2, the charge density increases in the 1(1′) position (α-position) and 3(3′) position (γ-position) of the pentamethine, but de-

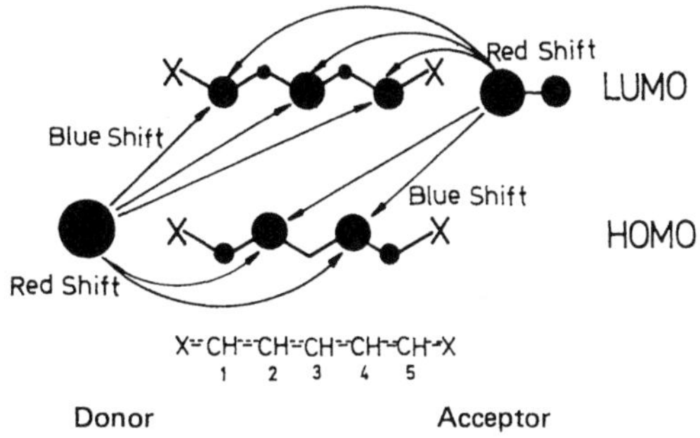

Fig. XII.2. Squared expansion coefficients of the frontier orbitals of pentamethinecyanine and spectral effects of external donor or acceptor substitution

creases in the 2(2′) position (β-position). In a more general labeling, it increases in odd-numbered positions, but decrease in even-numbered positions.

Hence, bathochromic shift occurs if methine groups are replaced by more electronegative elements in odd-numbered positions and, conversely, hypsochromic shift when methine groups are replaced in even-membered positions. The inverse effect is anticipated for heteroatom-replacement by electropositive elements in comparison with carbon. These conclusions about internal substitution effects are assembled in Fig. XII.3.

In order to discuss the effect of external substitution on polymethines we return to Fig. XII.2. The p_π orbital of the donor will efficiently overlap in positions which exhibit high expansion coefficients. For the HOMO this condition is fulfilled in even-numbered positions. Attachment of donor groups in even-numbered positions is expected to give red shift. Bent arrows in Fig. XII.2 indicate these positions. Donor substituents in odd-numbered positions give rise to a stronger overlap with the LUMO than with the HOMO, however. This may result in a hypsochromic shift. Corresponding considerations on the spectral effect of acceptor substitution lead to the prediction of red shift in odd-numbered positions, whereas blue shift should occur with substitution in even-numbered positions. The resulting spectral effects of external substitution by either donor or acceptor substituents are summarized in Fig. XII.4.

These color rules are well established by data for numerous substituted polymethine dyes (cf. Sect. XII.2 and XII.4). The spectral effect resulting from internal aza substitution at the central methine group was formerly known as Kendal's rule [782]. Kuhn first succeeded in explaining this rule in the framework of the free electron MO method [60]. A more general formulation of the color rules was given by Dewar [70] and Knott [49].

Dewar derived the regularities by perturbation theoretical analysis, considering polymethine-cyanines as iso-conjugated hydrocarbons. Although vinylogous carbanions may

Substitution of =CH–	X – CH(1) – CH(2) – CH(3)···CH–CH–X		
	1	2	3
More Electronegative Elements	bath.	hyps.	bath.
Less Electronegative Elements	hyps.	bath.	hyps.

Fig. XII.3. Spectral effects resulting from internal substitution along the polymethine chain

Attachment to =CH-	X–$\overset{1}{C}$H–$\overset{2}{C}$H–$\overset{3}{C}$H···CH–CH–X
Donor	hyps. bath. hyps.
Acceptor	bath. hyps. bath.

Fig. XII.4. Spectral effects resulting from external substitution of polymethines along the polymethine chain

represent poor reference structures for cationic polymethines in some respects, they nonetheless enable useful spectral predictions to be made. Since the vinylogous carbanions belong to the odd-membered alternant hydrocarbons, they have a non-bonding HOMO, which exhibits non-zero expansion coefficients only in positions designated above as even-numbered ones. This feature reflects the HOMO of the polymethine quite well (cf. Fig. XII.2) and the same conclusions can therefore be attained by starting from the wave function of the hydrocarbons. Since this approach is not restricted to straight chain structures, predictions can easily be made for more complex polymethine dyes, and also diphenylmethane and triphenylmethane dyes.

The perturbational approach has also proved to be highly useful in rationalizing the effect of departure from the molecular coplanarity [690]. Since all bond orders decrease upon excitation in the case of idealized polymethines, twisting around any bond should cause a bathochromic effect. This is indeed observed. Examples are given in Sect. XII.2.

More complex polymethine dyes can be considered, in the main, from two distinct points of view: either as derivatives of streptopolymethines (or their iso-π-electronic hydrocarbon) according to model *B*, or as vinylogous carbocations or carbanions which bear terminal substituents according to model *A* (cf. Fig. XII.5).

Whereas model *B* permits transfer of experience and rules which are valid for simple chain structures to more complicated ones, model *A* enables comparison of the distinct effects of end groups.

It should here only be mentioned that many of the dyes discussed in previous sections might be considered as being related to polymethine dyes, because polymethinic

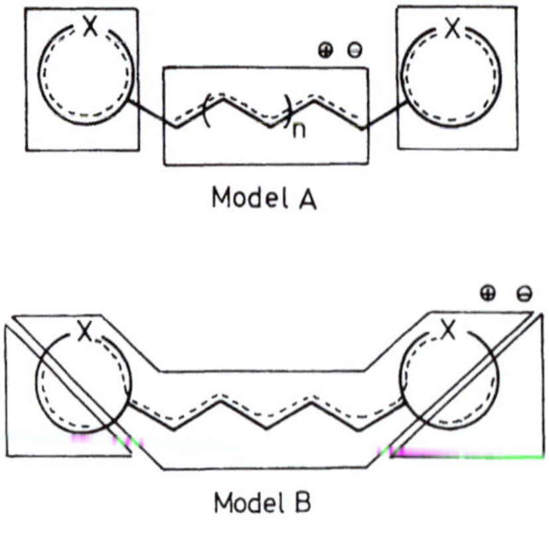

Fig. XII.5. Different fragmentation modes in describing heterocyclic cyanine dyes

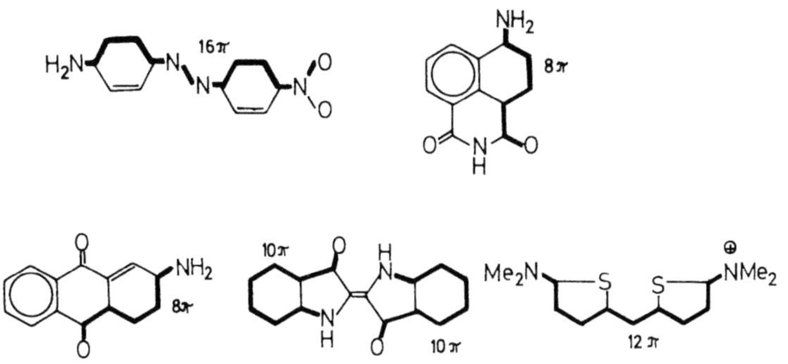

Fig. XII.6. Polymethine subchromophores in complex dye structures

structure are formally involved (Dähne's polymethine coupling conception [24]). Some examples are given in Fig. XII.6.

However, PPP configuration analyses have revealed that modelling in molecular building blocks according to model *B* becomes less useful if the polymethinic subchain is involved in an aromatic ring system [116, 134].

2. Symmetrical Polymethine Dyes

The absorption wavelengths of symmetrical polymethine dyes strongly depend upon the nature of the terminal groups. Alkyl as well as aryl substitution at the terminal heteroatoms cause red shift as to be expected from perturbation theoretical arguments. This feature is exemplified in Table XII.2 with streptopolymethine-cyanines *4* terminally sub-

Table XII.2. Longest-wavelength absorption maxima of simple streptopolymethine-cyanines

n		λ_{max} in nm						
		4a[a]	*4b*[b]	*4c*[c]	*4d*[d]	*4e*[e]	*5*[f]	*6*[g]
	R:	H	H	CH_3	H	CH_3		CH_3
	R':	H	CH_3	CH_3	C_6H_5	C_6H_5		
0				224				610
1		286	296	312.5	382	350	434	693
2		378	395	416	485		545	790
3			493	519	590	552	636	883
4				625	690	657	740	

[a] in water [881]
[b] in methanol [957]
[c] in methylene chloride [701]
[d] Ref. [958]
[e] in ethanol [959]
[f] in ethanol [720]
[g] in acetic acid [691]

stituted by methyl or phenyl groups. The blue shift observed in passing from *4d* to *4e* probably results from steric hindrance resulting in molecular non-coplanarity.

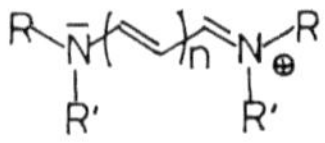

4

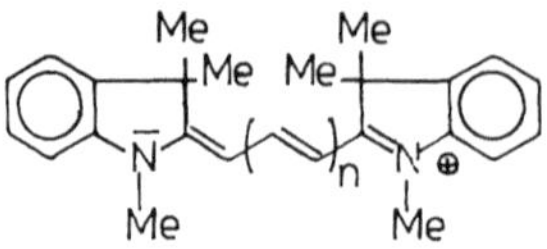

5

R,R′ cf. Table XII.2

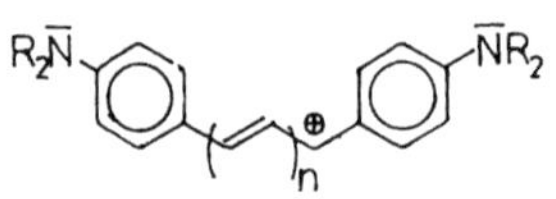

6

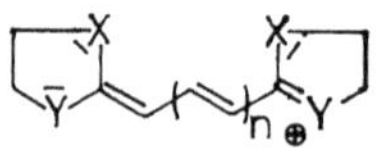

7

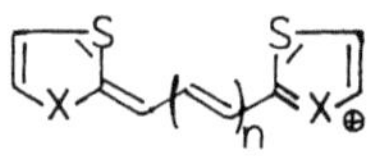

8

X,Y cf. Table XII.3

This explanation is in line with the fact that rigidized structures such as *5* display red-shifted color bands in comparison with *4d*.

The bathochromic substituent effect resulting from terminal substitution can easily be explained by considering streptopolymethines as ω,ω′-di-substituted vinylogous carbocations according to model *A*. As depicted in Fig. XII.7, HOMO substituent orbitals interact with the NBMO of the cationic chain and the lowest-energy transition occurs between the resultant delocalized orbitals. If substituents at the terminal nitrogen atoms raise the energy of the occupied MO (e.g. alkyl or phenyl groups), the excitation energy is lowered, as observed experimentally. This rough model is also useful for explaining other experimental findings, such as the bathochromicity of *6* in comparison with *4d*. The iso-π-electronic compounds of these series only differ by the linking of the aniline residues to the central carbocationic chain. The vinylogous diphenylmethane dyes *6* absorb at considerably longer wavelengths than N,N′-diphenyl streptocyanines *4d*, because the equal substituents distinctly interact with the central carbocation. Since the HOMO expansion coefficients of aniline are considerably larger at its nitrogen atom than at the carbon atom in *para* position, the resulting energy gap between the HOMO and

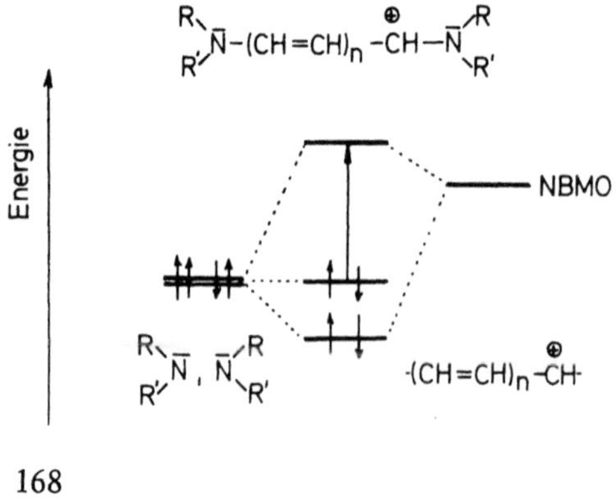

Fig. XII.7. Cyanine dyes considered as substituted vinylogous carbocations

Table XII.3. Longest-wavelength absorption maxima of simple streptopolymethine-cyanines with two heteroatoms in each group

n		λ_{max} in nm					
		7a[a]	*7b*[b]	*7c*[c]	*7d*[d]	*8a*[e]	*8b*[f]
	X:	O	NMe	NEt	S	NEt	S
	Y:	O	NMe	S	S		
0				335	420	409	521
1		383	407	442	548	556	681
2			499.5	540		640	

[a] 4,5,4′,5′-tetramethyl-*7a*, in chloroform [960]
[b] in methylene chloride [961]
[c] in methanol [962]
[d] in acetic acid [963]
[e] 4,4′-dimethyl-*8a*, in methanol [962]
[f] in methylene chloride, aliphatic residues in the 4,5- and 4′,5′-positions of the end group [696]

LUMO is larger for structures of the type *4d*. Consequently, *4d* exhibits higher excitation energies than *6*.

Fig. XII.7 also suggests that the bathochromic effect of terminal heteroatoms increases both with an increase in the donor capacity (decrease of the ionization potential) and a decrease in the bond strength (decrease of the interfragmental resonance integral). In good agreement, progressive red shift occurs in *7* in the sequence O < NMe < S. Examples are assembled in Table XII.3.

This sequence retains unaltered when the extrachromophoric aliphatic bridges are replaced by conjugative double bonds. These double bonds contribute to the chromophoric system [783]. A marked bathochromic shift occurs in passing from *7* to *8*.

Only one heteroatom is contained in the heterocyclic end groups of *9* to *12*.

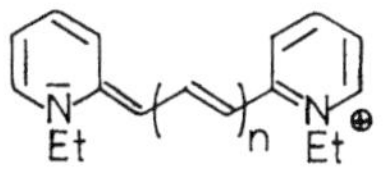

9

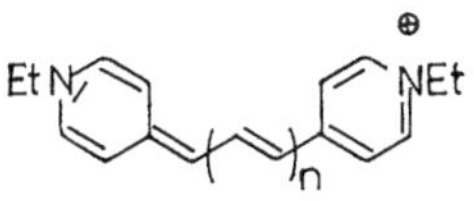

10

	9	10
n = 0	490 (4.63)	484 (5.30)
n = 1	558 (4.93)	603 (4.63)
	in ethanol [707, 784]	in ethanol [784]

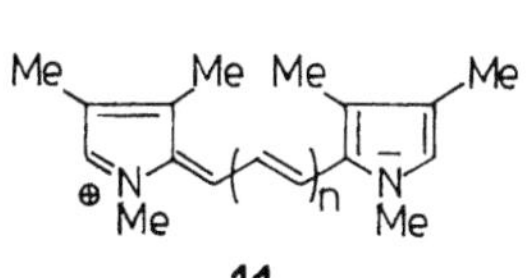

11

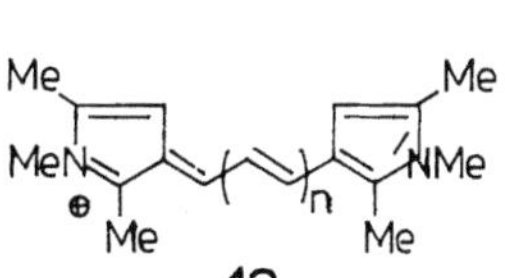

12

	11	12
n = 0	510 (4.75)	446 (4.54)
n = 1	581 (5.23)	536 (4.80)
	in chloroform [785]	in chloroform [785]

The absorption wavelengths depend, in that case, on the ring size, the position of linking and the steric situation. It is well known that bulky substituents adjacent to the methine group cause distortion around the bonds to the heterocyclic residues resulting in a bathochromic shift (Brunings-Corwin effect) [785], in good agreement with perturbation theoretical conclusions [690]. Like as predicted for simple streptopolymethines, the respective π bond orders of the heterocyclic polymethines decrease upon excitation. Consequently, distortion gives rise to a decrease of excitation energy. PPP calculations provided the same result [786].

The dyes *11* and *12* (n = 0) represent alkyl-substituted dipyrromethenes. The free-base dyes belong to the pyrrole pigments and are a constituent of bile pigments and the plant hormone Phytochrome (Bilatriene). The natural occurrence of these dyes led to extensive spectroscopic [787–789, 789a, b] and theoretical studies [788]. Alkyl-substituted dipyrromethenes of the structure *13* undergo protonation at the aza nitrogen. The conversion into the symmetrical polymethine chromophore is accompanied by a bathochromic shift of about 35 nm and increase in absorption intensity [787]. The probably all-*cis* configurated Bilatriene *14* [790] absorbs at considerably longer wavelengths than the dipyrromethene substructure. Theoretical calculations support a particular long-wavelength absorption in the all-*cis* configuration [791]. They also suggest marked environmental effects in biological systems [792].

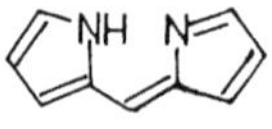

13

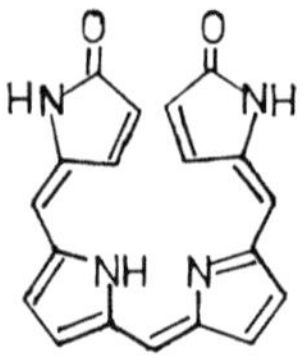

14

Entries in Tables XII.4 illustrate the spectral effect of terminal annelation in dyes of the structure *8*.

According to these spectral data the red shift is only very low when the terminal double bonds are replaced by benzo residues, but increases noticeably with the introduction of more extended conjugated ring systems. Pronounced spectral annelation effects have also been found in compounds of the structure *10*. For example, replacement of pyridinium by quinolinium or acridinium brings about color band shifts of about 100 [760] and 200 nm [793], resp.

Whereas terminal annelation of polymethine dyes can cause a considerable red shift, the color band is only weakly influenced by substituents attached at the terminal benzo residues, e.g. in the 5,5'- or 6,6'-positions of *15* (X = NEt, Y = NR, O, S; n = 1) [794–797]. The spectral shifts are nearly additive. Deviations from the additivity were attributed to sterically forcing the substituents out of the plane of the heterocyclic ring [798]. The deviations increased with increasing bulk of the substituents. The spectral effect of external substitution was also studied in other series, such as *4d* [799, 800], and was found to be moderate.

Table XII.4. Spectral effect of annelation on the color band of cyanine dyes *8* (X = NEt)

Derivatives of *8* (n = 1)

End Group	λ_{max} (in nm)	Ref.	End Group	λ_{max} (in nm)	Ref.
unsubstituted	550	[760]	phenanthreno[4,5-d]-	612,5	[964]
benzo[1,2-d]-	557	[760]	acenaphtheno[d]-	670	[795]
naphtho[1,2-d]-	597	[760]	azuleno[5,6-d]-[a]	630	[965]
naphtho[2,3-d]-	570	[760]	pyreno[4,3-d]-	600	[966]
naphtho[2,1-d]-	593	[760]	anthraceno[1,2-d]-	619	[967]
			anthraceno[2,1-d]-	621	[967]

[a] 3,8-dimethyl derivative

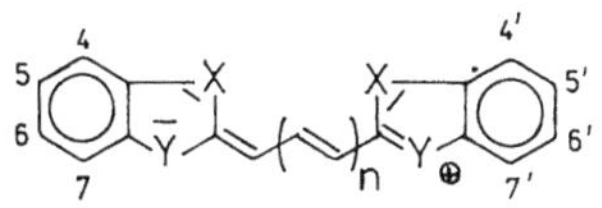

15

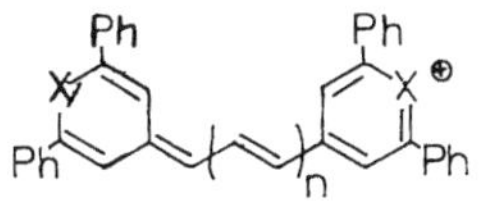

16

Pronounced spectral shifts are accompanied with heteroatomic substitution within the heterocyclic residues. Data tabulated in Table XII.5 demonstrate that sulphur- and selenium-containing polymethine dyes are always more strongly bathochromic then the heteroanalogous oxygen or nitrogen compounds. Similar results have been found in heteroanalogous series which differ from *16* by 2-substitution instead of 4-substitution of the terminal heterocyclic group [801].

The data collected in the Table XII.5, as well as those in the Tables XII.2 and XII.3, show that lengthening of the conjugated chain by one additional vinylene group has the effect of shifting the absorption wavelength by an amount which stays fairly constant at about 100 nm. Thus, the absorption wavelengths vary for streptocyanines approximately linearly with the number of double bonds (cf. Sect. XII.1).

If the absorption wavelengths are alternatively correlated with the number of methine groups of the polymethine chain, the "chain constant" A and the "color factor" B are obtained [802]. Whereas the constant A represents a characteristic value for the increase of the wavelength per methine group, the constant B defines the wavelength region of absorption discriminative for the end group. Chain constants and color factors of the series *15* and *16* are given in Table XII.5.

Irregularities in the vinylene shift have been observed in two cases. Both deviations concern the shift between the first two members of the vinylogous series. In the first case the vinylene shift is lower than expected, for example in the series *15f* in Table XII.5. In the second case, contrarily, the vinylene shift is larger than expected, for example in the series *15d*, *15e* and *15g* in Table XII.5 and *7c*, *7d*, *8a*, and *8b* in Table XII.3.

The first kind of anomaly results from steric hindrance in the monomethine cyanine, which prevents the molecules from being coplanar. Thus, *15f* is known to be distorted around the central carbon-carbon bonds [803, 804]. This distortion brings about red shift

Table XII.5. Influence of the heteroatom on the position of the color band in heterocyclic cyanine dyes. A and B are the chain constants and color factors of the series [802]

n		λ_{max} in nm							
		15a[a]	*15b*[a]	*15c*[b]	*15d*[b]	*15e*[b]	*15f*[c]	*15g*[a]	*15h*[d]
	X/Y:	NMe/ CMe_2	NEt/NEt	NEt/O	NEt/S	NEt/Se	NEt/ CH=CH	S/S	S/O
0		434	386	376	425	429	525.5	504	440
1		545	500	484	560	572	605	654	558
2		642	583[e]	583	655	660	708	758	
3		740[c]	685[e]	683	760	770	810		
4					870				
5					995				
A		49		50	54[f]	50[f]	51	52[f]	
B		397		332	410	416	452	498	

n		λ_{max} in nm			
		16a[g]	*16b*[g]	*16c*[g]	*16d*[g]
	X:	NEt	O	S	Se
0		512	552	627	672
1		638	676	755	795
2		748	798	879	910
3				1,000	1,035

[a] in acetic acid [752]
[b] in ethanol [968]
[c] taken from Ref. [760]
[d] in acetic acid [963]
[e] in methylene chloride [961]
[f] value calculated without the monomethine absorption
[g] in nitromethane [793]

for reasons discussed in Sect. XII.1. The "vinylene jump" from the monomethine to the trimethine cyanine is therefore lower than that between higher vinylogues of the same series.

The second kind of anomaly concerns exclusively sulphur- and selenium-containing polymethine dyes [805]. The enlarged vinylene shift is here due to the non-bonding sulphur-sulphur interaction. This effect can be explained by first-order perturbation theory if the sulphur atoms are presumed to overlap in a *cis* configuration. Since the HOMO of (4m + 2) π systems is symmetrical and the LUMO antisymmetrical, introduction of an additional bond parameter between the adjacent sulphur atoms increases the spacing between the frontier orbitals (left hand part of Fig. XII.8) [696]. In that case, monomethines show blue-shifted color bands and, consequently, enhanced vinylene shifts. Some oxonols belong, however, to the 4m π electron systems. The spectral effect of the non-bond-

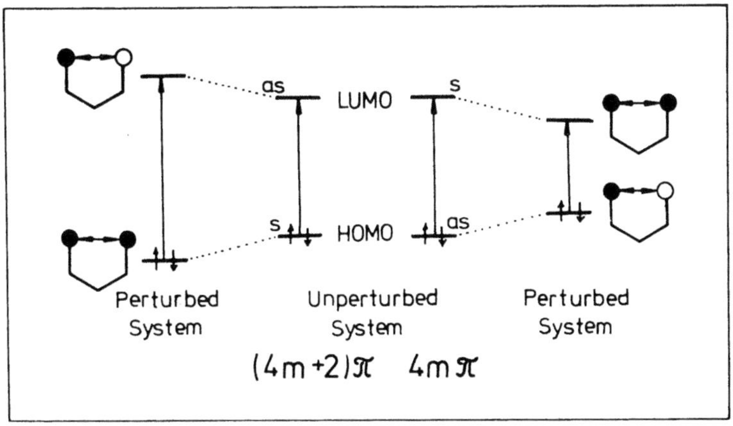

Fig. XII.8. Effect of the no-bond interaction between terminal atoms of symmetrical cyanine dyes (indicated by arrows) on the frontier orbitals in dependence on the number of π electrons

ing sulphur-sulphur interaction is now opposite (right hand part of Fig. XII.8). The absorption maxima of these dyes appear at longer wavelengths than expected, giving rise to reduced vinylene shifts. Examples of this type have been found in the dithienylmethane series [696].

Cyanine dyes generally have all-*trans* structure, but they may undergo photoisomerization [762, 806, 807]. *Cis* isomers absorb with lower intensity (lower oscillator strengths f) than the respective *trans* isomers. This relationship is supported, for instance, by the spectral properties of the rigidized structures *17–19* [808]. Bent mono-*cis* isomers are less bathochromic than the all-*trans* extended dyes.

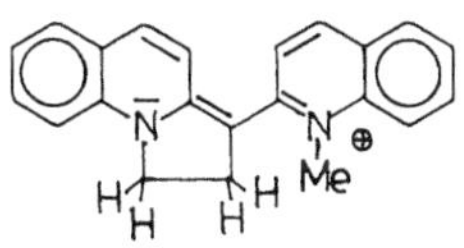

17

all-*trans*
571 (4.80)
(f 0.71)

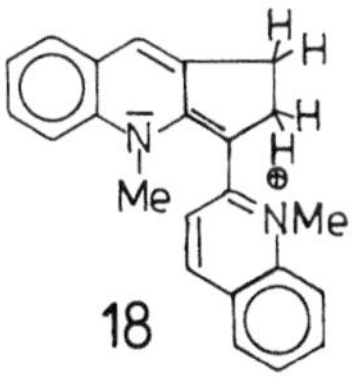

18

mono-*cis*
560 (4.69)
(f 0.55)

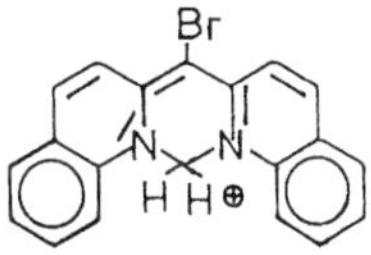

19

di-*cis*
557 (4.31)
(f 0.23)

in ethanol [808]

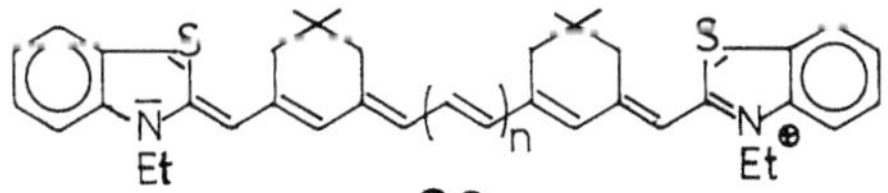

20

n = 1   1020 (5.00)
n = 2   1140 (4.54)
n = 3   1240 (3.85)

in pyridine [809]

There is some evidence for increasing tendency to *trans-cis* isomerization with increasing chain length [809]. Incorporation of the polymethine chain into cyclohexane [810], decalene [811] or tetradecahydroanthracene rings [812] increases the proportion of *trans* isomer over *cis*. But even for compounds of the series *20* a steady decrease of absorption intensity has been observed. The increasing hypsochromicity is accompanied by a flattening-out of the longest-wavelength absorption band. This conspicuous feature of cyanine dyes of very long chain length has been attributed to the presence of equilibrium mixtures of stereoisomers [809].

Methylene bridged compounds are also known in other series of cyanine dyes. Stable heptamethine pyrylium dyes, which absorb in the infrared region, have been recently advanced as lasing material [813].

Less spectral data are accessible about the negatively charged oxonols derived from streptooxonols *2*. Incorporation of the terminal CO-group into heterocyclic residues brings about pronounced red shifts, as exemplified with the dyes *21–26* in Table XII.6.

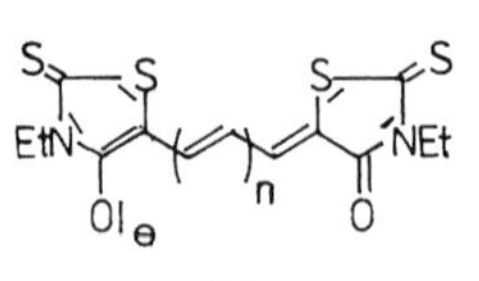
21

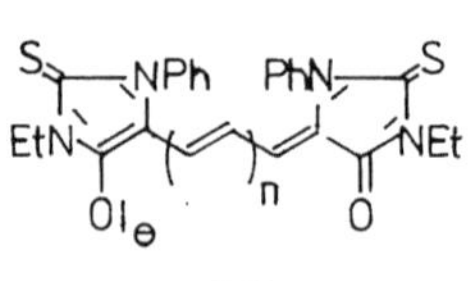
22

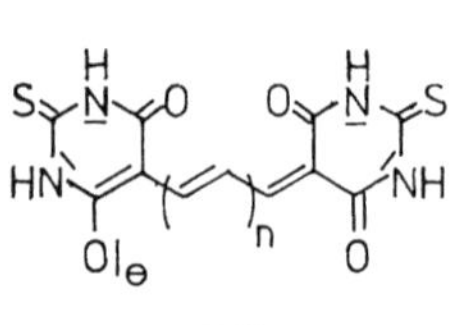
23

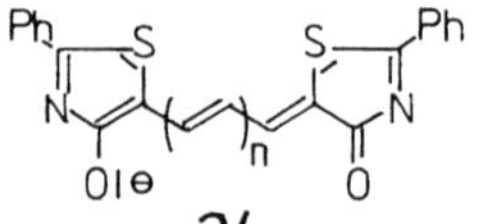
24

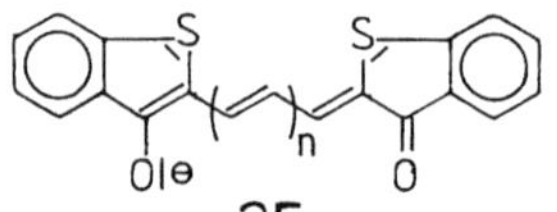
25

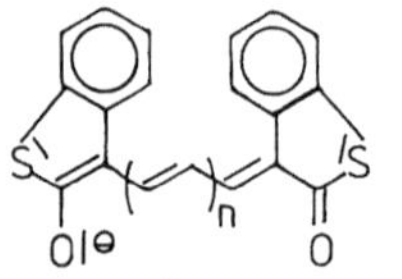
26

Owing to the instability of higher vinylogous oxonols only a few compounds have been synthesized hitherto.

Other series of negatively charged polymethine dyes contain terminal nitrile groups such as *27* [814, 815]. The absorption behaviour of *27* was studied by PPP calculations

Table XII.6. Longest-wavelength absorption maxima of heterocyclic oxonols

n	λ_{max} in nm					
	21[a]	*22*[b]	*23*[c]	*24*[d]	*25*[e]	*26*[f]
0	540	530	453	600	585	500
1	613		532	678	650	
2	714		630	781		

a in methanol [760]
b in methanol/piperidine [842, 969]
c in methanol [970]
d in methanol [971]
e in ethanol/Et_3N [760]
f in ethanol [972]

[81, 816]. Also the Meisenheimer-type complex *28* may be regarded as a *cis*-fixed anionic polymethine with terminal NO_2 groups [817, 818].

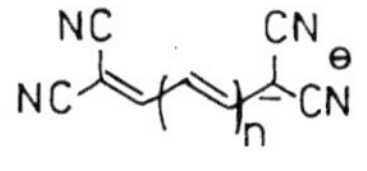

27

n = 0 346 (3.50)
n = 1 440 (3.94)
n = 2 540 (4.20)
n = 3 632

in methanol [814, 815]

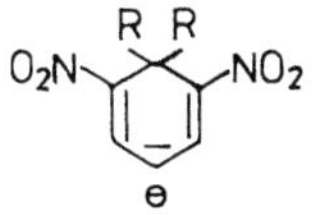

28

R = OCH_3 584 (4.37)
in methanol [817]

The cyanine and oxonol dyes suffer an appreciable spectral shift to either longer or shorter wavelengths if one or several of the methine groups are replaced by aza groups. The spectral data of *29* to *34* assembled in Table XII.7 afford striking examples.

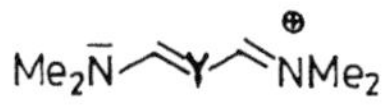

29

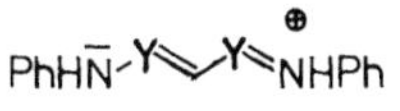

30

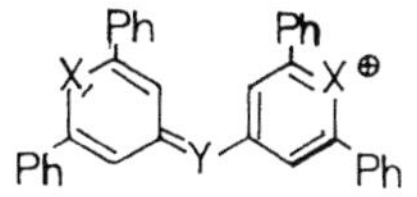

31

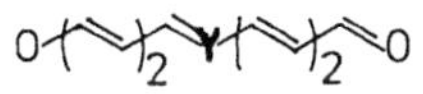

32

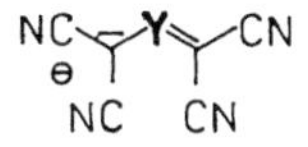

33

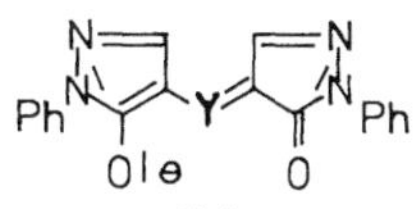

34

Table XII.7. Spectral influence of heteroatomic substitution of the chain methine groups by aza nitrogen

Y	λ_{max} in nm						
	29	*30*	*31*		*32*	*33*	*34*
			X = O	X = S			
CH	312.5[a]	382[c]	543[e]	627[e]	644[a]	346[h]	438[j]
N	268[b]	650[d]	436[f]	499[f]	552[g]	370[i]	528[j]

[a] in methylene chloride [701]
[b] Ref. [973]
[c] Ref. [958]
[d] in methanol [974]
[e] in nitromethane [793]
[f] in acetonitrile [975]
[g] Ref. [976]
[h] in methanol [815]
[i] Ref. [977]
[j] in methanol/Et_3N [978]

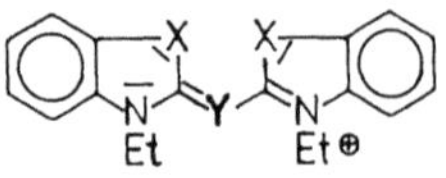

35

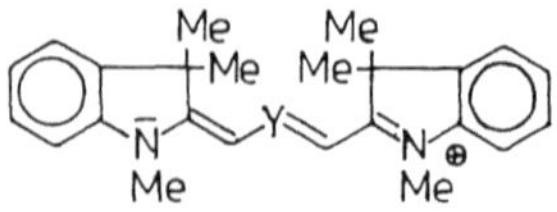

36

The direction of shift is completely in line with the Dewar-Knott rule [49, 70] and corresponds to that predicted for the polymethine subchain (cf. Sect. XII.1). The spectral shift is weak if nitrogen atoms are introduced in odd- as well as evennumbered positions of the polymethine chain. Experimental data of multiply aza-substituted cyanine dyes are compiled in Ref. [29, 819].

When the heteroatomic group is varied in the same position, the spectral shift reflects the electronegativity of the heteroatoms. This is illustrated in two series of the structures *35* in Table XII.8 and is in line with more recently published data [817a]. Whereas the nitrogen atom causes blue shift in comparison with the unsubstituted compound, red shift is observed on introduction of phosphorus or arsenic. The red shift is strengthened by the decrease in bond strength between the heteroatom and the adjacent carbon atoms. In the trimethine dye *36* electronegativity and bond effect are contrary. Since bathochromic shift has been observed by replacing the *meso* methine group by phosphorus, the bond effect obviously overcompensates the electronegativity effect.

In a more recent study the spectral effect of replacing a polymethine double bond by a triple bond has been discussed [820]. Dyes of the type *37* display visible absorption maxima hypsochromically displaced from those of the corresponding trimethine cyanine dyes by 30–40 nm (values of the respective parent compounds in round brackets):

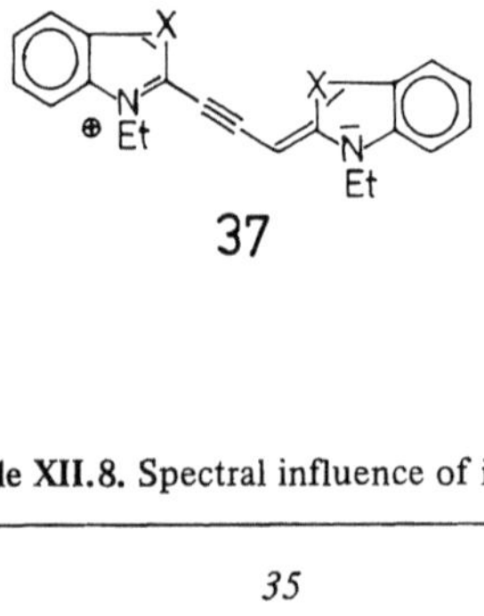

37

X = S	513 (5.00) [553 (5.20)]
X = CMe_2	513 (4.92) [541 (5.18)]
X = CH=CH	568 (4.94) [601 (5.18)]

in acetonitrile [820]

Table XII.8. Spectral influence of internal substitution by different heteroatoms

Y	*35*		*36*
	X = S λ_{max} (lg ε)	X = CH=CH λ_{max} (lg ε)	λ_{max} (lg ε)
N	375 (4.71)[a]	421.5 (4.84)[a]	602[d]
CH	422 (4.92)[b]	523 (4.93)[b]	553[d]
P	478 (4.71)[b]	605 (4.71)[b]	586[d]
As	491[c]	642[c]	

a in acetonitrile [979]
b in methanol [980]
c in ethylene chloride [981]
d in chloroform [982]

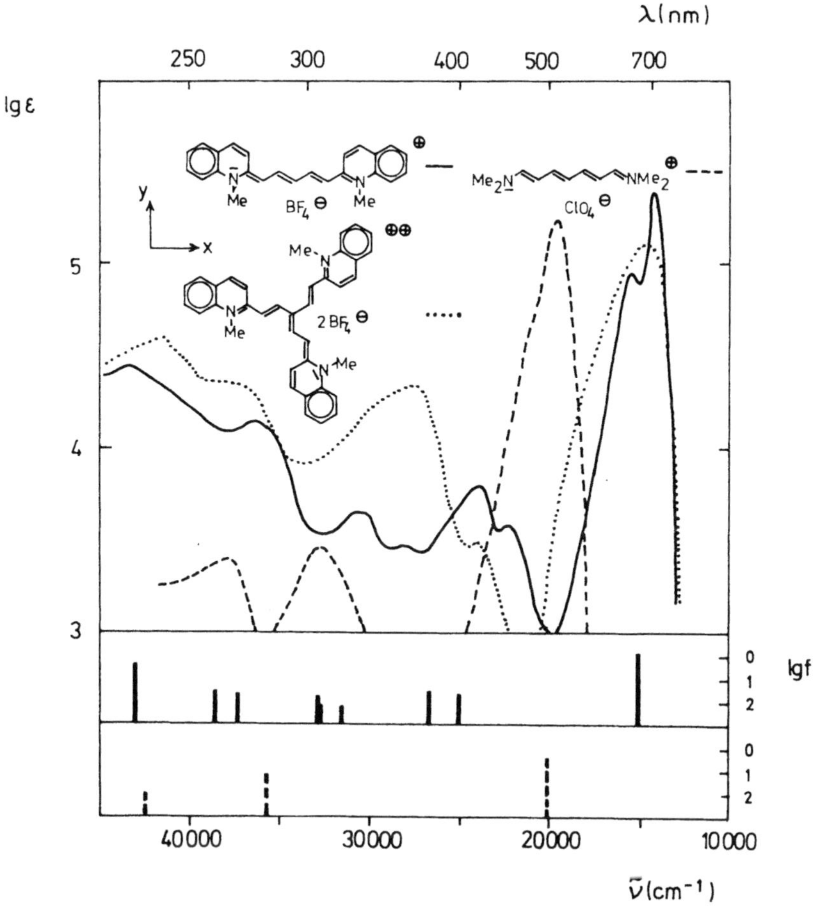

Fig. XII.9. Spectral absorption curves of the binuclear 2,2′-dicarbocyanine and the corresponding trinuclear [2.2.2]heptamethine in methanol [898], juxtaposed with the spectrum of streptopolymethine-cyanine in methanol [39]

The acetylenic cyanine dyes are no longer symmetrical [821]. Theoretical calculations suggest that the primary differences between acetylenic dyes and their symmetrical trimethine cyanine analogues result from the asymmetry in the conjugated π electron system [822].

In general, the spectral features of the symmetrical polymethine dyes have been successfully reproduced by PPP calculations [81, 86, 707, 789, 822–824]. For the sake of illustration both the spectral absorption curve and the theoretical results for 2,2′-dicarbocyanine are presented in Fig. XII.9 along with the spectrum of the parent streptoheptamethine-cyanine. The calculated lowest-energy transition of the dicarbocyanine is clearly separated from the higher-energy transitions in good agreement with the experimentally observed absorption features, and occurs at longer wavelength than that of the parent streptocyanine.

The lowest-energy transition of cyanine dyes is polarized along the longitudinal molecular axis (x-axis in Fig. XII.9). This result agrees with dichroic measurements [151, 825].

The absorption spectrum of 4,4′-carbocyanine (Cryptocyanine) has been discussed in more detail in connection with the fluorescence polarization spectrum [826].

The lowest-energy transition of symmetrical polymethine dyes is a nearly pure one-electron transition involving only the frontier orbitals. This fact may account for the good correlations obtained by plotting absorption wavenumbers of the color band against HMO transition energies [71, 74, 827–835].

Symmetric polymethines containing terminal methylene groups or carbocyclic rings instead of heteroatomic groups or heterocyclic ring have been excluded from this section. Spectral data of such compounds have been reviewed elsewhere [29, 836]. Theoretical results have been compared with experimental ones in Ref. [71] and [81].

3. Unsymmetrical Polymethine Dyes

Unsymmetrical polymethine dyes are encountered in unsymmetrical cyanines *38*, styrylcyanines *39* or merocyanines (neutrocyanines) *40*. These dyes contain carbocyclic or heterocyclic residues R_A and R_B, such as assembled in Table XII.9, or more simple acyclic heteroatomic groups. Their application for spectral sensitization prompted synthesis of a great variety of compounds and numerous spectral data have been recorded. Since representative data are tabulated and discussed elsewhere [13], only a few examples are given here.

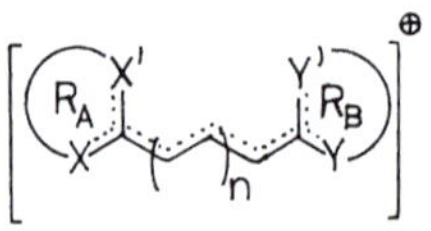

38

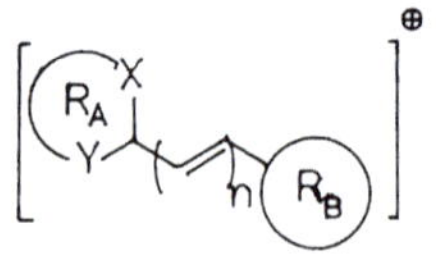

39

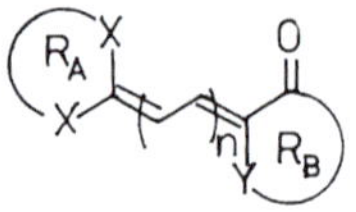

40

Unsymmetrical polymethine dyes of the general formula *41* display large vinylene shifts, but they are lower than those of symmetrical dyes. The vinylene shifts decrease with increasing chain length. On the other hand, the wavelength of an unsymmetrical polymethine dye λ_{AB} is, in general, lower than the arithmetic mean of the wavelengths of the corresponding two symmetrical dyes with the identical end groups ($R_B = R_A : \lambda_A$, $R_A = R_B : \lambda_B$). The wavelength difference $\Delta\lambda$, calculated according to Eq. (XII.3), is known as the "Brooker deviation" [837, 838] from Hamer's mean rule [839].

$R_A \cdots (CH \cdots CH)_n \cdots CH \cdots R_B$ **41**

$$\Delta\lambda = \lambda_{AB} - (\lambda_{AA} + \lambda_{BB})/2 \qquad \text{(XII.3)}$$

This deviation is generally negative in sign (negative or normal Brooker deviation). This is exemplified in Table XII.10. for dyes that contain an odd number of methine groups (such as *38*) as well as for dyes with an even number of methine groups (such as *39* and *40*).

Table XII.9. Some heterocyclic terminal groups R_A (R_B) involved in cyanine and merocyanine dyes of the structure *38* to *40*

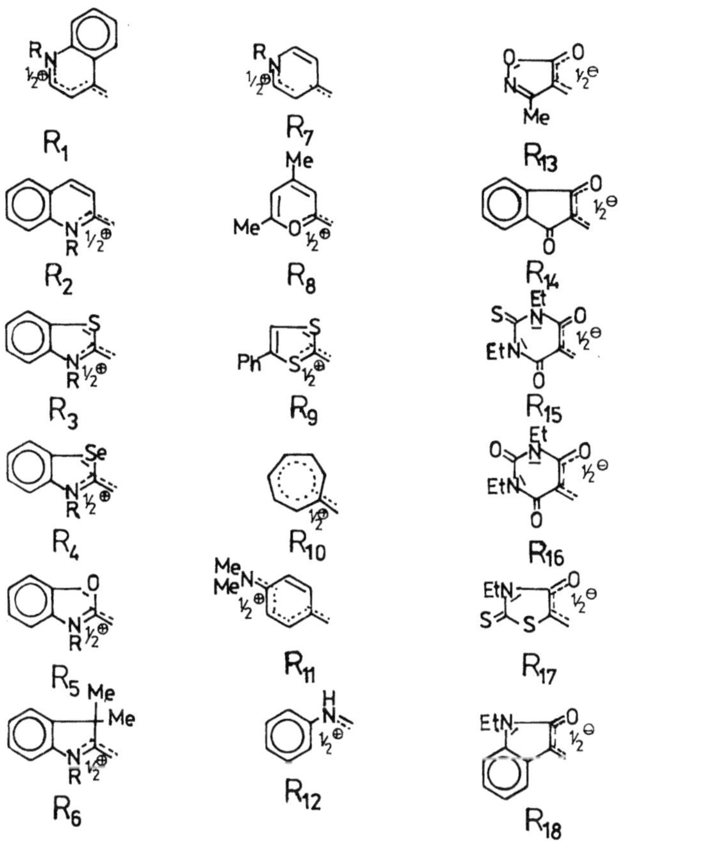

Their deviation can be calculated in a different manner referring to distinct sets of symmetrical parent compounds. This is illustrated schematically in Fig. XII.10.

Normally, the two calculated deviation values do not differ appreciably from one another, except in the presence of sulphur and selenium containing terminal groups, as discussed below. If the shape of the color band differs markedly, the maxima of the absorption bands can be replaced by their centres of gravity [840]. Experimentally derived shift values enable absorption wavelengths of unsymmetrical polymethine dyes to be calculated. As shown recently by Il'chenko [841] theoretical and experimental values satisfactorily agree within a limit of 10 nm.

According to Brooker [44] and Platt [51] the deviation is best explained by resonance theory. This approach suggests that the deviation increases with increasing non-equivalence of the resonance structures *42a* and *42b*, which, in turn, is determined by the different ability of the end groups to release electrons ("Brooker basicity" of the end groups).

Table XII.10. Longest-wavelength absorption maxima λ (in nm) of unsymmetrical polymethine dyes *38* to *40* (with n = 1) and the Brooker deviation Δλ (in nm). The residues R_A and R_B are specified in Table XII.9

			Normal Brooker Deviation				Abnormal Brooker Deviation		
Compound:	*38*		*39*		*40*		*38*		
R_B:	R_1		R_{11}		R_{17}		R_3		
R_A	$\lambda^{a,b}$	$\Delta\lambda$	λ^{c}	$\Delta\lambda^{d}$	λ^{c}	$\Delta\lambda^{e}$	$\lambda^{a,b}$	$\Delta\lambda_1{}^{f}$	$\Delta\lambda_2{}^{g}$
R_1	705	–	546.5	–110	616	–2.5	630	+13.5	–2
R_2	656	–1	525	–80	565	–3	578	+12.5	–3.0
R_3	630	–1	524	–58	526	–21	557.5	+21	±0
R_4	632[h]	–0.5	536	–52	526	–25	562[h]	+20.5	–2
R_5	596	–3	496	–48	490	–17.5	520	+18.5	±0
R_6	604	–21	546	–30	503.5	–36	542	+12.5	–10

[a] Ref. [760] [b] in methanol [983] [c] in methanol [842]
symmetrical reference compounds:
[d] *6* (n = 1), *15* (n = 0)
[e] *15* (n = 1), *21* (n = 0)
[f] *15*, X = S (n = 0), *15* X ≠ S (n = 1)
[g] *15*, X = S (n = 1), X ≠ S (n = 0)
[h] Ref. [616]

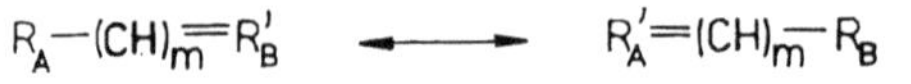

a 42 b

Experimental results led to the following sequence of Brooker's basicity [842] (R_A in Table XII.10).

$$R_{A(B)}: R_1 > R_2 > R_3 > R_4 > R_5 > R_6$$

The ^{13}C-NMR chemical shifts in β-position of pentamethine dyes support the relationship between the electron releasing effect of the terminal group and Brooker's basicity [843]. A relationship even exists to the true basicity as measured by pK_a values [844].

Whereas the normal Brooker deviation has found some foundation within the framework of the VB theory [44, 51], the first MO theoretical considerations were less successful. Herzfeld and Sklar [46] assumed that the deviation largely results from bond alternation which increases with the increasing non-equivalence of the end groups. This conclusion contradicts, however, with the fact that the deviation increases as the chain length increases. Later attempts aimed to reproduce the Brooker sequence of basicity numerically by HMO calculations [845], or made use of the perturbation theoretical approach in the framework of the one-electron approximation [123].

In the case of sulphur- and selenium-containing polymethine dyes positive deviation values have been calculated (anomalous Brooker deviation). This deviation has a com-

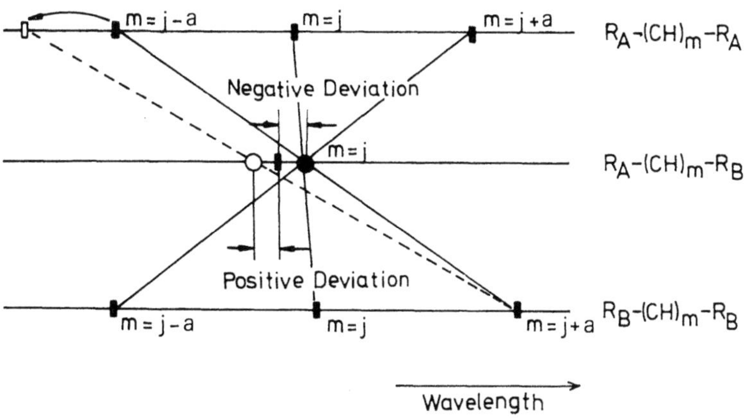

Fig. XII.10. Derivation of mean values of non-symmetrical cyanine dyes from absorption wavelengths of appropriate symmetrical dyes and deviation in respect to the experimental value (j = 3,5, ...; a = 2 or j = 2,4, ...; a = 1; the experimental values are indicated by full rectangles, the calculated mean value by a full cycles). Whereas negative deviation occurs in the case of regular vinylene shifts of the symmetrical parent compounds, positive deviation results, if one of the shorterchain reference compound absorbs at anomalous short wavelength (open rectangle)

pletely different origin as the normal deviation [696]. Due to the occurrence of the non-bonding sulphur-sulphur interaction, some sulphur- and selenium-containing heterocyclic monomethine dyes absorb at considerably shorter wavelengths than expected from the regular vinylene shift (cf. p. 171). This interaction is absent in unsymmetrical dyes which only have one sulphur- or selenium-containing residue. Therefore the reference mean value is too low, resulting in a positive deviation. The appearance of the positive deviation is also visualized in Fig. XII.10.

External substituent effects have been extensively studied for styryl dyes *39* which bear substituents in *para* position of the phenyl group. Some examples are given in Table XII.11. The longest-wavelength absorption maxima correlate excellently with ex-

Table XII.11. Longest-wavelength absorption maxima of styrylcyanines (*39*, n = 1, R_B = p-X–C_6H_4, R_A is specified in Table XII.9)

X		λ_{max} in nm			
	R_A:	R_7[a]	R_8[b]	R_9[b]	R_{10}[c]
H		335	409	460	440
Me			434	480	461
OH		375	466	519	
OMe		368	476	524	503
SMe			500	551	
NMe_2		454	605	652	659

[a] in water [984] [b] in methylene chloride [985] [c] in acetonitrile [986]

perimental ionization potentials of the respective substituted benzenes [700]. This fact indicates the CT character of the lowest-energy transition. However, a progressive bond equalization along the chain occurs in passing from the unsubstituted to the amino-substituted styryl dyes. Hence the data demonstrate, at the same time, the steady changeover from the polyene-like to the polymethine-like structure. This is accompanied with a progressive deepening in colour.

Similar spectral properties have been observed for compounds, in which heterocyclic cations are directly linked to the donor substituted benzene (*39*, n = 0). *Para*-hydroxy substituted *39* (n = 0, 1) undergo deprotonation and provide merocyanines, e.g. *43, 44* and *45a*.

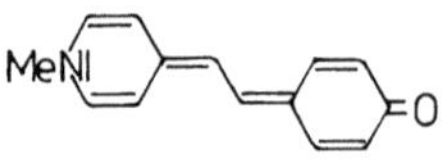

43

628

in chloroform [102]

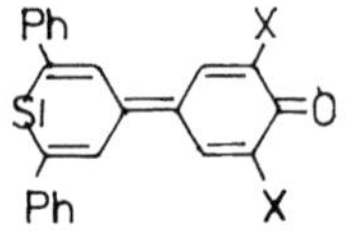

44

525 (4.58)

in methylene chloride [846]

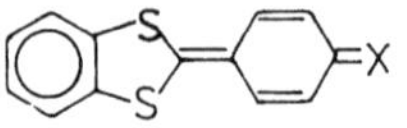

45 a: X=O
b: $X=C(CN)_2$

X = O: 450 (4.62)
$X = C(CN)_2$: 645 (4.75)
in chloroform [847]

The quinoid formulae *43* to *45* reflect the electronic structure better than alternative benzenoid formulae with charge separation. Blue shift occurs with benzo annelation in the quinoid part [846, 847]. The same behaviour has been observed for further merocyaninic quinone methides, such as *45b*.

The naturally occurring anthocyanidins are also aryl-substituted heterocyclic cations. The cyanidinium ion *46* and some related ions, which differ in the number of hydroxy groups, absorb consistently in the 550 nm region (in acidic ethanol) [440, 848, 849]. They are responsible for a variety of beautiful plant colours, particularly in flowers and fruits. Results of PPP calculations are in satisfactory agreement with experimental spectral data [149, 850, 851]. Calculations as well as luminescence spectra agree with the former prediction that the phenolic group is non-planar with respect to the benzopyrylium ring by an angle of 45° or more [149].

Betalains, for example Betaxanthin *47*, are another class of naturally occurring cyanine dyes. The chromophoric cation represents a slightly modified strepto-pentamethine and absorbs, as expected, at about 480 nm [852].

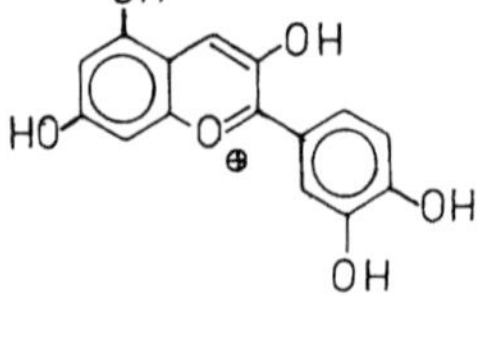

46

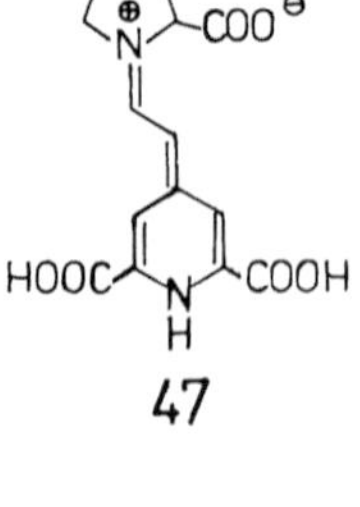

47

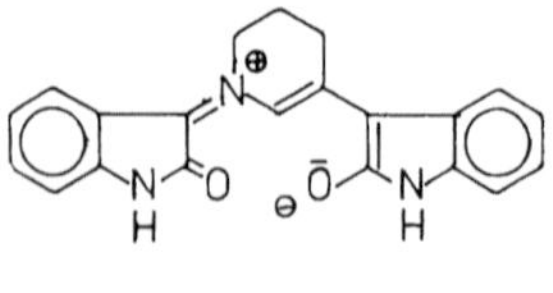

48

635 (4.60)
in dimethylformamide [860]

Numerous merocyanine dyes contain heterocyclic carbonyl compounds. Brooker and coworkers derived the following order of relative effectiveness of the residues R_B in formula *40,* which reflects the acceptor strength [842].

$$R_{13} > R_{14} > R_{15} > R_{16} > R_{17} > R_{18}$$

Long-wavelength absorptions are observed when donor and acceptor strength are sufficient to generate a polymethinic system with a high degree of bond equalization. Combination of both powerful donor groups (such as R_1) and powerful acceptor groups (such as R_9) bring about, however, very polar dye which may be more closely related to the conjugated betaine dyes rather than to the polymethine dyes (cf. Sect. XV).

Finally, aza substituted unsymmetrical polymethine dyes should be mentioned. Similar as described for symmetrical compounds red or blue shift occurs in dependence on the position of methine group substituted [853]. Since the formerly discussed color rules are also pertinent for that case, their longest-wavelength absorption feature is well understood. Mayer [331] has successfully rationalized the color of numerous disperse aza cyanine dyes in terms of their polymethinic subchains which undergo aza substitution. Numerous diazamerocyanines with nitrogen atoms in neighbouring position absorb in the same wavelength region as the parent compounds because of the opposite spectral shifts generated by aza substitution [331]. However, some diaza merocyanines absorb at considerably longer [331] or shorter wavelengths [335]. The diaza analogues of *39* belong to the first group.

Application in color photography [854] focussed attention to the electronic spectra of azomethine dyes derived from 5-pyrazolone [855, 856] and β-dicarbonyl compounds [857], which belong to aza merocyanines [858, 859]. The visible absorption spectra of pyrazolone azomethine dyes exhibit two maxima, the low intensity "y band" near 440 nm and the "x band" of higher intensity near 520 nm. In connection with these azomethine dyes the related indophenol dyes should be mentioned here [542] (cf. p. 97).

Isatin Blue *48* and related compounds also belong to the chain substituted oxonol dyes. They possess a betaine structure [860]. Due to replacing of one methine group by the azonium group in an odd-numbered position of the polymethine chain dyes of this structure display deep colors. PPP-CA calculations reproduce the position of the color band of *48* quite well and indicate a strong localization of the excitation in the polymethinic subchain [116].

4. Dyes with Substituted, Bridged and Coupled Chains

Three types of modification of the polymethine structure should be discussed and exemplified: external substitution, internal cyclization and coupling of the polymethine chain with cyclic or chain molecules.

The *external substituent effect* is governed by steric as well as by electronic effects. Steric effects may interfere with the coplanarity of the molecule or favour the *cis*-configuration of the dye molecule. This phenomenon complicates the interpretation of

Table XII.12. Spectral substituent effect on the longest-wavelength absorption maxima of the thiacyanine *49–51*

Compound	λ_{max} in nm									
X:	H	F	Me	OMe	Ph	Br	N=NPh	COR	CN	NO_2
49	556[a]	567[a]							523[a]	
50	556[a]	523[a]	543[b]	495[b]	560[b]			564[b] (R=OEt)	610[a]	
51	650[b]	666[c]	640[b]	662[d]	651[e]	645[b]	645[e]	597[f] (R=H)		559[g]

[a] in nitromethane [869]
[b] taken from Ref. [760]
[c] in MeOH [987, 987a]
[d] in EtOH [988]
[e] in MeOH [929]
[f] in MeOH [989]
[g] in MeOH [866]

shifts which result from electronic effects methyl [861–864], methoxy and methylmercapto substituents [865].

Well-defined and pronounced shifts arise from lone-pair donor as well as acceptor groups. Their attachment to the polymethine chain is characterized by the fact that the shifts differ in sign, in accord with arguments outlined in Sect. XII.1. Some examples are given in Table XII.12.

The data of the series *49* to *51* show, moreover, that the direction of the spectral shifts depends on the position of substitution (*49* and *50*) and becomes just reversed in passing from the *meso*(β)-substituted trimethines *50* to the *meso*(γ)-substituted pentamethines *51*. Similar effects have been found in other series [866]. In heptamethines, in turn, the direction of the spectral shift corresponds to that in trimethines [867, 867a]. The spectral shifts approximately follow Hammett substituent constants [866–868]. Unexpectately, cycloalkyl residues behave as electron acceptor groups [863].

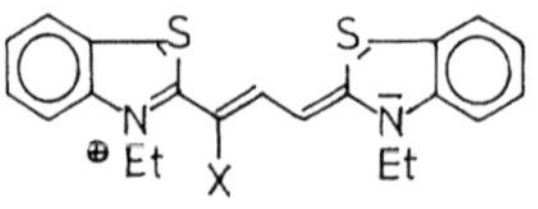

49

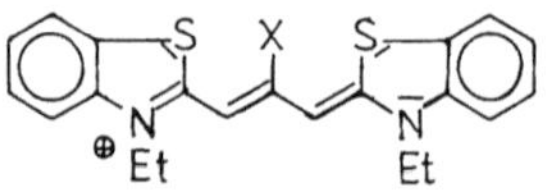

50

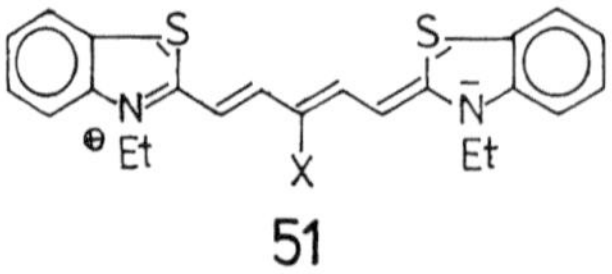

51

In the case of multiple substitution the substituent effect is nearly additive. This has been shown, for example, for disubstituted heterocyclic trimethines bearing substituents in the non-equivalent α,β-positions [869]. Interestingly, a dimethylene bridge in *52* acts as donor group and cause alternately bathochromic and hypsochromic shifts with increasing chain length (absorption wavelengths of the parent dyes in round brackets).

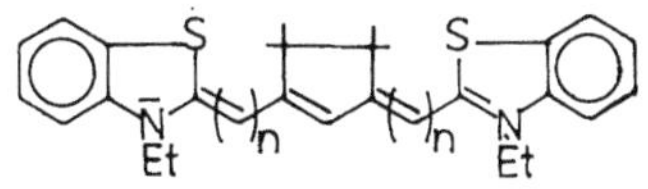

52

n = 0   599 (558)
n = 1   599 (650)
n = 2   802 (758)

in methanol [870]

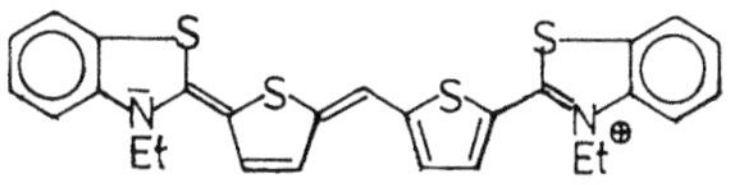

53

812 (860) [871]

The spectral effect is low when two donor substituents are attached, one at an odd- and the other one at an even-membered position. A special example of this type is the dye *53* where each σ-bivalent sulphur acts as twofold substituent leading to cyclic structures. Despite of the strong structural modification the dye still absorbs in the same spectral region as the parent dye.

The *effect of cyclization* is exemplified with the spectral absorption of the cyanine cations *54* to *56* and the oxonolanions *57* to *59* and their wavelength shifts with respect to the streptopolymethine structures (values in round brackets), which are visualized in the formulae by heavy lines.

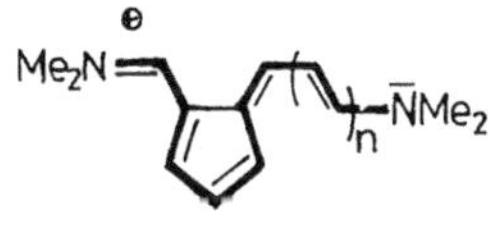

54

n = 0   409 (−110)
n = 1   505 (−120)
in methylene chloride [29, 872]

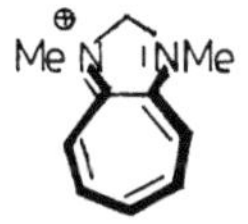

55

375 (−144)
in acetonitrile [873]

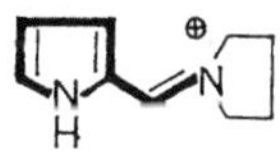

56

289 (−127)
in ethanol [874]

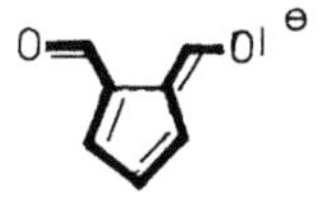

57

379 (−76)
in methanol [29]

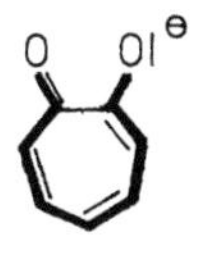

58

387 (−68)
in ethanol [875]

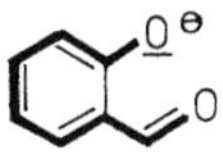

59

385 (−70)
in 0, 1 N NaOH [876]

In all above-mentioned cases blue shift has been observed in good agreement with conclusions obtained by first-order perturbation theory. These predictions are especially simple if the cyclized molecules retain a mirror plane of the parent structure which is perpendicular to the molecular plane, such as for *55*, *57* and *58*. These compounds contain ten π electrons. Since streptopolymethines with $(4m + 2)$ π electrons ($m = 1, 2, \ldots$) have a symmetrical HOMO and an antisymmetrical LUMO, the bond order between the joined chain position decreases upon excitation. Introduction of a π bond leads, consequently, to a blue shift. This consideration has been extended to cases of lower symmetry by taking explicitly into account the expansion coefficients of the parent compounds [877].

It should only be mentioned that numerous cyanine dyes with heterocyclic end groups, or cyanine dyes involving carbocyclic rings within the polymethine chain, could be considered as multiply-cyclized streptopolymethines, such as *60*, *61* and *62*.

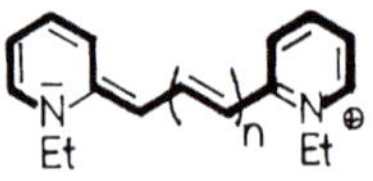

60

n = 0 480 (4.63)
n = 1 558 (4.93)
in ethanol [707, 784]

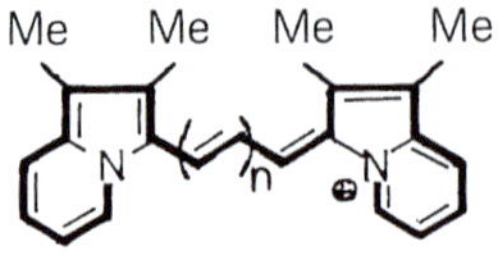

61

n = 0 629
n = 1 691
in ethanol [878, 991]

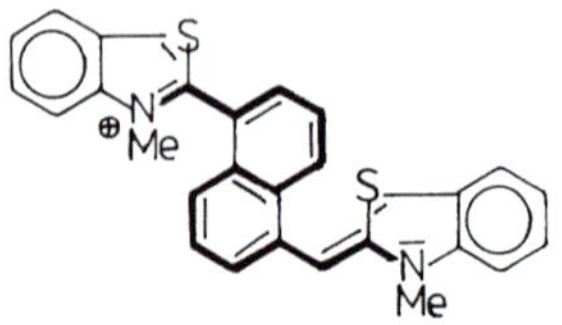

62

408 (3.97)
in ethanol [879]

Also in these cases the cyclized structures absorb at shorter wavelength than the hypothetical chain substructures, in good accordance with perturbation theoretical arguments. However, owing to the additional bonds the parent chain chromophore is now more strongly disturbed and the reference to the streptopolymethines is less close than in the above mentioned cases. This becames particularly apparent with *62*, where the individuality of the naphthalene fragment is essentially retained.

The *effect of coupling* of polymethines with chain and cyclic π fragments brings about mostly either noticeable red or blue shift. Strong red shifts have been observed, for example, for the cyanines *63*, *64* and *65* with respect to streptotrimethine-cyanine and streptoheptamethine-cyanine reference compounds, resp. (absorption wavelengths of the polymethine reference compounds in round brackets).

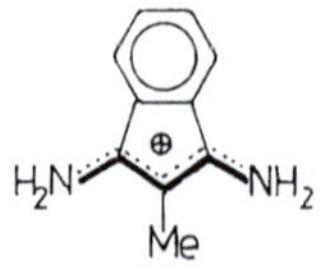

63

458 [880] (286) [881]
in ethanol

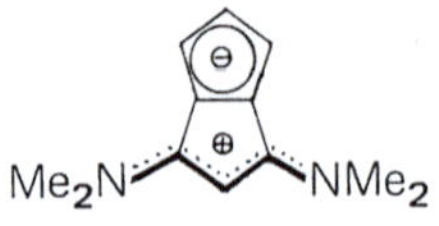

64

618 [882] (312.5) [701]
in methylene chloride

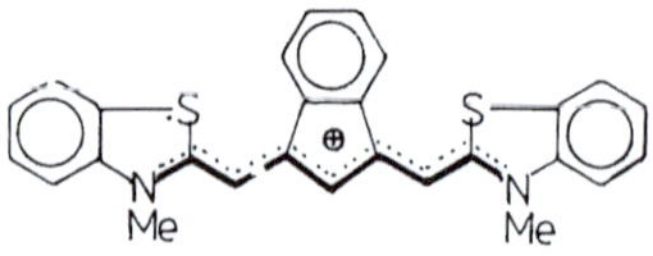

65

755 (652.5) [883]
in ethanol

On the other hand, pronounced blue shifts relative to streptopentamethine-cyanine results from the corresponding coupling to the dyes *66* and *67*.

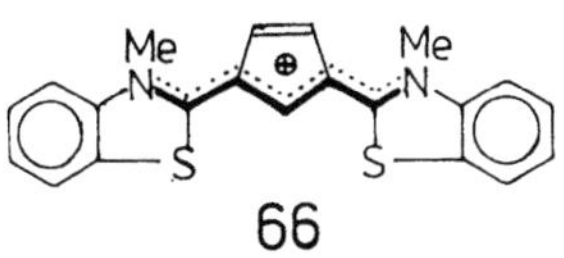

66

470 (558) [883]
in ethanol

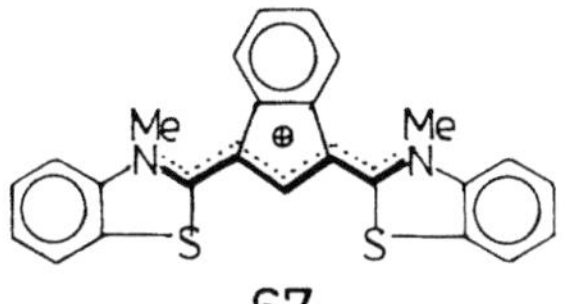

67

518 (558) [883]
in ethanol

The distinct spectral effect can be explained by considering the frontier orbital of the fragments [122, 129, 877]. The MO's of the symmetrically bivalent joined fragments are either symmetrical or antisymmetrical with respect to the common symmetry plane perpendicular to the molecular plane. Since in the first mentioned cases the HOMO-HOMO and LUMO-LUMO interaction of the fragments dominate for symmetry reasons, the lowest-energy excited state is more stabilized than the ground state and the transition energy is, consequently, lowered. In the second case the frontier orbitals undergo HOMO-LUMO and LUMO-HOMO interactions. Thus, the ground stabilization is larger than the excited state stabilization, which is reflected in an upturn of transition energy. Considerations along these lines have been extended to further symmetrically-joined structures, such as dyes resulting from coupling of polymethines with naphthalene [129]. These dyes have also been studied experimentally [832, 884].

Further examples of coupled polymethines are the compounds of the structure *68* and *69*. Whereas coupling to *68* is accompanied by a blue shift with respect to the uncoupled polymethine parent compounds (cf. Table XII.13), coupling to *69* gives rise, in the most cases, to a marked red shift (cf. Table XII.14). PPP-calculations have been performed to explain the red shift in *68* [884a].

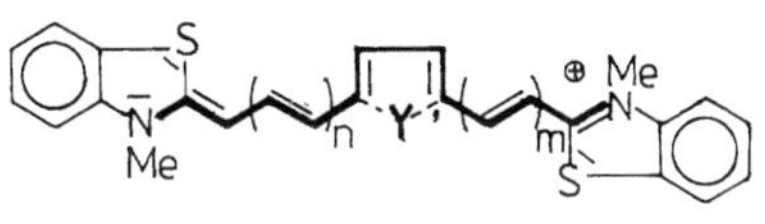

68

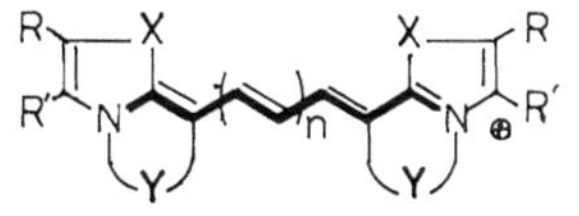

69

Table XII.13. Effect of coupling on the long-wavelength absorption maxima in thiacyanines *68* [990] (λ_{max} given in nm)

n	m	Y					
		H,H	O	S	CH=CH	Se	1,2-phenylene
0	0	650	596	604	520	620	514
1	0	762	676	695		730	
0	1	762	722	717	628	726	678

Table XII.14. Effect of twofold coupling on the longest-wavelength absorption maxima in symmetrical polymethines *69*

X	Y	R/R′	n	λ_{max} in nm	a	Ref.
CH=CH	Me–C=C–Me (with a vertical bond below each C)	H,H	0	587	588	[991]
			1	642		
			2	755		
MeN	Me–C=C–Me (with a vertical bond below each C)	1,2-benzo	0	569	500	[992]
			1	608	583	
S	1,2-phenylene	H, Me	0	552	557	[993]
			1	668	650	

a λ_{max} of the uncoupled parent dye

Drastic spectral effects are also encountered in compounds which arise from coupling of two polymethine moieties, as exemplified by *70* to *72*.

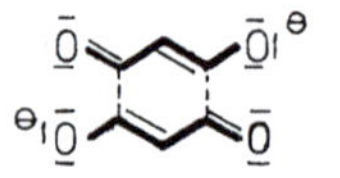

70

496 [24] (267.5) [701] in 0, 1 N NaOH

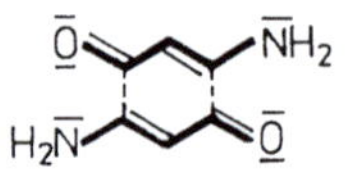

71

488 [24] (283) [701] in dimethylsulfoxide

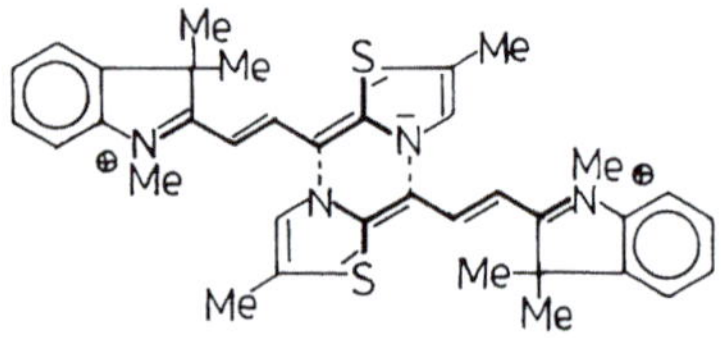

72

642 (552) [885]

For the sake of comparison the spectral data of the uncoupled parents are again given in round brackets.

The chromophoric *70* results from symmetrical junction of two oxonol trimethine chains. Because of the symmetry-allowed HOMO/HOMO and LUMO/LUMO overlap of the frontier orbitals, the large bathochromicity is readily understandable [129]. According to the symmetry conditions, σ-bivalent symmetrical junction of vinylogous chains leads only to deep-colored compounds, if the total number of π electrons of the composite molecule amounts to $4N\pi$ electrons ($N = 2, 3, \ldots$). In that case, moreover, the relative intensity of two lowest-energy electronic transitions is approximately described by molecular excition theory. Along this reasoning, for example, the lowest-energy transition of *70* is forbidden, for the parallel oriented transition moments compensate each other in the out-of-phase combination of the lowest-energy excited state.

The molecular excition model accounts for the splitting of the color band occurring with coupling of polymethine moieties via methylene bridges [886–888]. As might be expected, the absorption spectra of approximately σ-covalently bonded cyanines resemble those of dye dimers [889]. The molecular excition model has also been invoked to rationalize color bands of conjugatively-linked cyanine chains. Band splitting also occurs with joining cyanines via ethylene [890], benzene [891], naphthalene [892] or other conjugated residues. The more the angle ϕ between the transition moments of the sub-chromophores approaches to 180°, the more the color band is determined by the longer wavelength component. This is demonstrated, in the following, with *73* (n = 1, m = 0) [891].

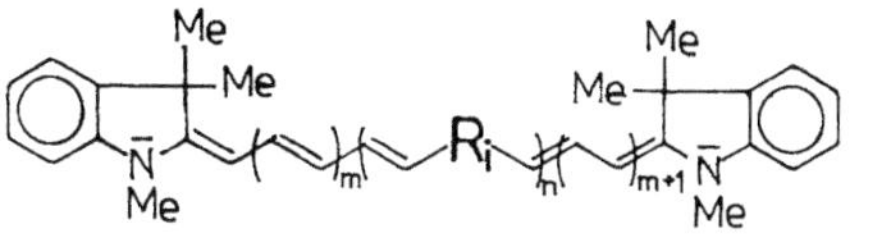

73

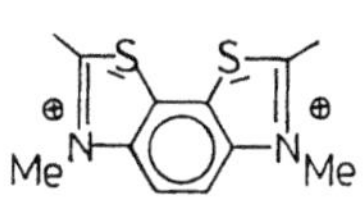

R_1

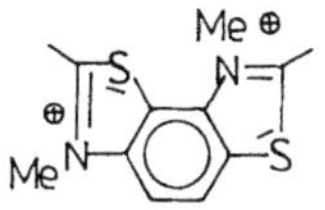

R_2

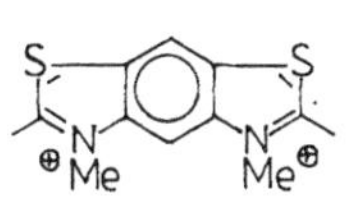

R_3

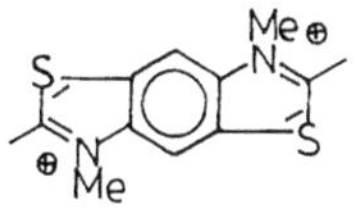

R_4

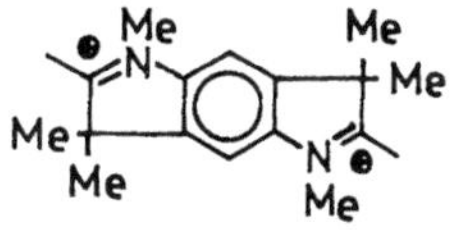

R_5

R_1: (ϕ 95°)	R_2: (ϕ 128°)	R_3: (ϕ 147°)	R_4: (ϕ 180°)
507 (5.24)	512 (3.99)	510 (3.58)	503 (–)
623 (5.22)	594 (5.54)	631 (5.54)	636 (5.59)

Theoretical studies of model type compounds have revealed the scope and limitations of the molecular excition model for these conjugated coupled cyanine dyes [133, 892a].

If the coplanarly arranged chromophores are inclined to each other, the two components of the color band are, as to be expected, polarized mutually orthogonally [893]. For a definite geometrical structure the splitting of the absorption depends on the nature of the terminal groups [894, 894a]. The splitting is largest when the Brooker's basicities of the terminal groups correspond to that of the central unit. As also exemplified in Table XII.15 with the series *73* (R_3, m = 0, 1, 2, . . .), the energy splitting (in cm^{-1}) hardly depends on the length of the polymethine chain [896].

Table XII.15. Longest-wavelength absorption maxima of biscyanines *73* of different chain length

R_i	m	n	λ_{max} in nm (lg ϵ) in nitromethane		$\Delta\lambda$ in nm	$\Delta\tilde{\nu}$ in cm^{-1}	Ref.
R_3	0	1	509 (4.57)	630 (5.56)	121	3,770	[896]
	1	1	595 (4.72)	760 (5.61)	165	3,630	
	2	1	685 (4.71)	894 (5.49)	209	3,400	
R_5	0	1	507 (4.34)	646 (5.55)	139	4,240	[895]
	0	2	495 (4.60)	699 (5.64)	204	5,890	
	0	3	529 (4.90)	729 (5.74)	200	5,180	

Tris-and tetrakis-cyanines of the series *73* (R_5, n = 2, 3) have also been prepared [895] (cf. Table XII.15). Due to the multiple splitting the long wavelength absorption pattern of these dyes is more complicated than that of the bis-cyanines.

5. Branched Polymethine Dyes

Many of the previously mentioned polymethines may be attributed to the class of branched polymethines. The term branched polymethine dyes was originally used by Dähne [23] to indicate dye molecules with two terminal heteroatoms which are connected by more than one polymethine chain, e.g. Hydrol blue-type dyes *6* or 4,4'-pyridocyanines *10*.

A more general definition of this term might also involve compounds which exhibit more than two heteroatomic end groups. According to this generalization the aforementioned substituted polymethines *50* (X = lone pair substituent groups) are branched polymethines.

In this subsection we confine the discussion to polynuclear polymethine dyes of the type *74* [897–899] and *75* [900].

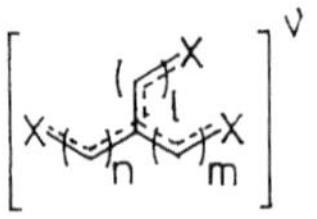

74

l, m, n = 0, 1, 2, . . .

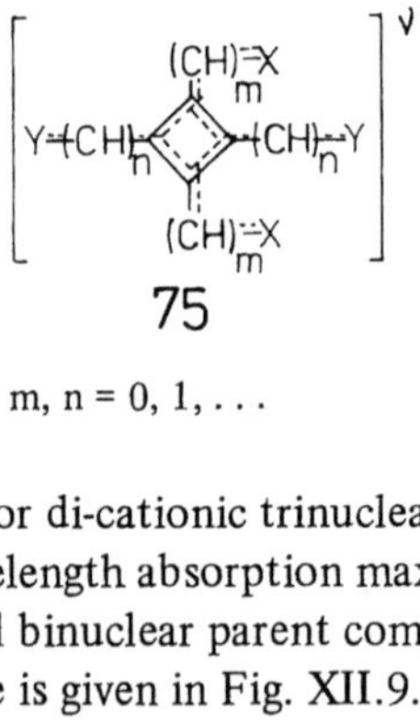

75

m, n = 0, 1, . . .

Representatives of *74* are the mono- or di-cationic trinuclear cyanines *76* or the neutral merocyanines *77*. Their long-wavelength absorption maxima appear at nearly the same position as those of the symmetrical binuclear parent compounds (values given in round brackets) [901]. A further example is given in Fig. XII.9.

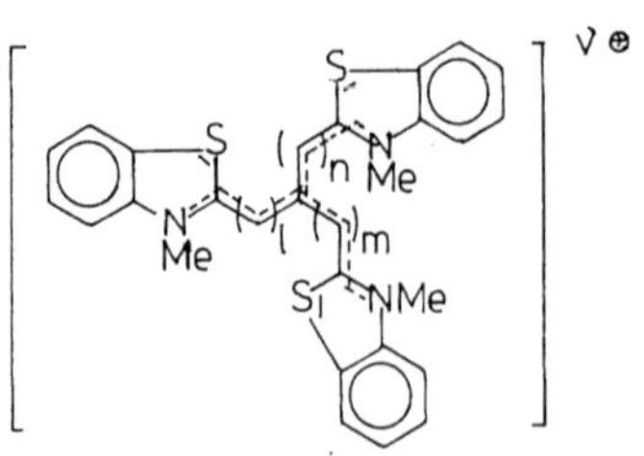

76

l	m	n	
1	1	1	555 (556) [760]
2	2	0	601.5 (650) [760]
2	2	2	633 (650) [898]

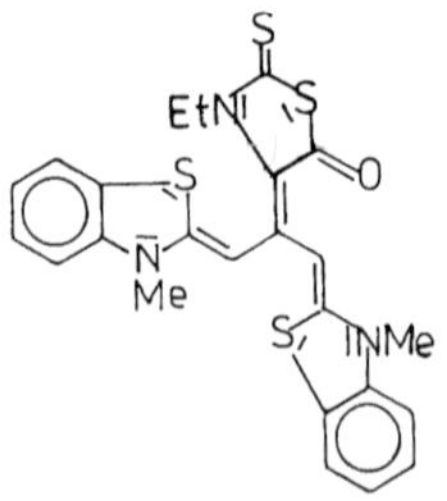

77

552 (556)
in methanol [902]

Table XII.16. Spectral data of several squaric acid polymethine dyes *78* and *79* (R_A is specified in Table XII.9)

R_A	n	*78* λ_{max} in nm (lg ε)	Open Chain Reference Dyes λ_{max} in nm	*79* λ_{max} in nm (lg ε)
R_1	1	–	–	670 (4.94)[b]
R_2	1	730[a]	708	600 (4.73)[b]
R_3	1	670[a]	660	520 (5.09)[b]
R_6	1	630[a]	642	–
R_{11}	0	628[c]	692	–
R_{12}	0	388 (4.82)[d]	382	342 (4.67)[b]

[a] in chloroform [994]
[b] in dimethylformamide [994]
[c] in chloroform [995]
[d] in 1 M H_2SO_4 [996]

A close resemblance of the color band to that of the parent chain chromophore has also been observed for dyes of the structure *78*, which contain a central four-membered ring. The isomeric *79* absorbs, however, at markedly shorter wavelengths (cf. Table XII.16).

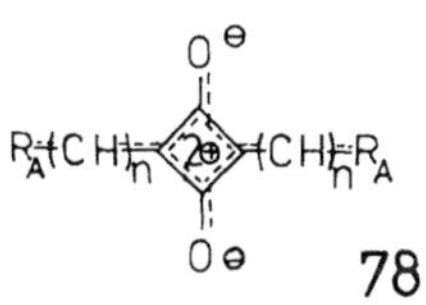
78

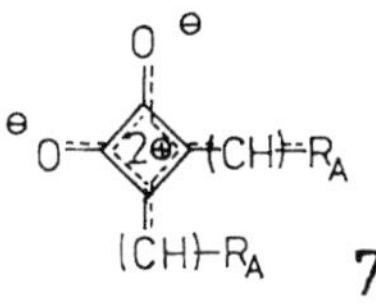
79

Although the branched dyes *78* represent particular chromophoric systems, the electronic excitation has been proved to be partly localized [133], for one of their frontier orbitals has a nodal plane through the oxygen atoms. If *78* is considered as substituted chain cyanine $R_A(CH)_{2n+3}R_B$, the approximate constancy of the absorption wavelenth upon substitution results from the simultaneous stabilization of its ground and excited state. Since for symmetry reason only the HOMO is affected by substitution in that case, the qualitative effect of the two substituents O^- and CO is easy to assess. Whereas the *meso*-donor substituent O^- raises the HOMO (stabilization of the excited state), the CO acceptor substituent lowers this orbital by virtue of the predominant HOMO-LUMO interaction (ground state stabilization). Thus, the excitation energy of *78* remains in the same order of magnitude as for the cyanine subchain, in spite of the strong substituent effect.

The shorter wavelength absorption of *79* in comparison with *78* can be rationalized by alternatively dissecting the structures *78* and *79* into two equal radical polymethine units. Along these lines the colour-determining MO's of the composite molecule result both from the singly occupied fragment MO. Due to the different positions of linkage, these MO's undergo a distinct energy splitting. The stronger overlap in *79* relative to *78*, arising from coupling in equivalent positions, accounts for the shorter wavelength absorption of *79*.

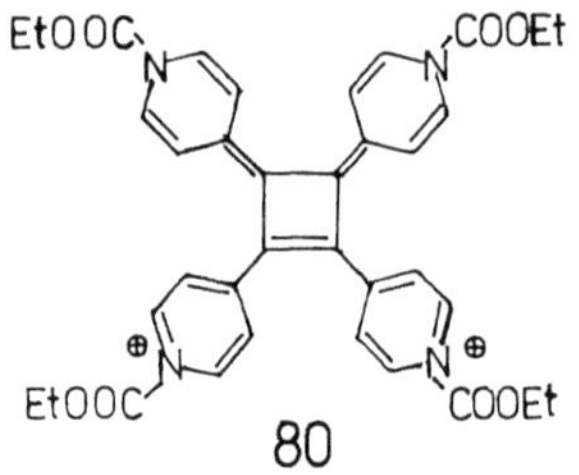

80

676 (> 5.03)
in chloroform/acetonitril [903]

Finally the highly symmetrical tetra-nuclear dye *80*, which has been prepared recently, should be mentioned [903]. Also in this case the position of the color band is similar to that of the 4,4'-pyrido-carbocyanine reference dye (~ 600 nm [784]).

5. Medium Effect

The solvent effects on the absorption wavelength of cationic polymethine dyes is small [904–906]. Surprisingly, the wavenumber shifts are roughly proportional to $(n^2 - 1)/(2n^2 + 1)$, as generally found for less polar compounds dissolved in non-polar solvents [904, 905]. However, the fundamental origin of this solvent effect is surely more complicated. NMR studies on some symmetrical cationic cyanines provide evidence for a strong solute-solvent interaction [907]. Protic solvents lead to specific solute-solvent interactions. Such interactions have been discussed in connection with calculated charge distributions [906, 908].

More pronounced spectral shifts arise from molecular aggregation of dyes which may either cause an hypsochromically shifted color band (formation of H-aggregates) or a bathochromically shifted color band (J-type aggregates). The tendency to aggregate does not only depend on the dye structure, but also on environmental effects [889, 909, 910]. Nature and origin of the aggregation have been reviewed in Refs. [889, 911]. In a few cases the spectra of dimers could be resolved [912]. In that case the color band splits into two components. They are the more strongly displaced from each other the longer the chain length.

The spectra of the polar merocyanines display a more pronounced solvent effect than the cationic cyanines [913]. This was first reported by Brooker and coworkers who studied the spectra of merocyanines dissolved in pyridine-water mixed solvents [914]. They found that variation of the percentage water content of the solvent gives rise to pronounced changes of both the maximum molar absorptivity, ϵ_{max}, and the corresponding wavelength, λ_{max}, at the visible absorption maximum. For very highly polar dyes the progressive addition of water to the solution of the dye in pyridine leads to a shift of λ_{max} to lower wavelengths, with a concomitant diminution of ϵ_{max}. The opposite behaviour was found for dyes which are comparatively more weakly polar. For dyes of intermediate polarity the initial addition of water to the pure pyridine solvent has an effect qualitatively similar to that observed with weakly polar dyes. However, upon further progressive addition of water to the solvent λ_{max} and ϵ_{max} pass through extreme values and the subsequent differential solvent effect is qualitatively similar to that observed with very highly polar dyes. The reversal point of the ϵ_{max} versus λ_{max} plot indicates a "reversal of solvatochromism". A similar behaviour has been observed when merocyanine dyes are dissolved in other mixed solvents [915] or in pure solvents of different polarity [916, 917].

The study of solvent shift within vinylogous series revealed that the spectral displacement increases with the chain length [842]. Furthermore, the vinylene shifts are non-convergent for λ_{max} in pyridine but strongly convergent in water. A similar spectral behaviour has been observed for simple streptopolymethine-merocyanines, as shown in Fig. XII.11.

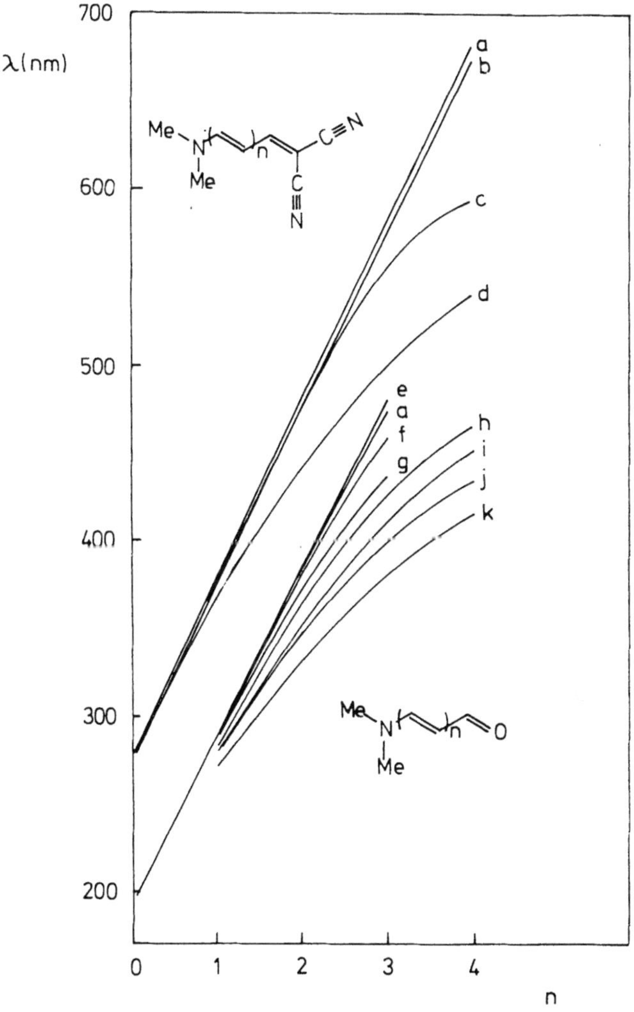

Fig. XII.11. Position of the lowest wavelength absorption of two series of streptopolymethine-merocyanines in different solvents in dependence on the number of double bonds along the chain [923, 997] (solvents: a water, b acetonitrile, c 2-Me-tetrahydrofuran, d Me-cyclohexane, e saturated LiCl solution, f formamide, g ethanol, h dimethylformamide, i tetrahydrofuran, j tetrachloromethane, k hexane)

Some merocyanines belong to the negatively solvatochromic dyes [204]. For example, *43* undergoes blue shift on passing from chloroform (628 nm) to methanol (482 nm). Whereas negative solvatochromism remains with mono-benzoannelation of the quinoid part of *43*, positive solvatochromism is observed with appropriate di-benzoannelation. This finding caused Hünig and Rosenthal to conclude that benzoannelation favours the quinoid merocyanine structure over the charge-separated stilbenoid structure. Aza merocyanines display, in general, a positive solvatochromism even then when the carbon analogues are negatively solvatochromic. This has been demonstrated, for example, for diaza analogues of *43* by Hünig and coworker [335].

There are several proposals to describe the solvent effect of merocyanines empirically. Thus, solvent parameters have been advanced for this purpose, such as the Z [918], the E_T [919] and the PM values [920]. Good linear correlations have been found with E_T values [919]. Solvent parameters characterize the solvent polarity including all different kind of solvent-solute interactions.

The solvent effect of merocyanines results both from specific and non-specific solute-solvent interactions. The non-specific interactions determine the solvent effect in non-protic solvents. In accordance with the continuum approach [921], either bathochromic or hypsochromic shift can be expected. Roughly speaking, the displacement towards longer wavelengths occurs when the dipole moment increases upon excitation (positively solvatochromic merocyanines), and displacement towards lower wavelength occurs, conversely, when the dipole moment decreases (negatively solvatochromic merocyanines). According to this approach, polarization effects on the solute molecule by surrounding solvent molecules are assumed to be of minor importance. On the other hand, Dähne and coworker [101] placed special emphasis on the solvent-induced change of the electronic structure of the solute molecule. This change is indicated by NMR data [922]. According to these authors, positively solvatochromic dyes undergo a change of the electronic structure towards the symmetric polymethine structure in strongly polar or protic solvents. This assumption explains the constant vinylene shift of merocyanine dyes in polar solvents. Some theoretical models has been advanced to establish the alteration of π-electron structure upon solvatation [101–103, 498, 923, 924].

In alkaline and acidic media spectral shifts can also be arise from protonation or deprotonation processes. These processes can proceed at the heteroatomic groups of the polymethinic chain, as shown for compound *81a* [925]. Addition of a proton leads to the polymethinic cation *81b*, and removal of a proton to the polymethinic anion *81c*. Both ions are iso-π-electronic to the parent compound. Although different in charge, both ions absorb at nearly the same wavelength, which exceeds that of the parent compound *81a* (about 400 nm):

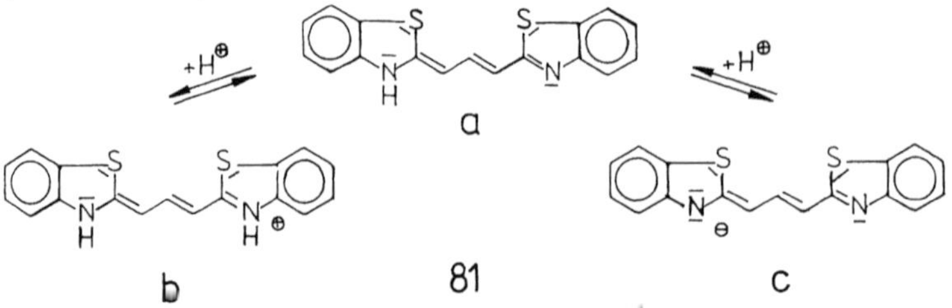

81

548
in ethanol/HCl [925]

536
in ethanol/KOH [925]

Interestingly, a bathochromic shift of the same magnitude is observed in propanolic solutions which contain transition metal ions (e.g. N-methyl-*81* and $Co(NO_3)_3$: 555 nm) [926].

Spectral shifts may also arise from deprotonation or protonation'at heteroatoms which may be regarded as internal or external substituents of the polymethine chain. The shifts are easily to interpret. Thus, the dye *82a* undergoes fading in acids [927], because protonation occurs at the aza group in *meso*-position:

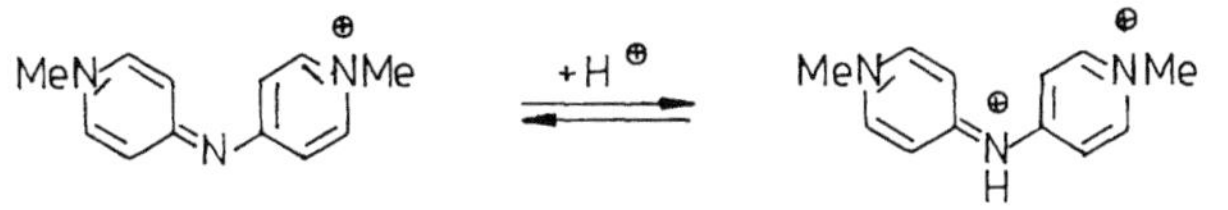

a 82 b

392 in ethanol

328 in acid

Although the π-electron system of *82* is evidently not reduced upon protonation, a remarkable blue shift occurs. An explanation can be achieved in terms of the formerly discussed substituent effects on the polymethines. The protonation at the central nitrogen atom enhances the electronegativity of this heteroatom, which replaces a methine group at an even-numbered position of the polymethine chain. Thus, the observed blue shift is in accordance with the color rules (cf. Fig. XII.3).

Pronounced spectral effect have also been observed when deprotonation or protonation occurs at external substituents of the polymethine chain, such as at the *meso*-hydroxyl group of *83a* [928] and at *meso*-phenylazo group of *84a* [929], resp. The absorption maxima undergo blue shift in both cases.

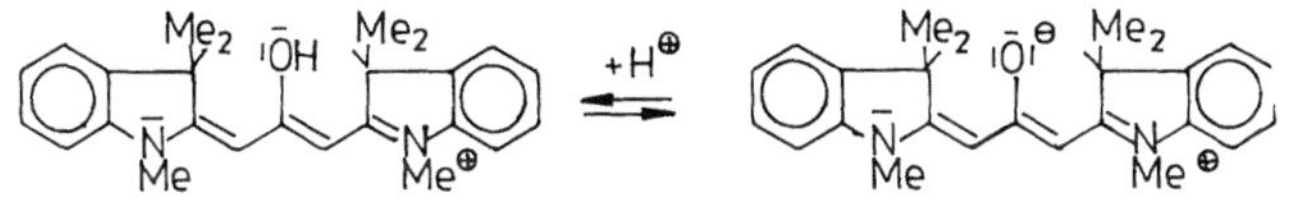

a 83 b

500 (3.50) in methanol/0.01 n HCl

425 (4.74) in methanol [928]

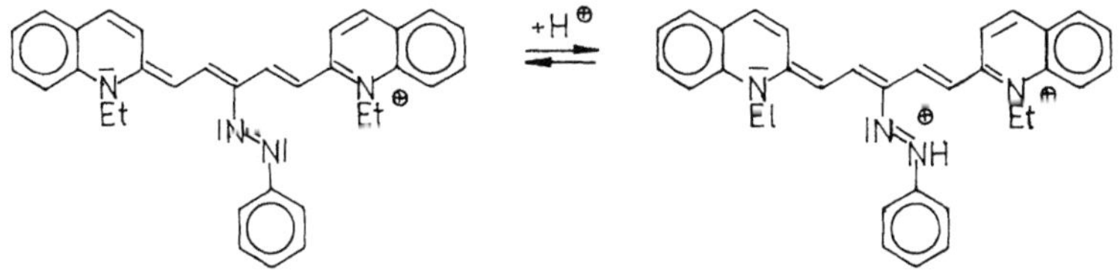

a 84 b

688 (5.08) in methanol

498 (4.40) in methanol/H_2O, pH 3.75 [929]

In the first case removal of the proton from the OH-group strengthens the donor capacity of the substituent. As linked at on odd-numbered position of the polymethine chain, the observed blue shift agrees with the prediction according to the aforementioned color rules (cf. Fig. XII.4). Attachment of a proton to the azo group strengthens, in the second case, the acceptor capacity of the substitutent, which is linked at an even-numbered position of the polymethine chain. Thus, the observed blue shift is completely in line with the color rules. Since protonation of the azo group in azo dyes mostly causes red shift of the color band, the observed blue shift of *84b* is quite remarkable.

Merocyanines are generally protonated at the terminal oxygen. Either blue shift or red shift has been observed for positively solvatochromic dyes [847, 930], the color band of the negatively solvatochromic *43* is blue-shifted upon protonation from 442 nm in water/base mixtures to 372 nm in water/acid mixtures [931].

A dramatic change in color is observed when protonation destroys the polymethine chromophore. This reaction mostly leads to a complete decoloration of the dye solution, such as observed for *85* [932] and many other dyes [722, 933, 934].

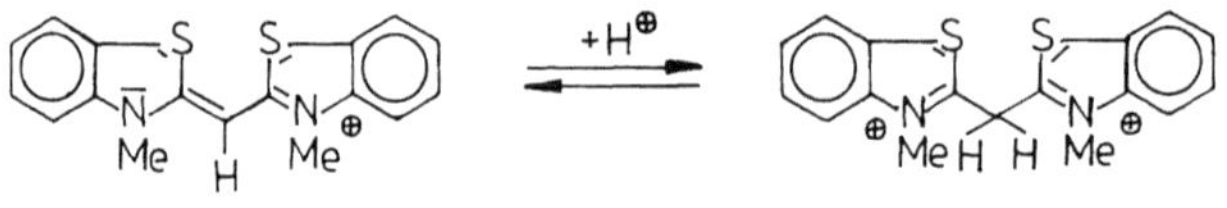

a 85 b

429 (4.78)
in methylene chloride [932]

296 (3.98)
in perchloric acid [932]

Cyanine dyes are, in general, the more easily protonated the longer the chain length, but there are exceptions [844, 935, 935a]. Protonation occurs at methine groups which bear a negative charge (even-numbered positions of the polymethine chain according to the enumeration in Fig. XII.3), and preferably at that methine group adjacent to the heterocyclic ring [936]. The experimentally determined pK_a values [844, 937–939] correlate with the electrophilic superdelocalizabilities [940].

Another fading process consists in addition of a nucleophilic species to the molecule, such as an amine [722, 933, 934] or alcohol [941]. This addition occurs at positively-charged odd-numbered positions of the polymethine chain and is accompanied again with the destruction of the basic chromophore. Anthocyanines add the hydroxyl group in 2-position of the flavylium cation [942].

Colorless structures may also result from intramolecular processes. Thus, several di- and tri-quinoylmethanes, such as *86,* exist in two prototropic forms [943–945]. One of them (*86b*) is colorless. The position of the prototropic equilibrium depends on the solvent, the temperature, and the size of the aromatic residue. The existence of a prototropic equilibrium has, however, given rise to doubt concerning the rigidized N,N′-methylene derivative of *86a* [946].

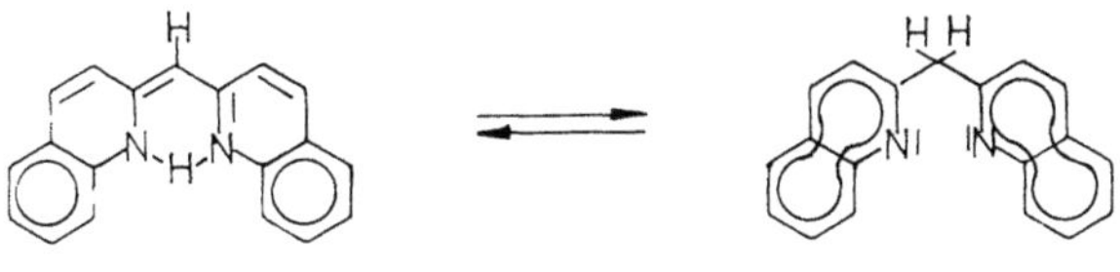

a 86 b

blue colorless

Similarly, merocyanines of the type *87a* are known to be in equilibrium with the isomeric spiro compounds *87b* [947–949]. The equilibrium can also be influenced by the solvent, the temperature and the structure of the compounds, and moreover, by exposure to light [950]. The merocyanines are either negatively or positively solvatochromic dyes [951]. Due to the formation of "photomerocyanines", spiro compounds of the structure *87* belong to the photochromic materials.

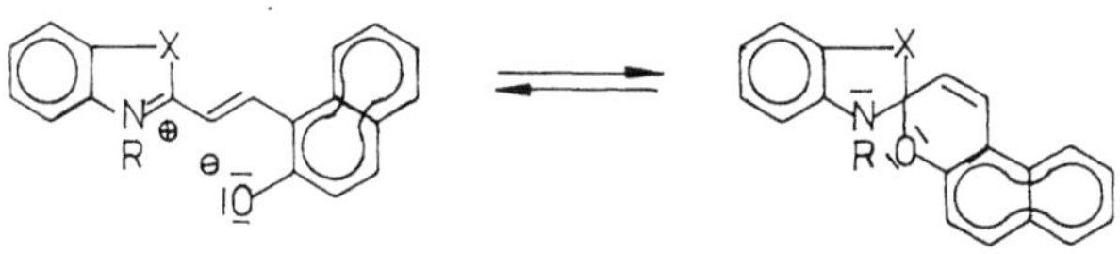

a 87 b

In order to rationalize the formation and the spectral properties of photomerocyanines several theoretical studies have been performed in the framework of the PPP and all-valence-electron approximations [952–956, 956a].

The recently synthesized ketocyanines belong to the strongly positively solvatochromic dyes, for their absorption wavelengths shift from 460 to 500 nm in heptane to 660 to 740 in acetic acid [956b].

XIII. Porphyrins and Phthalocyanines

Important naturally occurring pigments such as haemin and chlorophyll and synthetic pigments such as metal phthalocyanines owe their color to large cyclic conjugated systems. They represent tetradentative ligands which are chromophorically derived from porphin.

Porphins have a tetrapyrrolic structure in which the rings are linked by methine bridges at the α-positions of pyrrole. They constitute an important group of porphyrins, and include the natural pigment haemin. The parent compound *1* comprises nine conjugated double bonds and further two double bonds which are branched-conjugated with the main cyclic conjugated system. According to X-ray diffraction analyses of *1* and simple derivatives [998], and to ab-initio calculations [999] the molecule is planar and the two inner hydrogen atoms are attached to opposite nitrogens. The two independent pairs of pyrrole rings possess small but significant differences in their molecular geometry.

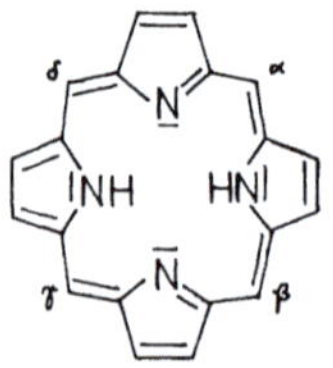

1

Based on the average carbon-carbon and carbon-nitrogen bond lengths displayed in Fig. XIII.1a, a 18-membered ring is considered as main path of conjugation, but the large bond distances of the pyrrolic α,β-carbon-carbon bonds and the short pyrrolic β,β-carbon bonds alternatively suggest the predominance of an inner 16-membered ring [1000] (cf. Fig. XIII.1b). Calculated π-bond orders obtained in a recent ab initio calculation were interpreted in term of both structural models [1001]. The two pathways of conjugation corresponds to stable aromatic systems of 18 π electrons, and are consistent with the Hückel 4N + 2 rule [1002].

Another large group of porphyrins are chlorins (including chlorophyll) and bacteriochlorins (including bacteriochlorophyll). The parent compounds are the dihydroporphyrin *2* and the tetrahydroporphyrin *3* with reduced rings in opposite positions.

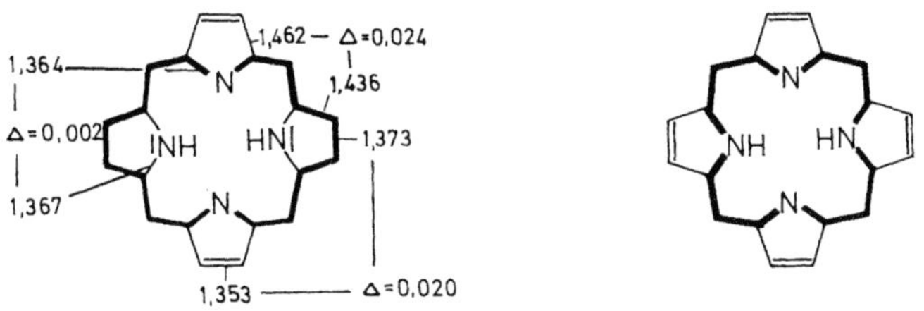

a b

Fig. XIII.1. Averaged bond lengths in Å, differences in bond lengths (a) and two different main conjugated systems indicated by heavy lines (b)

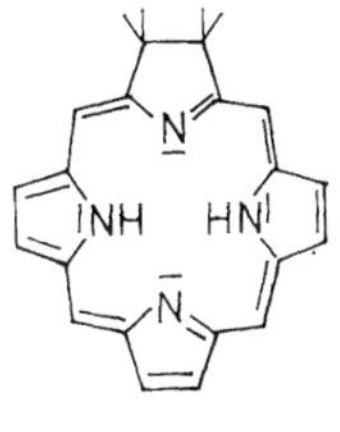

2

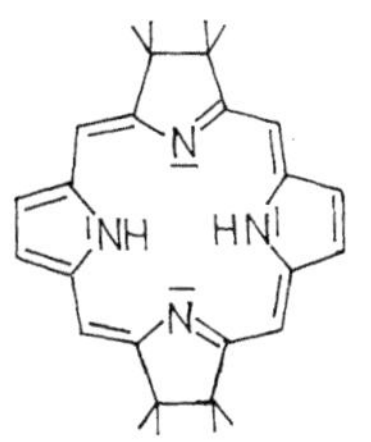

3

The electronic absorption spectra of free-base porphyrins consist of four bands in the 500–700 nm region, an extremely intense band at about 400 nm (Soret band), and several bands at shorter wavelengths. The absorption spectrum of octaethyl porphin is shown in Fig. XIII.2 (dotted curve).

Alkyl- and aryl-substitution as well as benzoannelation of porphin bring about bathochromic displacement of the absorption bands. Furthermore, characteristic spectral shifts in the long-wavelength region have been observed with successive reduction of the exocyclic double bonds (cf. Table XIII.1). Pronounced red shift occurs in passing from porphins via chlorins to bacteriochlorins.

Since octaethyl-octahydro-porphin still absorbs at about 500 and 345 nm according to Table XIII.1, the 18 π electron system of *4* may be regarded as parent chromophore

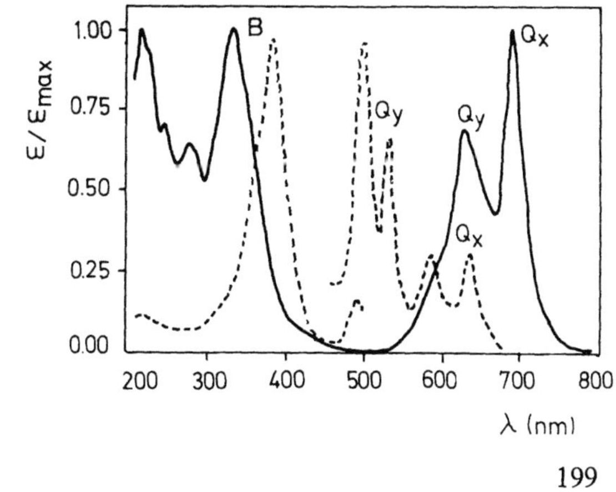

Fig. XIII.2. Absorption spectral curves of octaethyl-porphin [1045] (dotted line) and phthalocyanine [1012] (full line) in the vapour phase and assignment of the absorption bands. The intensities of the UV and visible bands are separately normalized to unity

Table XIII.1. Long-wavelength absorptions of some porphyrins (in benzene)

Compound	λ_{max} in nm (lg ϵ)									Ref.
	Q-Bands							B-Bands		
1	563.5 (3.70)	489.5 (4.20)						396.5 (5.42)		[1036]
Octaethyl-*1*	622 (3.76)	569 (3.18)	568 (3.83)	532 (4.03)	498 (4.16)			4.00 (5.20)		[1037]
$\alpha,\beta,\gamma,\delta$-Tetraphenyl-*1* [a]	650 (3.6)	595 (3.6)	550 (3.9)	515 (4.3)	485 (3.5)			420 (5.6)	395 (4.9)	[1038]
Tetrabenzo-*1* [b]	668 (4.80)	621 (4.90)	612 (4.90)	604 (4.80)	590 (4.21)	576 (4.25)	566 (4.22)	438 (5.34)	422 (5.21)	[1039]
Dihydro-octaethyl-*1*	645.5 (4.84)	618 (3.74)	591 (3.71)	545 (3.40)	522 (3.72)	496 (4.12)	489 (4.16)	392 (5.26)		[1037]
Tetrahydrooctaethyl-*1* (adjacent)	634 (3.95)	586 (4.51)	546 (4.25)	511 (3.97)	482 (3.72)	401 (4.72)		379.5 (4.97)	370 (5.01)	[1040]
Tetrahydrooctaethyl-*1* (opposite)	721 (5.18)	685 (4.15)	661.5 (3.79)	604 (3.40)	491 (4.69)	463 (4.31)		434 (4.96)	374 (5.26)	[1040]
Hexahydrooctaethyl-*1*	636 (3.89)	557.5 (4.26)	527 (4.12)	480 (3.81)	440–436 (3.66)			357–355 (4.64)		[1040]
Octahydrooctaethyl-*1*	505–500 (4.64)							349–339 (4.58)		[1040]

[a] in pyridine [b] in chloronaphthalene

of porphyrins. The iso-π-electronic [16]annulenyl dianion *5* absorbs at 600 nm (lg ϵ 4.20) and 560 nm (lg ϵ 4.14) and very intensively at 412 nm (lg ϵ 5.24) in dimethylformamide [1003].

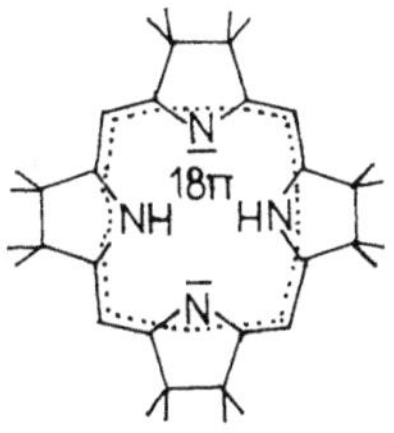

4

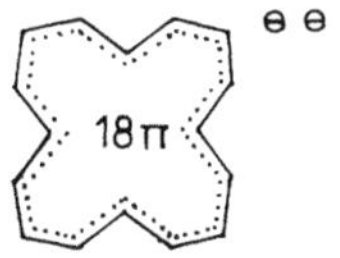

5

The sharp long-wavelength absorptions between 400 and 700 nm have been extensively used as "finger print" region in order to characterize the porphyrins. The inspection of the long-wavelength absorptions enables porphyrins to be classed as porphin or chlorin. Furthermore, the presence of electron-accepting moieties attached to the perimeter of the ring system may be discerned by the intensity change of the absorption bands. The numerical spectral data for a great variety of substituted porphyrins has been compiled elsewhere [1004–1008].

Replacement of the methine groups in α-, β-, γ- or δ-position of *1* by nitrogen atoms leads to aza porphins, the first visible band changing enormously in intensity but hardly in position. The ultraviolet band changes considerably in position but hardly in intensity, however (cf. Table XIII.2).

Fusion of four benzene rings to the pyrrole nuclei in passing from tetraazaporphine *6* to phthalocyanine *7* gives rise to a marked change of the spectral absorption features. Whereas the Soret band nearly persists, the first band appears now at about 700 nm (cf. Table XIII.3). The spectral feature of *7* is shown in Fig. XIII.2.

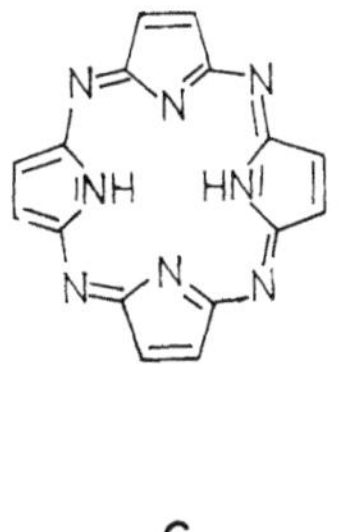

6

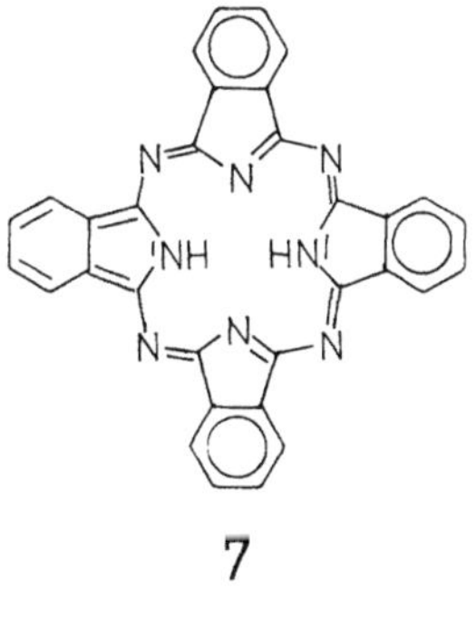

7

According to X-ray [1009] and neutron diffraction studies [1010] of phthalocyanine *7*, the four benzene fragments are weakly joined to the 16-membered inner ring system. This may account for the fact that the longest-wavelength absorptions hardly change with exchanging the annelated ring (cf. Table XIII.3). Also, the spectral effect of substituents at the benzo groups of phthalocyanine is weak [1008].

Table XIII.2. Most intense absorption maxima of the longest-wavelength absorptions of tetramethyl-tetraethyl-azaporphins in pyridine [1041]

Substitution Pattern	λ_{max} in nm (lg ϵ)		
Assignment [1015]:	Q_x	Q_y	B_x
none	619 (3.66)	529 (3.88)	398 (5.05)[a]
monoaza	610 (4.42)	534 (4.38)	376 (5.06)[a]
α,γ-diaza	620 (4.67)		375 (5.16)
α,β,γ,δ-tetraaza	624 (4.91)	556 (4.67)	340 (4.99)

a shortest-wavelength absorption measured in dioxane

Table XIII.3. Longest-wavelength absorptions of annelated α,β,γ,δ-tetraaza porphyrines (Q-bands)

Annelation Pattern	λ_{max} in nm (lg ϵ)						Ref.
none[a]	617 (4.75)	545 (4.60)					[1041]
tribenzo[b]	663	626	590				[1044]
tetrabenzo[c]	703 (5.23)	668 (5.22)	636 (4.75)	604 (4.52)	576 (4.02)	556 (3.88)	[1042]
tetra-[1,2]-naphtho[d]	720	677	645	606			[1042]
tetra-[2,3]-naphtho[d]	720	678	650	606			[1042]
tetra-[2,3]-pyridino[d]	700	673	634	608	576	554	[1043]

a in chlorobenzene
b in pyridine
c in chloronaphthalene
d in chlorobenzene

Extensive spectral studies on *7* in the solid state [1011] and in the gaseous state [1012] have revealed the absorption feature in the UV region.

The two central hydrogen atoms of phthalocyanine can be easily replaced by various metals and metalloids. Similarly to the metal free parent compound they are intensively colored. With a few exceptions, the variation in the absorption spectrum is less marked. This is particularly true of the position of the 700 nm band, which give rise only to a single sharp band, whereas the Soret band varies slightly with the nature of the metal ion.

If one or two opposite pyrrole or isoindoline groups in *6* or *7* are replaced by benzene or pyridine moieties, compounds of the structure *8* result. They do not longer possess the characteristic perimeter of the porphyrins. In contrast to porphyrins, *8a* and *8b* are not colored.

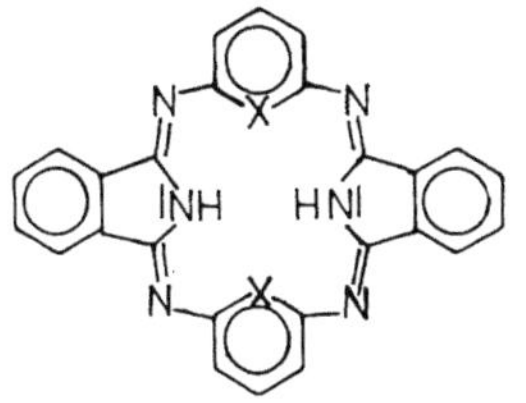

8

a: X = CH 343 (4.23)
b: X = N 343
in dimethylformamide [1013]

Porphyrins have been extensively studied MO-theoretically [1002]. The first interpretation of the spectral absorptions of porphyrins and phthalocyanines was advanced in the framework of the free electron method, considering porphin as a hypothetical 18-membered or 16-membered ring. The notation of the excited states given by Platt [135] has been generally accepted. According to him the long-wavelength absorptions of porphyrins are determined by electron excitations to the Q- and B-states, resp. In the free base porphin the Q- and B-transitions are split in x- and y-components, which are polarized parallel and perpendicular to the H–H axis, resp. Platt also presented a rationale of the substituent effects on the long-wavelength Q-transitions. Formulae were given which describe the intensity pattern depending on the kind of substitution.

After the failure of attempts to describe the spectral feature of porphyrins by Hückel-type orbitals, Gouterman and coworkers emphasized the importance of the configuration interaction [1014, 1015]. Within Gouterman's "four orbital model" the longest-wavelength visible and near ultraviolet absorption bands are interpreted as arising from electronic excitation from the highest two occupied orbitals 1 and 2 to the lowest two empty orbitals 1′ and 2′ (cf. Fig. XIII. 3a). The resulting configurations interact pairwise giving electronic states, in which the transition dipoles cancel or reinforce each other. This mixing gives rise to weak Q-transitions (Q_x and Q_y) and intense B-transitions (B_x and B_y), cf. Fig. XIII.3b.

Within the "four orbital model" the four visible absorption maxima of the free base porphyrins have been assigned to vibronic subbands of the Q_x- and Q_y-transitions, whereas the Soret band mainly results from the B-transitions (cf. Fig. XIII.1). The Q_y-transition shows increasing intensity in passing from porphin *1* to chlorin *2* and bacteriochlorin *3* and migrates to lower energy than the Q_x-transition.

PPP [1016–1022] and CNDO/S calculations [1023–1025] have confirmed the validity of the four-orbital model in describing the lowest energy transitions of porphyrins.

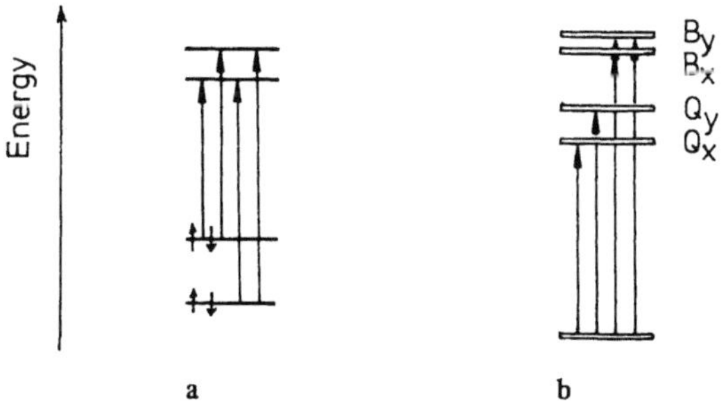

Fig. XIII.3. MO energy level diagram (a) and term level diagram (b) of porphin (four-orbital four-transition model)

However, in a varying degree non-four-orbital configurations contribute to the B-transitions [1016, 1021–1023]. The most of these calculations overestimate the energy gap between the two Q- and the two B-transitions, and do not reproduce the experimentally found intensity ratio between the respective absorption bands. The same deficiencies are apparent from results of calculations which have been performed within an improved free electron model including electron interaction [1026]. Better agreement between the experimental and theoretical results have been achieved within the PPP approach by a new approximation for electron repulsion [1027]. Recently, both photoelectron spectra and electronic spectra have been satisfactorily analysed by using the spectroscopically parametrized CNDO/S3 method [1028].

The lowest-energy $n \rightarrow \pi^*$ transition was located in the region of the intense Soret band according to extended PPP [1021, 1022] and ab initio calculations [1029], whereas this transition appears at higher energies by CNDO/S calculations [1023, 1025]. No experimental data are available so far.

There are also several theoretical studies about derivatives of porphins, for example for dianions or dications, which result from the porphyrin parent compounds upon deprotonation or protonation, resp. [1021, 1023, 1028, 1030, 1030a]. These ions have a higher symmetry than the parent compounds [1030b]. This accounts for the coincidence of their Q_x- and Q_y-absorption bands. Since metal porphyrins behave in a spectroscopically similar way to the ionic porphyrins, most of the theoretical studies on metal porphins have been performed by only presuming the equivalence of the four nitrogen atoms [1016–1020, 1027]. Calculations of iron-(II) porphyrins by refined Hückel methods showed that the color determining molecular orbitals are hardly affected by the 3d, 4s and 4p orbitals on iron [1031, 1032]. Later extended Hückel calculations, however, did not reproduce the small variation of the spectrum observed between different porphyrins [1033]. Ross and Sundbom [1034] calculated copper(II) porphin by incorporating the copper atom explicitly into an extended PPP formalism. According to them, $d \rightarrow d^*$ type transitions and metal-to-ligand charge transfer transitions affect the electronic transition energies of this porphin.

More recent PPP studies have aimed at an interpretation of spectral properties of bacteriochlorophyll and related naturally occurring dyes [1035].

Excitonic coupling of the electronic transition occurs in cofacial diporphyrins, resulting in a relatively large hypsochromic shift of the Soret band and a prominent tail at longer wavelengths [1035a]. Covalently linked tetraarylporphyrin dimers show a splitting of the Soret band [1035b].

XIV. Conjugated Betaine Dyes

Although colored conjugated betaines have not yet found application in textile dyeing, they have attracted much attention. Their strong spectral solvent effects have rendered them useful as indicators for solvent polarity [919, 921].

In principle, some of the former mentioned dyes could be attributed to the conjugated betaines. Illustrative examples are the "stilbene dye" *14a* in Chap. VI or the "azo dye" *22a* in Chap. VII. However, the zwitterionic formulae of these dyes reflect only in part their real electronic structure. As indicated by the respective second mesomeric formulae *14b* and *22b*, some intramolecular compensation of the charges takes place towards a merocyanine-like structure.

In truly zwitterionic compounds the charges are overwhelmingly fixed at different molecular fragments. These fragments are two rings as in the case of the colored pyridinium N-cyclopentadienide *1* [1046].

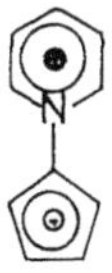

1

Formula *1* is a good representation of its electronic structure, for 71% of the π-charges are localized at each of the two rings according to CNDO/S and PPP calculations [1047]. The occurrence of long-wavelength absorptions in donor-acceptor complexes between cations and anions suggested that the colour of the betaine *1* arises from an intramolecular charge-transfer transition [1048]. This assumption agrees with results of quantum chemical calculations. According to PPP-CA and CNDO/S-CA calculations two intramolecular CT transitions determine the longest-wavelength absorption of *1* [1047, 1049] (Fig. XIV.1). Only one of them is of high intensity. This intense electronic transition occurs between the ground state and second excited state. In terms of the composite molecule approach, the excited state mainly results from the mixing of the NB ground state configuration with the lowest-energy CT configuration of the same symmetry. This configuration mixing is the necessary condition for sufficiently high transition probability [1050].

In the one-electron picture, the color determining electronic transition is roughly described by an electronic excitation from one of the degenerate HOMO's of the cyclopentadiene anion fragment to the LUMO of the pyridine fragment (cf. Fig. XIV.2).

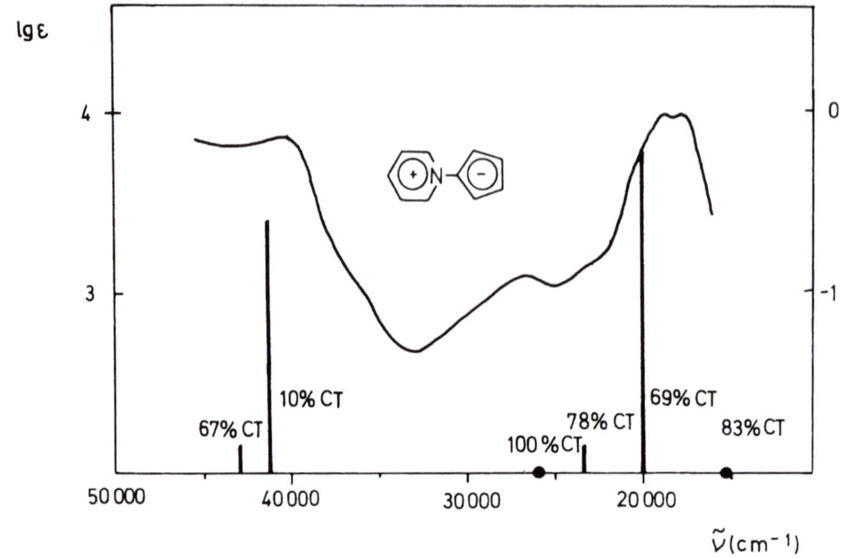

Fig. XIV.1. Spectral absorption curve of pyridinium N-cyclopentadienide in heptane [1048] and results of PPP calculations and OKK analyses of the $\pi \rightarrow \pi^*$ transitions. The figures indicate the cyclopentadienide-to-pyridinium charge transfer contributions to the electronic excitations

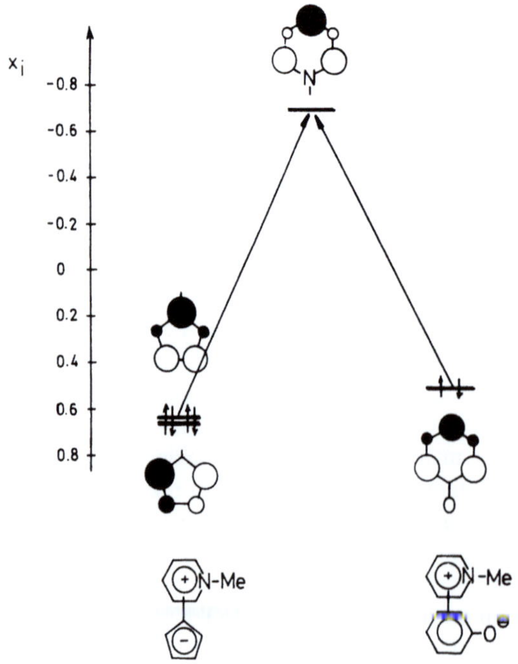

Fig. XIV.2. Pictorial representation of the expansion coefficients of donor and acceptor fragments which form zwitterionic molecules (cf. caption to Fig. VII.5)

Since the HOMO and LUMO energies corresponds to the experimental ionization potential I of the donor and electron affinity EA of the acceptor, resp., the CT excitation energy is approximately given by Eq. (XIV.1).

$$h\nu_{CT} = I - EA - J + 2K \qquad \text{(XIV.1)}$$

where J and K represent electronic repulsion terms.

The K term is relatively small or even zero for zwitterionic donor-acceptor systems, thus giving rise to low excitation energies. In agreement with Eq. (XIV.1) the CT absorption wavelength increases with increasing donor or with increasing acceptor capacity. In accordance with the different donor capacity, pyridinium N-cyclopentadienide *1* (dissolved in chloroform) exhibits a red color while a blue color has been observed for the benzologous pyridinium N-indenylide [1046] and pyridinium N-fluorenide [1051]. Whereas the latter betaines are not stable, compounds of the series *2* [1052, 1053], *3* [1053] and *4* [1054] were isolated and spectroscopically measured.

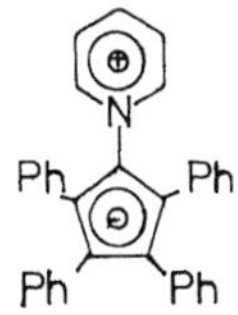

2

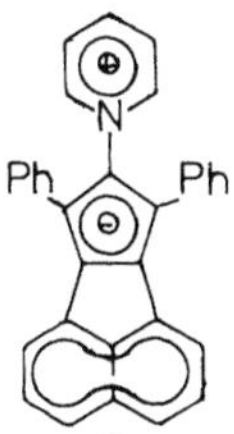

3

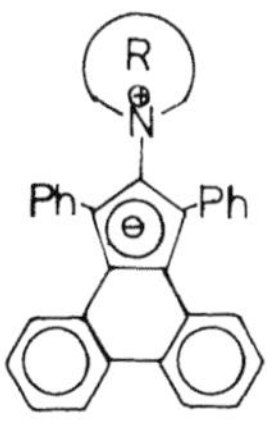

4

Due to the higher donor capacity of their anionic moieties in comparison with cyclopentadienide, they absorb at longer wavelengths than *1* (cf. Table XIV.1). On the other hand, bathochromic shift has been observed when pyridinium is replaced by the stronger acceptors quinolinium or *para*-cyanpyridinium [1053, 1054].

Related compounds are heteronium ylids which result from substituting the pyridinium residue in *2* by XPh_3^+ (X = Bi, Sb, As, P) or YPh_2^+ (Y = Te, Se, S) [1055]. However

Table XIV.1. Longest-wavelength absorption maxima of pyridinium N-cyclopentadienide betaines and related compounds in benzene

Compound	R	λ_{max} in nm (lg ϵ)	Ref.
1	–	511.2 (4.34)[a]	[1048]
2	–	590 (3.97)	[1053]
3	–	630 (4.15)	[1053]
4	pyridinium	650 (3.88)	[1054]
	1-chinolinium	700 (3.81)	[1054]
	2-chinolinium	710 (3.71)	[1054]
	p-cyanopyridinium	727 (4.26)	[1054]

[a] in chloroform

marked hypsochromic shift occurs in passing from the triphenylbismuthonium to the triphenylphosphonium, and from the diphenyltelluronium to the diphenylsulfonium ylid. This shift has been attributed to a considerable structural change from the zwitterionic ylid towards the non-polar ylene structure [1055, 1056].

Blue shift likewise occurs if the cyclopentadiene anion is linked at the *ortho*- or *para*-position of the pyridinium nucleus instead at the nitrogen atom, as demonstrated with *5* and *6* [1057].

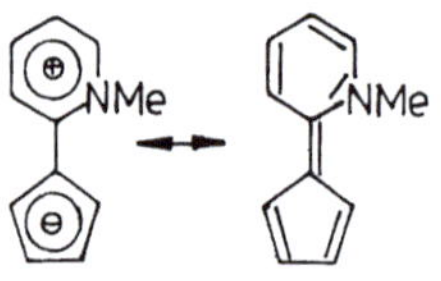

a 5 b

369 (f 0.14)
429 (f 0.34)
in dioxane [1057]

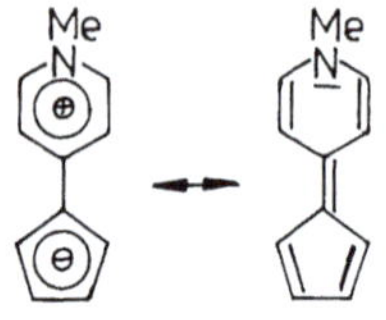

a 6 b

429 (f 0.66)

in dioxane [1057]

In contrast to *6*, the absorption band of *5* splits up into two components. PPP calculations reproduce this spectral behaviour, but suggest additional weak absorptions at longer wavelengths both for *5* and *6*. The blue shift of the first intense absorptions of *5* and *6* as compared to that of *1* results partly from the lower acceptor capacity of the pyridinium residue due to methyl substitution, and partly from the higher ground state stabilization in *5* and *6*, as reflected by the mesomeric formulae *5b* and *6b*.

Since ground state stabilization is determined by the HOMO-LUMO overlap, inspection of the orbital expansion coefficients permits a rough prediction about the degree of stabilization in the different positions of the pyridinium ion. According to the pictorial representation of the acceptor LUMO in Fig. XIV.2, larger ground state stabilization occurs in the *ortho*- or *para*-position than in the *meta*-position. The isomer *7* should have, consequently, a more pronounced zwitterionic electronic structure than *5* and *6* and is expected to absorb at longer wavelengths [116]. No experimental results are yet known.

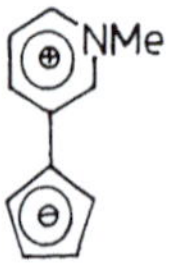

7

Other series of conjugated betaines differ from the above mentioned compounds by the donor fragment. The anionic hydrocarbon can be replaced by enolic [1058] or oxonolic anions [1059]. The resultant compounds are of yellow, red or violet color. Results of HMO calculations were presented to establish the CT-type long-wavelength absorption [1060]. The experimental CT excitation energies were found to correlate with the difference between the ionization potential of the donor and the electron affinity of the acceptor part [1061, 1061a].

The betaine *8* [1062–1064] displays a similar position of the color band as found for the iso-π-electronic *9* [1064–1067].

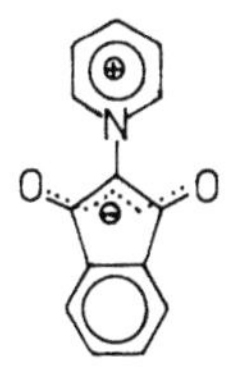

8

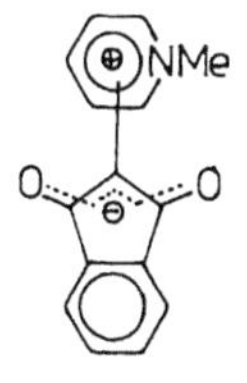

9

388 (4.25)

in ethanol [1064]

ortho: 394 (4.33) [1067]
meta: 386 (3.87) [1064]
para: 393 (4.67) [1064]
in ethanol

According to the above considerations the *meta*-substituted *9* should belong to the zwitterionic structures rather than the *ortho*- and *para*-isomers. This particular feature of *meta-9* is less reflected in the absorption wavelength than in the absorption intensity. The PPP calculation of *meta-9* predicts a longer wavelength component, which has not been observed experimentally. The transition may be hidden if the energy of this transition is increased due to negative solvatochromism.

Junction of pyridinium with phenoxide fragments leads to the pyridinium N-phenoxide betaines *10*. They exhibit long-wavelength absorptions of moderate intensity [1068, 1069] (cf. Table XIV.2).

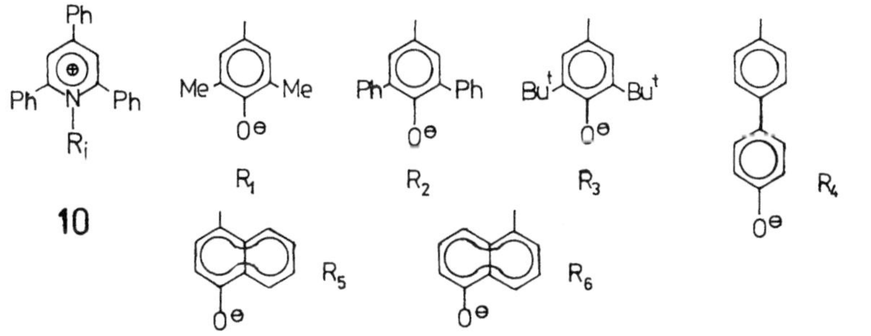

Table XIV.2. Longest-wavelength absorption maxima of pyridinium N-phenoxide betaines *10* [1068]

R_i	λ_{max} in nm (lg ε)	
	in Methanol	in Chloroform
R_1	487 (3.39)	653.5 (3.67)
R_2	515 (3.55)	730 (3.92)
R_3	563 (3.53)	792 (4.10)
R_4	422.5 (3.47)	452 (3.13)
R_5	508 (3.23)	–
R_6	478 (2.71)	–

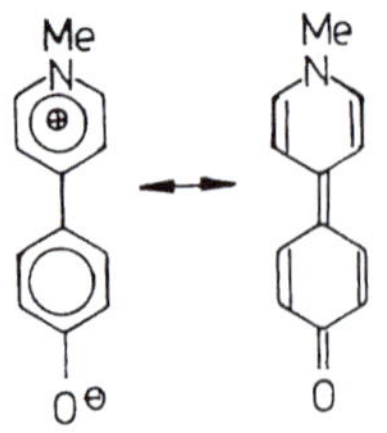

a 11 b

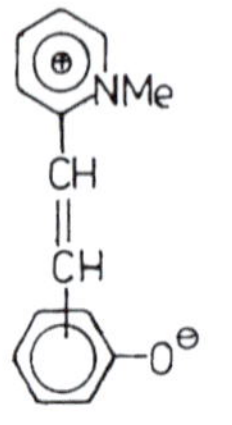

12

ortho: 494 (4.36)
meta: 425–440s (3.6)
para: 494–496 (4.44)
in alkaline ethanol [1072]

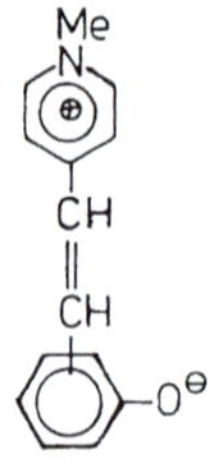

13

ortho: 520 (3.62)
meta: –
para: 515 (3.99)
in ethanol [1071]

In contrast to *1* an single intense $\pi \rightarrow \pi^*$ transition causes the longest-wavelength absorption of pyridinium N-phenoxides *10* [116, 1049], characterized by a strong intramolecular charge transfer according to the PPP configuration analysis [116]. The transition was calculated to be polarized in the long molecular axis. Accordingly, considerable red shift and doubling of the molar absorptivity has been observed by linking two equal betaine chromophores through a *para* phenylene bridge [1069]. Substituents attached at the *para* phenyl group of the pyridinium cause spectral shifts which obey a straight-line relationship with Hammett substituent constants [1070]. As expected, electron withdrawing substituents bring about red shift. Conversely, blue shift has been observed with electron releasing substituents.

Whereas junction of phenoxides with pyridinium in the *ortho-* or *para-*position of both fragments leads to quinone methides rather than pyridinium phenoxide betaines, as pictured by the mesomeric formula *11b*, for example, the zwitterionic structure is retained for *meta* junction. This conclusion can be drawn from the frontier orbitals of the fragments, which should weakly overlap in the *meta-*position (cf. Fig. XIV.2). In this connection spectral data about the vinylogous compounds *12* and *13* should be mentioned [1071–1074]. The spectrum of *meta-12* clearly differs from those of the *ortho-* and *para-*isomers in that the color band is hypsochromically and hypochromically shifted [1072]. However, the blue shift is in contradiction with the results of PPP calculations, which predict higher absorption wavelengths for *meta-12* [116]. The discrepancy may derive, in turn, from the large negative solvatochromism, which is generally observed for true zwitterionic structures.

Whereas the solvatochromic blue shift amounts to less than 30 nm for *1* and *6* in passing from chloroform to ethanol, a shift of about 100 nm has been found for *para-13* and more than 200 nm for *10* ($R_i = R_2$). The extremely strong negative solvatochromism of the pyridinium N-phenoxide betaines brings about a striking alteration of the visual color of the dissolved dye upon change of the solvent. The betaine *10* (R_2) is green in acetone (677 nm), blue in 3-methyl-butanol-1 (608 nm), violet in ethanol (550 nm) and red in methanol (515 nm). Therefore, the excitation energy of this compound has been measured in a great variety of organic solvents in order to define solvent polarity parameters (E_T-values) [1068, 1068a]. The lowest excitation energies of many zwitterionic or polar dyes, such as *1* [919], *3* [1053], *4* [1054], *ortho-13* [1074] and *para-13* [919] give straight-line relationships when plotted against E_T-values. This subject has been extensively reviewed by Reichardt and Dimroth [919].

The negative solvatochromism of betaine dyes is well described by an improved version of the continuum theory [1049]. Within this conception the blue shift is mainly due to the decrease of the dipole moment upon excitation, which was estimated to about 9 Debye for *10* ($R_i = R_3$) in dioxane [1069]. Evidence has been furnished that solvation of strongly polar solvents also causes an appreciable change in the electronic structure of *para-13* [102, 498].

XV. Multiple Chromophore Dyes

If dye molecules encompass two different chromophoric systems, their classification becomes somewhat arbitrary. In many cases these dyes reflect the spectral characteristics of the constituent chromophoric systems. The position of their longest-wavelength absorptions are then only slightly shifted relative to those of the distinct parent subchromophores. This feature may even be retained when the π-electron systems of the subchromophores undergo a moderate π-type overlap.

As long as the overlap between the two chromophores is very weak, the spectral shift and the intensity pattern of the lowest energy transitions are sufficiently described by the molecular exciton theory [109]. Significant spectral shifts are only to be expected if the subchromophores have a nearly equal excitation energy and if they display a head-to-tail arrangement of the pertinent transition moments. With increasing overlap between the subchromophores CT excitations between the chromophoric fragments cannot longer be ignored and the components of the color band, which result from excitonic splitting, are appreciably displaced to longer wavelengths.

Examples of mixed-chromophoric dyes which contain two distinct subchromophores in non-negligible π-type overlap are afforded by the compounds *1* [1075], *2* and *3* [1076]:

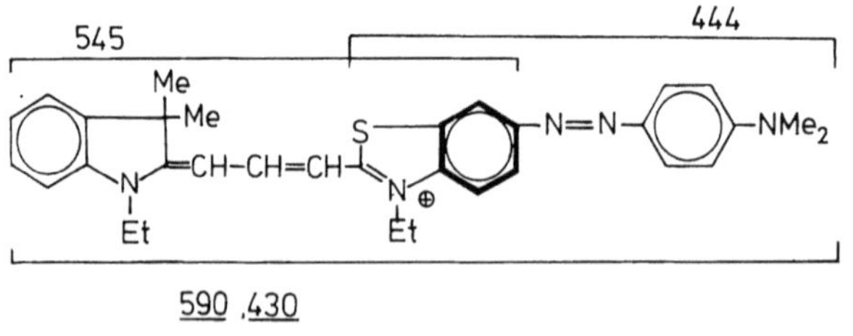

1

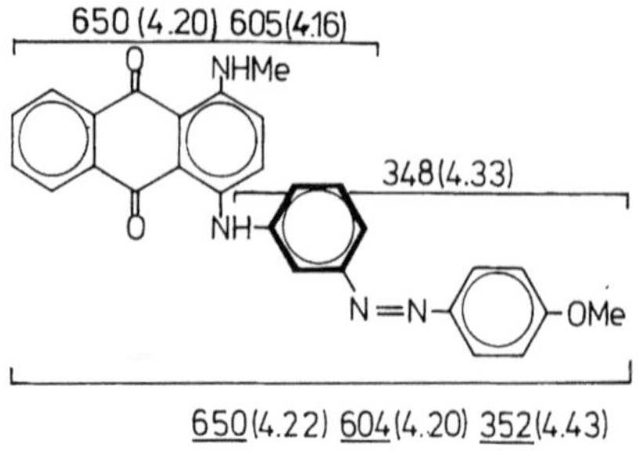

2

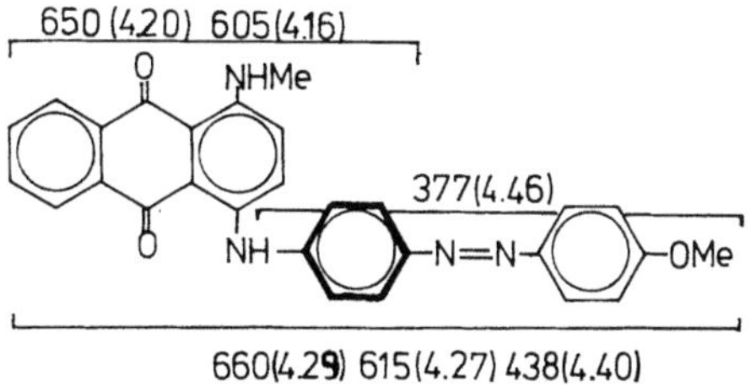

3

In these cases, the numerical agreement between the absorption wavelengths of the bichromophoric dyes and those of the chromophoric parts is less close. However, more appropriate reference subchromophores are defined if they include some electron delocalization. The subchromophores now involve an additional part of the π-electron system of the bichromophoric dye. For the above mentioned dyes *1* to *3* this can achieved by using the interchromophoric benzene rings twice in defining the subchromophores, as indicated with the formulae. The absorption wavelength of the bichromophoric dyes and those of the subchromophores then nearly coincide. This finding reflects a certain degree of localization of the excitation rather than a localization of the electronic states [116]. Kiprianov considered numerous bichromophoric dyes according to this modelling in subchromophores [110].

Meso-phenylazo-2,2′-dicarbocyanine *4* is exceptional in that the longest-wavelength absorption appears at shorter wavelength than that found for unsubstituted cyanine dyes [929]. According to the nodal properties of symmetrical dicarbocyanines the HOMO but not the LUMO is affected by the *meso*-substituent. Since the phenylazo group acts as acceptor rather than as donor substituent, the interaction between the acceptor LUMO and the cyanine HOMO should be considered. Owing to this interaction the cyanine HOMO is lowered in energy and, consequently, the color-determining energy of thc transition between the frontier orbital of the cyanine dye increases. Since the orbital interaction is weak because of the large energy gap between the interacting MO's, the hypsochromic effect is also weak and the location of the colour band of the dye *4* is similar to that dicarbocyanine reference dye.

The recognition of any localization of the electronic excitation in multiple chromophore dyes is greatly facilitated by analyzing the electronic transition with the aid of OKK- or LSU-analysis, mentioned in Sect. IV.1. However, a high degree of localization does not necessarily result in a coincidence of the absorption wavelength with that of the respective local part.

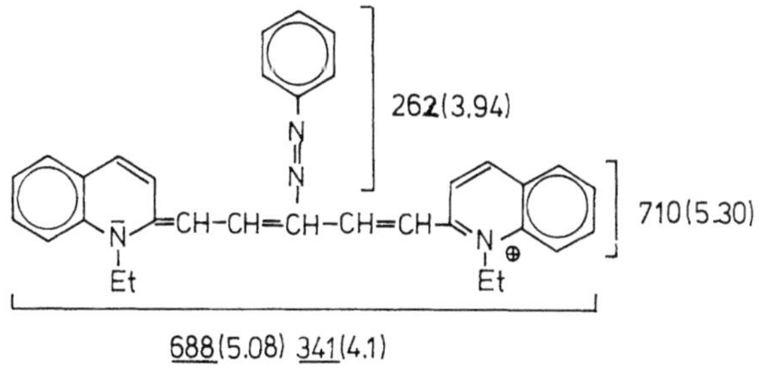

4

Linking of two chromophores may markedly influence the visual color of the dye. This fact has been utilized to obtain dyes of appropriate color properties. Thus, green dyes result from composing yellow and blue chromophores. Suitable parent structures for the yellow components can be azo dyes and quinoid dyes for the blue components, such as exemplified with the green dyes *2* and *3*. Along these lines, the variety of infrared lasers should be enriched by appropriate junction of dyes which both absorb at long wavelengths. Studies on UV-absorbing bichromophoric compounds have revealed interesting pecularities in their photophysical properties relative to those of the parent compounds [1077]. Less is known so far about the properties of colored bi- and polychromophoric dyes. The appearence of two luminescence maxima of the mixed-chromophoric *5* under certain conditions shows that the individuality of the molecular parts can be retained even in electronically excited states [1078].

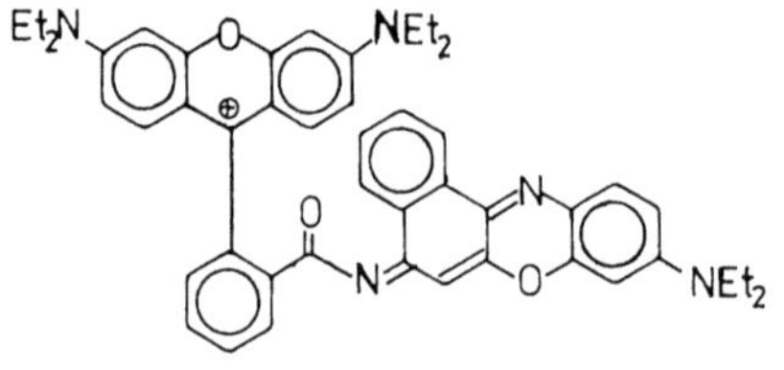

5

References

1. Dähne, S.: Z. Chem. *10,* 133, 168 (1970)
2. Lewis, G. N., Calvin, M.: Chem. Rev. *25,* 273 (1939)
3. Ferguson, L. N.: Chem. Rev. *43,* 385 (1948)
4. Maccoll, A.: Quart. Rev. *1,* 16 (1947)
5. Graebe, C., Liebermann, K.: Ber. dtsch. chem. Ges. *1,* 106 (1868)
6. Witt, O. N.: Ber. dtsch. chem. Ges. *9,* 522 (1876)
7. Kauffmann, H.: Sammlg. chem. u. chem.-techn. Vorträge *9,* 277 (1907); *12,* 1 (1907)
8. Dilthey, W.: J. prakt. Chemie [2] *109,* 273 (1925)
9. Dilthey, W., Wizinger, R.: J. prakt. Chem. [2] *118,* 321 (1928)
10. Wizinger, R.: Organische Farbstoffe, Berlin, Bonn: Dümmler 1933
11. Wizinger, R.: Chimia *15,* 89 (1961)
12. Mohler, H.: Chimia *4,* 284 (1950)
13. Griffiths, J.: Colour and constitution of organic molecules. London: Academic Press 1976
14. Nietzky, R.: Chemie d. organ. Farbstoffe. Berlin: Springer Verlag 1889
15. Armstrong, H.: Proc. Chem. Soc. *4,* 21 (1888)
16. von Baeyer, A.: Liebigs Ann. Chem. *354,* 152 (1907)
17. Willstätter, R., Piccard, J.: Ber. dtsch. chem. Ges. *41,* 1458 (1908)
18. Iz'mailskii, V. A.: Thesis, Technical College Dresden 1913
19. Eistert, B.: Sammlg. chem. u. chem.-techn. Vorträge, N.F. *40,* 1 (1938)
20. Bury, C. R.: J. Amer. Chem. Soc. *57,* 2115 (1935)
21. Haller, R., von Kostanecki, S.: Ber. dtsch. chem. Ges. *30,* 2947 (1897)
22. König, W.: J. prakt. Chem. [2] *112,* 1 (1926)
23. Dähne, S.: Z. Chem. *5,* 441 (1965)
24. Dähne, S., Leupold, D.: Angew. Chem. *78,* 1029 (1966); Int. Edit. *5,* 984 (1966)
25. Dähne, S., Leupold, D.: Ber. Bunsenges. Phys. Chem. *70,* 618 (1966)
26. Dähne, S.: Wiss. Z. Techn. Univers. Dresden *20,* 671 (1971)
27. Dähne, S., Kulpe, S.: Structural principles of unsaturated organic compounds, Abh. Akad. Wiss. DDR Nr. 8; Berlin: Akademie Verlag 1977
28. Dähne, S.: Science *199,* 1163 (1978)

28a. Dähne, S., Kulpe, S.: J. prakt. Chem. *320,* 395 (1978)

29. Hünig, S., Quast, H.: In: Opt. Anregung organ. Systeme (ed. W. Foerst), p. 184. Weinheim: Verlag Chemie 1966
30. Ishchenko, A. A., Derevyanko, N. A., Kudinova, M. A., Dyadyusha, G. G., Tolmachev, A. I.: Teor. Eksp. Khim. *13,* 828 (1977)
31. Allen, R. L. M.: Colour Chemistry. London: Nelson 1971
32. Rao, C. N. R.: Ultraviolet and visible spectroscopy. London: Butterworth 1961
33. Jaffé, H. H., Orchin, M.: Theory and application of ultraviolet spectroscopy. New York: Wiley 1962
34. Murrell, J. N.: Theory of the electronic spectra of organic molecules. London: Methuen 1964
35. Suzuki, H.: Electronic absorption spectra and geometry of organic molecules. New York: Academic Press 1967

36. Formánek, J.: Untersuch. u. Nachweis organ. Farbstoffe auf spektroskop. Wege, Vol. 1–2. Springer 1908, 1926, 1927
37. Lasareff, P. P.: Atlas des spectres des substances colourants. Publ. Acad. Sc. URSS, Leningrad 1927
38. Gurr, E.: Synthetic dyes in biology, medicine and chemistry. London: Acad. Press 1971
39. DMS-UV Atlas of organic compounds, Vol. 1–5. London: Butterworth; Weinheim: Verlag Chemie 1966–1971
40. Organic electronic spectral data, Vol. 1–13 (1946–1971). New York: Intersc. Publ. 1960–1977
41. Commerford, T. R.: Textile Chem. Col. *6,* 39 (1974)
42. Pauling, L.: Proc. Nat. Acad. Sci. USA *25,* 577 (1939)
43. Lewis, G. N., Calvin, M.: Chem. Rev. *25,* 273 (1939)
44. Brooker, L. G. S.: Rev. Mod. Phys. *14,* 275 (1942)
45. Förster, Th.: Z. Physikal. Chem. B *47,* 245 (1940), B *48,* 12 (1941)
46. Herzfeld, K. F., Sklar, A. L.: Rev. Mod. Phys. *14,* 294 (1942)
47. Moffitt, W. E.: Proc. Phys. Soc. *63A,* 700 (1950)
48. Förster, Th.: Z. Elektrochem. *45,* 548 (1939)
49. Knott, E. B.: J. Chem. Soc. *1951,* 1024
50. Shibata, K.: Bull. Chem. Soc. Japan *26,* 105 (1953)
51. Platt, J. R.: J. Chem. Phys. *25,* 80 (1956)
52. Simpson, W. T.: Theoret. Chim. Acta *18,* 354 (1970)
53. Murrell, J. N.: J. Chem. Soc. *1959,* 296
54. Grinter, R., Heilbronner, H.: Helv. Chim. Acta *45,* 2496 (1962)
55. Heilbronner, E.: In: Molecular orbital in chemistry, physics and biology (ed. P.-O. Löwdin, B. Pullman), p. 329. New York: Academic Press 1964
56. Kuhn, H.: J. Chem. Phys. *16,* 840 (1948); Helv. Chim. Acta *31,* 1441 (1948)
57. Bayliss, N. S.: J. Chem. Phys. *16,* 287 (1948), Quart. Rev. *6,* 319 (1952)
58. Platt, J. R.: J. Chem. Phys. *17,* 484 (1949)
59. Simpson, W. T.: J. Chem. Phys. *16,* 1124 (1948)
60. Kuhn, H.: Fortschr. Chem. Org. Naturstoffe *16,* 170 (1958), *17,* 404 (1959)
61. Försterling, H. D., et al.: In: Optische Anreg. organ. Systeme (ed. W. Foerst), p. 55. Weinheim: Verlag Chemie 1966
62. Ham, N. S., Rüdenberg, K.: J. Chem. Phys. *25,* 13 (1956)
63. Olszewski, S.: J. Chem. Phys. *44,* 3297 (1966)
64. Streitwieser, A.: Molecular orbital theory for organic chemists. New York: Wiley 1962
65. Heilbronner, E., Bock, H.: Das HMO-Modell, Bd. 1–3. Weinheim: Verlag Chemie 1968, 1970
66. Streitwieser, A., Brauman, J. I., Coulson, C. A.: Supplement tables of molecular orbital calculations, Vol. 1 and 2. Oxford: Pergamon Press 1965
67. Kutzelnigg, W.: Einführ. in d. theoret. Chemie, Vol. *2,* Chap. 11. Weinheim: Verlag Chemie 1978
68. Fabian, J., Hartmann, H.: Theoret. Chim. Acta *36,* 351 (1975)
69. Gutman, I.: Croat. Chem. Acta *48,* 97 (1976)
69a. Gutman, I.: Chem. Phys. Letters *50,* 488 (1977); Theoret. Chim. Acta *50,* 287 (1979); Match *5,* 161 (1979)
70. Dewar, M. J. S.: J. Chem. Soc. (London) *1950,* 2329; *1952,* 3532, 3544
71. Zahradník, R.: Topics Curr. Chem. *10,* 1 (1968)
72. Götz, H., Heilbronner, E.: Helv. Chim. Acta *44,* 1365 (1961)
73. Gołębiewski, A., Nowakowski, J.: Acta Phys. Polen *25,* 647 (1964)
74. Sturmer, D. M., Gaugh, W. S.: Photogr. Sci. Eng. *17,* 146 (1973)
75. Kakitani, T., Kakitani, H.: Theoret. Chim. Acta *46,* 259 (1977)
76. Parr, R. G.: The quantum theory of molecular electronic structure. New York: Benjamin 1964
77. Fratev, F., Hiebaum, G., Gochev, A.: Chem. Phys. Letters *16,* 349 (1972)

78. Downing, J. W., Michl, J., Jørgensen, P., Thulstrup, E. W.: Theoret. Chim. Acta *32*, 203 (1974)
79. Schulten, K., Ohmine, I., Karplus, M.: J. Chem. Phys. *64*, 4422 (1976)
80. Nishimoto, K., Forster, L. S.: Theoret. Chim. Acta *3*, 407 (1965), *4*, 155 (1966)
81. Fabian, J., Zahradník, R.: Wiss. Z. Techn. Univers. Dresden *26*, 315 (1977)
82. Brown, R. D., Heffernan, M. L.: Trans. Faraday Soc. *54*, 757 (1958)
83. Ferré, Y., Vincent, E.-J., Larivé, H., Metzger, J.: Bull. Soc. Chim. France *1972*, 3862
84. Pancíř, J., Matoušek, I., Zahradník, R.: Collect. Czechoslov. Chem. Commun. *38*, 3039 (1973)
85. Hammond, H. A.: Theoret. Chim. Acta *18*, 239 (1970)
86. Ferré, Y., Larivé, H., Vincent, E.-J.: Photogr. Sci. Eng. *18*, 457 (1974)
87. Griffiths, J., Lockwood, M.: J. Chem. Soc. Perkin. Trans. 1, *1976*, 48
88. Del Bene, J., Jaffé, H. H.: J. Chem. Phys. *48*, 1807 (1968).
89. Yonezawa, T., Konishi, H., Kato, H.: Bull. Chem. Soc. Japan *42*, 933 (1969)
90. Kuhn, J. E.: Thesis, Univers. Zürich 1971
91. Lipari, N. O., Duke, C. B.: J. Chem. Phys. *63*, 1748 (1975)
92. Warshel, A., Karplus, M.: Chem. Phys. Letters *17*, 7 (1972); Warshel, A.: In: Modern theoretical chemistry (ed. G. A. Segal), Vol. *7*, Part A, p. 133. New York: Plenum Press 1977
93. Momicchioli, F., Baraldi, I., Bruni, M. C.: J. Chem. Soc., Faraday Trans. 2, *68*, 1556 (1972)
94. Fujimura, Y., Onda, M., Nakajima, T.: Bull. Chem. Soc. Japan *46*, 2034 (1973)
95. Kakitani, T.: Progr. Theoret. Physics *52*, 1721 (1974)
96. Pullman, A.: In: The world of quantum chemistry (ed. R. Daudel, B. Pullman), p. 239. Dortrecht: D. Reidel Publ. Comp. 1974
97. Del Bene, J. E.: J. Amer. Chem. Soc. *95*, 6517 (1973); J. Chem. Phys. *62*, 666, 1961 (1975)
98. Morokuma, K., Iwata, S., Lathan, W. A.: In: The world of quantum chemistry (ed. R. Daudel, B. Pullman), p. 277. Dortrecht: Reidel Publ. Comp. 1974; Iwata, S., Morokuma, K.: J. Amer. Chem. Soc. *97*, 966 (1975)
99. De Jeu, W. H.: Chem. Phys. Letters *7*, 153 (1970)
100. Rao, C. N. R., Murthy, A. S. N.: Theoret. Chim. Acta *22*, 392 (1971)
101. Nolte, K.-D., Dähne, S.: Adv. Mol. Relax. Interact. Proc. *10*, 299 (1977)
102. Benson, H. O., Murrell, J. N.: J. Chem. Soc., Faraday Trans. 2, *68*, 137 (1972)
103. Scheibe, P., Schneider, S., Dörr, F., Daltrozzo, E.: Ber. Bunsenges. Phys. Chem. *80*, 630 (1976)
104. Germer, H. A.: Theoret. Chim. Acta *34*, 145, 273 (1974)
105. Klessinger, M.: Theoret. Chim. Acta *5*, 251 (1966)
106. Csizmadia, I. G.: Progr. theoret. organic chem., Vol. 1, Theorie and practice of MO calculations on organic molecules p. 2. Amsterdam: Elsevier 1976
107. Platt, J. R.: J. Chem. Phys. *19*, 101 (1951)
108. Klessinger, M., Lüttke, W.: Tetrahedron *19*, Suppl. 2, 315 (1963)
109. Kasha, M., Rawls, N. R., El-Bayoumi, M. A.: Pure Appl. Chem. *11*, 371 (1965)
110. Kiprianov, A. I.: Usp. Khim. *40*, 1283 (1971)
111. Longuet-Higgins, H. C., Murrell, J. N.: Proc. Phys. Soc. A *68*, 601 (1955)
112. Fabian, J.: J. Signalaufzeichnungsmaterialien *6*, 307 (1978); *7*, 67 (1979)
113. Mulliken, R. S., Person, W. D.: Molecular complexes. New York: Wiley 1969
114. Heilbronner, E., Weber, J.-P., Michl, J., Zahradník, R.: Theoret. Chim. Acta *6*, 141 (1966)
115. Baba, H., Suzuki, S., Takemura, T.: J. Chem. Phys. *50*, 2078 (1969)
116. Fabian, J.: unpublished 1976
117. Polansky, O. E., Derflinger, G.: Int. J. Quantum Chem. *1*, 379 (1967)
118. Dewar, M. J. S., Dougherty, R. C.: The PMO theory of organic chemistry. New York: Plenum Press 1975
119. Dewar, M. J. S.: J. Amer. Chem. Soc. *74*, 3341, 3345, 3350, 3353, 3355, 3357 (1952)
120. Mason, S. F.: J. Soc. Dyers Colour. *84*, 604 (1958); The chemistry of synthetic dyes (ed. E. Venkataraman), Vol. *3*, p. 169. New York: Academic Press 1970
121. Hallas, G.: J. Soc. Dyers Colour. *84*, 510 (1968), *86*, 237 (1970)

122. Dyadyusha, G. G.: Ukr. Khim. Zh. *30,* 929, 1186 (1964)
123. Dyadyusha, G. G.: Ukr. Khim. Zh. *31,* 1171, 1293 (1965)
124. Dyadyusha, G. G., Kachkovskii, A. D.: Ukr. Khim. Zh. *44,* 948, 1069 (1978)
125. Ohta, T., Kuroda, H., Kunii, T. L.: Theoret. Chim. Acta *19,* 167 (1970)
126. Lusanov, A. V., Sukhorukov, A. A., Umanskii, V. E.: Teor. Eksp. Khim. *10,* 456 (1974), Lusanov, A. V., ibid. *13,* 579 (1977)
127. Giovanelli, K. H., Hohlneicher, G., Straub, P. A.: Ber. Bunsenges. Phys. Chem. *76,* 857 (1971)
128. Fabian, J.: J. prakt. Chem. *322,* 1 (1980)
129. Fabian, J.: J. prakt. Chem. *320,* 361 (1978)
130. Fabian, J., Hartmann, H.: Z. Chem. *18,* 145 (1978)
131. Heilbronner, E., Straub, P. A.: Hückel molecular orbitals. Berlin: Springer Verlag 1966
132. Fabian, J., Hartmann, H.: J. Mol. Structure *27,* 67 (1975)
133. Fabian, J., Zimmer, A.: J. prakt. Chem. *321,* 449 (1979)
134. Fabian, J., Tröger-Naake, G.: Int. J. Quantum Chem. *11,* 259 (1977)
135. Platt, J.: In: Radiation biology (ed. Hollaender), Vol. 3, p. 71. New York: McGraw-Hill 1956
136. Christensen, R. L., Kohler, B. E.: J. Phys. Chem. *80,* 2197 (1976)
137. Ovchinnikov, A. A., Ukrainskii, I. I., Kventsel, G. F.: Usp. Fiz. Nauk *108,* 81 (1972)
138. Simmons, H. E.: In: Progr. physical organic chem. (ed. A. Streitwieser, R. W. Taft), Vol. 7, p. 1. New York: Interscience 170
139. Merz, J. H., Straub, P. A., Heilbronner, E.: Chimia *19,* 302 (1965)
140. Plotnikov, V. G.: Opt. Spektr. *23,* 39 (1967)
141. Inuzuka, K., Becker, R. S.: Bull. Chem. Soc. Japan *47,* 88 (1974), Das, P. K., Becker, R. S., J. Phys. Chem. *82,* 2081 (1978)
142. Allinger, N. L., Tai, J. C.: J. Amer. Chem. Soc. *99,* 4256 (1977)
143. Isler, O.: Carotenoids. Basel: Birkhäuser 1971
144. Bauernfeind, J. C., Brubacher, G. B., Kläui, H. M., Marusich, W. L.: In: Carotenoids (ed. by O. Isler), p. 743. Basel: Birkhäuser 1971
145. Zechmeister, L.: Fortschr. Chem. Org. Naturstoffe *18,* 223 (1960)
146. Weedon, B. C. L.: Fortschr. Chem. Org. Naturstoffe *27,* 81 (1969). In: Carotenoids (ed. by O. Isler), p. 267. Basel: Birkhäuser Verlag 1971
147. Vetter, W., Englert, G., Rigassi, N., Schwieter, U.: ibid. p. 189.
148. Bloor, J. E., Gilson, B. R., Brearley, M.: Theoret. Chim. Acta *8,* 35 (1967)
149. Song, P.-S., Moore, T. A., Sun, M.: In: The chemistry of plant pigments. New York: Academic Press 1972
150. Bart, J. C. J., MacGillavry, C. H.: Acta Cryst. B *24,* 1587 (1968)
151. Eckert, R., Kuhn, H.: Z. Elektrochem. *64,* 356 (1960)
152. Moore, T. A., Song, P.-S.: J. Mol. Spektrosc. *52,* 209 (1974)
153. Hiryama, K.: J. Amer. Chem. Soc. *77,* 373 (1955)
153a. Sugimoto, T., Nakano, K., Kitajima, H., Suzuki, H.: J. Phys. Soc. Jap. *46,* 1307 (1979)
154. Honig, E., Warshel, A., Karplus, W.: Acc. Chem. Res. *8,* 92 (1975)
155. Becker, R. S., Hug, G., Das, P. K., Schaffer, A. M., Takemura, T., Yamamoto, N., Waddell, W.: J. Phys. Chem. *80,* 2265 (1976)
156. Inoue, Y., Tokito, Y., Chûjô, R., Miyoshi, Y.: J. Amer. Chem. Soc. *99,* 5592 (1977)
157. Pitt, G. A. J.: In: Carotenoids (ed. by O. Isler), p. 717. Basel: Birkhäuser Verlag 1971
158. Favrot, J., Leclercq, J.-M., Roberge, R., Sandorfy, C., Vocelle, D.: Chem. Phys. Letters *53,* 433 (1978)
159. Nash, H. A.: J. Theor. Biol. *22,* 314 (1969)
160. Pullman, B., Langlet, J., Berthod, H.: J. Theor. Biol. *23,* 492 (1969)
161. Langlet, J., Pullman, B., Berthod, H.: J. Mol. Structure *6,* 139 (1970)
162. Langlet, J., Giessner-Prettre, C.: J. Mol. Structure *13,* 317 (1972)
163. Wiesenfeld, J. R., Abrahamson, E. W.: Photochem. Photobiol. *8,* 487 (1968)
164. Honig, B., Karplus, M.: Nature *229,* 558 (1971)
165. Rowan, R., Warshel, A., Sykes, B., Karplus, M.: Biochemistry *13,* 970 (1974)
166. Warshel, A., Karplus, M.: J. Amer. Chem. Soc. *96,* 5677 (1974)

167. Gilardi, R. D., Karle, I. L., Karle, J.: Acta Crystallogr. B *28,* 2605 (1972)
168. Becker, R. S., Inuzuki, K., Balke, D. E.: J. Amer. Chem. Soc. *93,* 38 (1971)
169. Langlet, J., Berthod, H., Pullman, B.: J. Chim. Phys. Physicochim. Biol. *66,* 566 (1969)
170. Burke, M. J., Pratt, D. C., Faulkner, T. R., Moscowitz, A.: Exp. Eye Res. *17,* 557 (1973)
171. Weimann, L. J., Maggiora, C. M., Blatz, P. E.: Int. J. Quantum Chem., Quantum Biol. Symp. *2,* 9 (1975)
172. Birge, R., Schulten, K., Karplus, M.: Chem. Phys. Letters *31,* 451 (1975)
173. Schaffer, A. M., Waddell, W. H., Becker, R. S.: J. Amer. Chem. Soc. *96,* 2063 (1974)
174. Irving, C. S., Byers, G. W., Leermakers, P. A.: Biochemistry *9,* 858 (1970): J. Amer. Chem. Soc. *91,* 2141 (1969)
174a. Das, P. K., Becker, R. S., Hannak, D., Bayer, E.: J. Amer. Chem. Soc. *101,* 239 (1979)
175. Suzuki, T., Kito, Y.: Photochem. Photobiol. *15,* 275 (1972)
176. Blatz, P. E.: Photochem. Photobiol. *15,* 1 (1972)
177. Navangul, H. V., Blatz, P. E.: J. Amer. Chem. Soc. *95,* 1508 (1973), *100,* 4340 (1978)
178. Suzuki, H., Komatsu, T., Kitajima, H.: J. Phys. Soc. Japan *37,* 177 (1974)
179. Waleh, A., Ingraham, L. L.: Arch. Biochem. Biophys. *156,* 261 (1973)
180. Mantione, M. J., Pullman, B.: Int. J. Quantum Chem. Symp. *5,* 349 (1971)
181. Moore, T. A., Song, P.-S.: J. Mol. Structure *52,* 224 (1974)
182. Muthukumar, M., Weimann, L. J.: Chem. Phys. Letters *53,* 436 (1978)
183. Kriebel, A. N., Albrecht, A. C.: J. Chem. Phys. *65,* 4575 (1976)
184. Hausser, K. W., Kuhn, R., Smakula, A.: Z. Physikal. Chem. B *29,* 384 (1935)
185. Blair, H. S., Boyd, N. L.: J. Soc. Dyers Colour. *89,* 245 (1973), *92,* 14 (1976)
186. Traetteberg, M., Frantsen, E. B., Mijlhoff, F. C., Hoekstra, A.: J. Mol. Structure *26,* 57 (1975)
187. Traetteberg, M., Frantsen, E. B.: J. Mol. Structure *26,* 69 (1975)
188. Beveridge, D. L., Jaffé, H. H.: J. Amer. Chem. Soc. *87,* 5340 (1965)
189. Ljunggren, St., Wettermark, G.: Theoret. Chim. Acta *19,* 326 (1970)
190. Bally, T., Haselbach, E., Lanyiova, S., Marschner, R., Rossi, M.: Helv. Chim. Acta *59,* 486 (1976)
191. Bromberg, A., Muszkat, K. A.: Tetrahedron *28,* 1265 (1972)
192. Gołebiewski, A., Parczewski, A.: Acta Phys. Polon. A *41,* 727, 735 (1972)
193. Momicchioli, F., Bruni, M. C., Baraldi, I., Corradini, G. R.: J. Chem. Soc., Faraday Trans. 2, *70,* 1325 (1974)
194. Warshel, A.: J. Chem. Phys. *62,* 214 (1975)
195. Beringhelli, T., Cavezzotti, A., Simonetta, M.: J. Mol. Structure *12,* 333 (1972)
196. Wettermark, G., Tegmér, L., Mertensson, O.: Arkiv Kemi *30,* 185 (1968)
197. Tanazaki, Y., Inoue, N., Hoshi, T., Shiraishi, J.: Z. Physikal. Chem. *74,* 45 (1971)
198. Hoffmann, H.-J., Birner, P.: J. Mol. Structure *39,* 145 (1977)
199. Wolf, A., Schmidtke, H.-H., Knop, J. V.: Theoret. Chim. Acta *48,* 37 (1978)
200. Momicchioli, I., Corradini, G., Bruni, M. C., Baraldi, I.: J. Chem. Soc., Faraday Trans. 2, *71,* 215 (1975)
201. Popov, K. R., Smirnov, L. V.: Opt. Spectrosc. *30,* 178, 628 (1971)
201a. Bobrovich, V. P., Sarzhevskii, A. M., Senyuk, M. A.: Zh. Prikl. Spektrosk. *30,* 1022 (1979)
202. Ciohodaru, L., Meghea, A.: Rev. Roum. Chim. *21,* 1473 (1976)
203. Ciohodaru, L.: Rev. Roum. Chim. *22,* 1125 (1977)
204. Hünig, S., Rosenthal, D.: Liebigs Ann. Chem. *592,* 161 (1955)
205. Schulze, J., Gerson, F., Murrell, J. N., Heilbronner, E.: Helv. Chim. Acta *44,* 428 (1961)
206. Takeuchi, Y., Akiyama, S., Nakagawa, M.: Bull. Chem. Soc. Japan *45,* 3183 (1972)
207. Yasuhara, A., Akiyama, S., Nakagawa, M.: ibid. *45,* 3638 (1972)
208. Takeuchi, Y., Yasuhara, A., Akiyama, S., Nakagawa, M.: ibid. *46,* 909 (1973)
209. Mandal, K., Misra, T. N.: ibid. *49,* 198 (1976)
210. Hudson, B. S., Kohler, B. E.: Chem. Phys. Letters *14,* 299 (1972)
211. Birks, J. B., Birch, D. J. S.: ibid *31,* 608 (1975)
211a. Tavan, P., Schulten, K.: Chem. Phys. Letters *56,* 200 (1978)
212. Moore, T. A., Song, P.-S.: ibid. *19,* 128 (1973)

213. Gavin, R. M., Risemberg, S., Rice, S. A.: J. Chem. Phys. *58,* 3160 (1973)
214. Karplus, M., Gavin, R. M., Rice, S. A.: J. Chem. Phys. *63,* 5507 (1975)
215. Nikitina, A. N., Ponomareva, N. A., Y'anovskay'a, L. A., Dombrovskii, V. A., Kucherov, V. F.: Opt. Spectrosc. *40,* 251 (1976)
216. Stachelek, T. M., Pazoha, T. A., McClain, W. M., Drucker, R. P.: J. Chem. Phys. *66,* 4540 (1977)
217. Baraldi, I., Momicchioli, F., Bruni, M. C.: J. Chem. Soc., Faraday Trans. 2, *68,* 1571 (1972)
218. Chae, Qu., Song, P.-S., Johansen, J. E., Liaaen-Jensen, S.: J. Amer. Chem. Soc. *99,* 5609 (1977)
219. Sease, J. W., Zechmeister, L.: J. Amer. Chem. Soc. *69,* 270 (1947)
220. van Reijendam, J. W., Heeres, G. J., Janssen, M. J.: Tetrahedron *26,* 1291 (1970)
221. Wachters, A. J. H., Davies, D. W.: Tetrahedron *20,* 2841 (1964)
222. Kiprianov, A. I., Kornilov, M. Yu., Mikhailik, S. K.: Zh. Org. Khim. *2,* 552 (1966)
223. Kornilov, M. Yu., Ruban, E. M.: Ukr. Khim. Zh. *35,* 824 (1969)
224. Kornilov, M. Yu., Ruban, E. M.: Zh. Org. Khim. *9,* 2188 (1973)
225. Karrer, P., Eugster, C. H.: Helv. Chim. Acta *34,* 1805 (1951)
226. Feichtmayr, F., et al.: Tetrahedron *25,* 5383 (1969)
226a. Salares, V. R., Young, N. M., Carey, P. R., Bernstein, H. J.: J. Raman Spectrosc. *6,* 282 (1977)
227. Liptay, W.: In: Opt. Anreg. Organ. Systeme (ed. W. Foerst), p. 263. Weinheim: Verlag Chemie 1966
228. Lippert, E.: Z. Elektrochem. *61,* 962 (1957)
229. Körösy, F.: Acta Chim. Acad. Sc. Hung. *15,* 35 (1958)
230. Wassermann, A.: J. Chem. Soc. *1959,* 979
231. Blatz, P. E., Pippert, D. L.: J. Amer. Chem. Soc. *90,* 1296 (1968)
232. Blatz, P. E., Pippert, D. L., Balasubramaniyan, V.: Photochem. Photobiol. *8,* 309 (1968)
233. Sorensen, T. S.: J. Amer. Chem. Soc. *87,* 5075 (1965)
234. Dähne, S., Schob, F.: Prakt. Chem. *315,* 810 (1973)
235. Levisalles, J.: Bull. Soc. Chim. France *24,* 997 (1957)
236. Nussbaum, A. L., Mancara, O., Daniels, R., Rosenkranz, G., Djerassi, C.: J. Amer. Chem. Soc. *73,* 3263 (1951)
237. Inhoffen, H. H., von der Bey, G.: Liebigs Ann. Chem. *583,* 100 (1953)
238. Yankovskaya, L. A., Kovalev, A. C., Kucherov, V. F.: Izv. Akad. Nauk SSSR, Ser. Khim. *1965,* 684.
239. Blout, R. E., Fields, M.: J. Amer. Chem. Soc. *70,* 189 (1948)
240. Berry, R. W. H., Brocklehurst, P., Burawoy, A.: Tetrahedron *10,* 109 (1960)
241. Splitter, J. S., Calvin, M.: J. Org. Chem. *20,* 1086 (1955)
242. Henseke, G., Binte, H.-J.: Liebigs Ann. Chem. *612,* 205 (1958)
243. Bohlmann, F., Mannhardt, H. J.: Chem. Ber. *89,* 1307 (1956)
244. Zollinger, H.: Azo and diazo chemistry. New York: Interscience, 1961
245. Venkataraman, K.: The chemistry of synthetic dyes, Vol. 1–7. New York: Academic Press 1952–1974
246. Schweizer, H. R.: Künstl. Organ. Farbstoffe u. ihre Zwischenprodukte. Berlin: Springer Verlag 1964
247. Robin, M. B., Hart, R. R., Kuebler, N. A.: J. Amer. Chem. Soc. *89,* 1564 (1967)
248. Wong, D. P., Fink, W. H., Allen, L. C.: J. Chem. Phys. *52,* 6291 (1970)
249. Tinland, B.: Spectrosc. Lett. *3,* 51 (1970)
250. Baird, N. C., Swenson, J. R.: Canad. J. Chem. *51,* 3097 (1973)
251. Hsu, H.: Chem. Phys. *7,* 187 (1975)
252. Vasudevan, K., et al.: Chem. Phys. *7,* 187 (1975)
253. Winter, N. W., Pitzer, R. M.: J. Chem. Phys. *62,* 1269 (1975)
254. Haselbach, E., Schmelzer, A.: Helv. Chim. Acta *54,* 1575 (1971)
255. Haselbach, E., Heilbronner, E.: Helv. Chim. Acta *53,* 684 (1970)
256. Houk, K. N., Chang, Y.-M., Engel, P. S.: J; Amer. Chem. Soc. *97,* 1824 (1975)

257. Kobayashi, T., Yokota, K., Nagakura, S.: J. Electron Spectrosc. Relat. Phenom. *6*, 167 (1975)
258. Hochstrasser, R. M., Lower, S. K.: J. Chem. Phys. *36*, 3505 (1962)
259. Robin, M. B., Simpson, W. T.: J. Chem. Phys. *36*, 580 (1962)
260. Perić, M., Buenker, R. J., Peyerimhoff, S. D.: Canad. J. Chem. *55*, 1533 (1977)
261. Kortüm, G., Rau, H.: Ber. Bunsenges. Phys. Chem. *68*, 973 (1964)
262. Robin, M. B.: Higher excited states of polyatomic molecules, Vol. 2, p. 68. New York: Acad. Press 1975
263. Mosher, O. A., et al.: J. Chem. Phys. *62*, 3424 (1975)
264. Rau, H.: Angew. Chem. *85*, 248 (1973), Int. Edit. *12*, 224 (1973)
265. Bock, H.: Angew. Chem. *77*, 469 (1965), Int. Edit. *4*, 457 (1965)
266. Bock, H., Wittel, K., Veith, M., Wiberg, N.: J. Amer. Chem. Soc. *98*, 109 (1976)
267. Tomasik, P.: Rocz. Chem. *44*, 1369 (1970)
268. Bisle, H., Römer, M., Rau, H.: Ber. Bunsenges. Phys. Chem. *80*, 301 (1976)
268a. Kroner, J., Bock, H.: Chem. Ber. *101*, 1922 (1968)
269. Hohlneicher, G., Sänger, W.: In: Quantum aspects of heterocyclic compounds in chemistry and biochemistry (ed. E. D. Bergmann, B. Pullman), Vol. 2, p. 193. New York: Academic Press 1970
270. Brown, C. J.: Acta Cryst. *21*, 146 (1966)
271. Robertson, J. M.: J. Chem. Soc. *1939*, 232
272. Traetteberg, M., Hilmo, I., Hagen, K.: J. Mol. Structure *39*, 231 (1977)
273. Beveridge, D. L., Jaffé, H. H.: J. Amer. Chem. Soc. *88*, 1948 (1966)
274. Gondo, Y.: Mem. Fac. Sci., Kyushu University, Ser. C *6*, 63 (1968)
275. Bonchev, D., Rachin, E.: Monatsh. Chem. *101*, 1454 (1970)
276. Sühnel, J.: Thesis, Techn. Univers. Dresden 1975
277. Maiburov, S. P., Popov, K. P., Smirnov, L. V.: Opt. Spectrosc. *37*, 362 (1974)
278. Inoue, H.; Kanagawa Daigaku Kokakubu Hokoku *1969*, (7) 90
279. Haselbach, E., Heilbronner, E.: Helv. Chim. Acta *51*, 16 (1968)
280. Hofmann, H.-J., Birnstock, F.: J. Mol. Structure *44*, 231 (1978)
281. Bürgi, H. B., Dunitz, J. D.: Helv. Chim. Acta *54*, 1255 (1971)
282. Traetteberg, M., Hilmo, I., Abraham, R. J., Ljunggren, S.: J. Mol. Structure *48*, 395 (1978)
283. Skrabal, F., Steiger, J., Zollinger, H.: Helv. Chim. Acta *58*, 800 (1975)
284. Nurmukhametov, P. N., Lipasova, V. A.: Opt. Spectrosc. *39*, 652 (1975)
285. Goursot, A., Jacques, P., Faure, J.: J. Chim. Phys Physicochim. Biol. *73*, 694 (1976)
286. Hampson, G. C., Robertson, J. M.: J. Chem. Soc. *1941*, 409
287. Dähne, S.: Z. Wiss. Photographie, Photophys., Photochem. *62*, 183 (1968)
288. Fabian, J.: Chimia *32*, 323 (1978)
289. Kikot, G. S., Chauser, M. G., Cherkashin, M. I.: Izv. Akad. Nauk SSSR, Ser. Khim. *1976*, 2150
290. Castelino, R. W., Hallas, G.: J. Chem. Soc. B *1971*, 793
291. Hallas, G., Ho, W. L., Todd, R.: J. Soc. Dyers Colour. *90*, 121 (1974)
292. Peters, A. T.: Text. Res. J. *44*, 719 (1974)
293. Kiprianov, A. I., Zhmurnova, I. N.: Zh. Obshch. Khim. *23*, 626 (1956)
293a. Dix, J. P., Vögtle, F.: Angew. Chem. *90*, 892 (1978); Int. Edit. *17*, 857 (1978)
294. Griffiths, J.: J. Soc. Dyers Colour. *88*, 106 (1972)
295. Hanson, W. T.: J. Photogr. Sci. *25*, 189 (1977)
296. Griffiths, J.: J. Chem. Soc. B *1971*, 801
296a. Gregory, P., Thorp, D.: J. Chem. Soc., Perkin Trans 1, *1979*, 1990
297. Kokkinos, K., Wizinger, R.: Helv. Chim. Acta *54*, 330, 335, 338, 2744 (1971)
298. Hoyer, E., Schickfluß, R., Steckelberg, W.: Angew. Chem. *85*, 984 (1973). Int. Edit. *12*, 926 (1973)
299. Hallas, G., Ng, K. L.: J. Soc. Dyers Colour. *93*, 284 (1977)
300. Dickey, J. B., Towne, E. B., Hedberg, D. G., Wallace, D. J., Weayer, M. A., Straley, J. M.: Am. Dyestuff Rep. *1965*, 596

301. Skulski, L.: Bull. Acad. Pol. Sci. *12,* 719 (1964)
302. Linke, H. A. B., Pramer, D.: Z. Naturforsch. B *24,* 997 (1969)
303. Mustroph, H., Epperlein, J., Ebert, D.: J. Signalaufzeichnungsmaterialien *5,* 439 (1976)
304. Holt, P. F., Smith, A. E.: J. Chem. Soc. *1965,* 5245
305. Turner, H. S.: J. Chem. Soc. *1949,* 2282
306. Zhubova, N. A., Chekalin, M. A.: Zh. Org. Khim. *3,* 908 (1967)
307. Dahn, H., v. Castelmur, H.: Helv. Chim. Acta *36,* 638 (1953)
308. Ciotto, A.: Thesis, Univers. Freiburg, Schweiz 1965.
309. Piper, J. D., Brode, W. R.: J. Amer. Chem. Soc. *57,* 135 (1935)
310. Brode, W. R., Piper, J. D.: J. Amer. Chem. Soc. *63,* 1502 (1941)
311. Morris, R. J., Jensen, F. R., Lusenbrink, T. R.: J. Org. Chem. *19,* 1306 (1954)
312. Funkhouser, J., Brode, W. R.: J. Amer. Chem. Soc. *56,* 2172 (1934)
313. Kiprianov, A. I., Buryak, V. Yu.: Ukr. Khim. Zh. *34,* 1016 (1968)
314. Kiprianov, A. I., Buryak, V. Yu.: Ukr. Khim. Zh. *35,* 179 (1969)
315. Morris, R. J., Brode, W. R.: J. Amer. Chem. Soc. *70,* 2487 (1948)
316. Randall, J., Lewis, C. E., Slagan, P. M.: J. Org. Chem. *27,* 4098 (1962)
317. Wippel, H. G.: Melliands Textilber. *50,* 1090 (1969)
318. Hünig, S., Kießlich, G., Linhart, P., Schlaf, H.: Liebigs Ann. Chem. *752,* 196 (1971)
319. Bailey, A. S., Merer, J. J.: J. Chem. Soc. *1966* C, 1345
320. Hünig, S., Steinmetzer, H.-Chr.: Liebigs Ann. Chem. *1976,* 1060
321. Mikhailenko, P. A., Shevchuk, L. I., Kiprianov, A. I.: Khim. Geterots. Soed. *1973,* 923
322. Mikhailenko, F. A., Shevchuk, L. I.: Khim. Geterots. Soed. *1974,* 1325
323. Polevshikov, P. F., Iz'mailskii, V. A.: Uch. Zap. Kirov. Gos. Pedagog. Inst. *1972,* No. 36, 118
324. Pentimalli, L.: Tetrahedron *9,* 194 (1960)
325. Fabian, J., John, H.: unpublished 1974
326. Quast, H., Schmitt, E.: Liebigs Ann. Chem. *732,* 43 (1970)
327. Voltz, J.: Chimia *15,* 168 (1961)
328. Hünig, S.: Angew. Chem. *70,* 215 (1958)
329. Alberti, G., Cerniani, A., Seu, G.: Chim. Ind. *55,* 801; *56,* 495, 600, 684 (1974)
330. Hünig, S., Nöther, H.: Liebigs Ann. Chem. *628,* 84 (1959)
331. Mayer, U.: Textilveredlg. *7,* 492 (1972)
332. Baumann, H., Dehnert, J.: Chimia *15,* 163 (1961)
333. Hünig, S., Kießlich, G., Oette, K.-M., Quast, H.: Liebigs Ann. Chem. *754,* 46 (1971)
334. Jacobi, J.: Thesis, Techn. Univers. Dresden 1974
335. Hünig, S., Hermann, H.: Liebigs Ann. Chem. *636,* 32 (1960)
336. Bershtein, I. Ya., Ginsburg, O. F.: Usp. Khim. *41,* 177 (1972)
337. Cox, R. A., Buncel, E.: In: Chemistry of the hydrazo, azo and azoxy groups (ed. S. Patai), Part. 2, Chap. 18, p. 775. London: Wiley 1975
337a. Nurmukhametov, R. N., Betin, O. I., Khachaturova, G. T., Shigorin, D. N.: Zh. Prikl. Spektrosk. *27,* 464 (1977)
338. Reeves, R. L., Kaiser, R. S.: J. Org. Chem. *35,* 3670 (1970)
339. Juvvik, P., Sundby, B.: Acta Chem. Scand. *27,* 3625 (1973)
340. Kishimoto, S., Hirashima, T., Manaba, O., Hiyama, H.: Kogyo Kagaku Zaisshi *70,* 1379 (1967)
341. Kuder, J. E.: Tetrahedron *28,* 1973 (1972)
342. Jacques, P., Goursot, A., Faure, J.: Chem. Phys. *26,* 301 (1977)
343. Goursot, A., Jacques, P., Faure, J.: Chem. Phys. *20,* 319 (1977)
344. Schreiber, J., Socha, J., Rothschein, K.: Collect. Czechoslov. Chem. Commun. *35,* 857 (1970)
345. Geidysh, L. S., Nikoforov, G. A., Ershov, V. V.: Izv. Akad. Nauk SSSR, Ser. Khim. *1970,* 1045
346. Manda, E.: Bull. Chem. Soc. Japan *44,* 1620 (1971)
346a. Kishimoto, S., Kitahara, S., Manabe, O., Hiyama, H.: J. Org. Chem. *43,* 3882 (1978)
347. Pilipenko, A. T., Savranskii, L. I.: Zh. Prikl. Spektrosk. *13,* 918 (1970)

348. Dedkov, Yu. M., Kotov, A. V.: Izv. Akad. Nauk SSSR, Ser. Khim. *1973,* 1334
349. Zaitsev, B. E., Sheban, G. V., Smirnov, L. D., Dyumaev, K. M.: Khim. Geterots. Soed. *1974,* 1364
350. Zaitsev, B. E., Sheban, G. V., Dyumaev, K. M.: Zh. Prikl. Spektrosk. *20,* 546 (1974)
351. Parent, R. A.: J. Soc. Dyers. Colour. *92,* 368 (1976)
352. Yasuda, M.; Bull. Chem. Soc. Japan *40,* 1239 (1967)
353. Andrisano, R.: Ann. Chim. Rome *41,* 545 (1951)
354. Arriau, J., Campillo, J. P., Deschamps, J., Terrago, G., Jaquier, R.: Bull. Soc. Chim. France *1973,* 1398, 1403
355. Arriau, J., Campillo, J. P., Deschamps, J., Parmentier, P.: Tetrahedron *30,* 1353 (1974)
356. Arriau, J., Campillo, J. P., Elguero, J., Pereillo, J. M.: ibid. *30,* 1345 (1974)
356a. Sheban, G. V., Zaitsev, B. E., Tambieva, O. A., Lisitsyna, E. S., Dyumaev, K. M.: Zh. Prikl. Spektrosk. *29,* 178 (1978)
357. Yagi, Y.; Bull. Chem. Soc. Japan *36,* 487, 492 (1963)
358. Dehari, Ch., Matsunaga, Y., Tani, K.: ibid. *43,* 3404 (1970)
359. Matsunaga, Y., Miyajima, N.: ibid. *44,* 361 (1971)
360. Matsunaga, Y., Shimane, M.: ibid. *45,* 295 (1972)
360a. Whitaker, A.: J. Soc. Dyers Colour. *94,* 431 (1978)
361. Monahan, A. R., Flannery, J. B.: Chem. Phys. Letters *17,* 510 (1972)
362. Lutskii, A. E., Bocharova, V. V., Kreslawskii, M. R.: Zh. Obshch. Khim. *45,* 2276 (1975)
363. Hünig, S., Requardt, K.: Liebigs Ann. Chem. *592,* 180 (1955)
364. Irick, G., Pacifici, J. G.: Text. Res. J. *42,* 391 (1972)
365. Hünig, S., Bernhard, G., Liptay, W., Brenninger, W.: Liebigs Ann. Chem. *690,* 9 (1965)
366. Monahan, A. R., Blossey, D. F.: J. Phys. Chem. *74,* 4014 (1970)
367. Monahan, A. R., Germano, N. J., Blossey, D. F.: J. Phys. Chem. *75,* 1227 (1971)
368. Yeh, S.-J., Jaffé, H. H.: J. Amer. Chem. Soc. *81,* 3279, 3283 (1959)
369. Gerson, F., Heilbronner, E., VanVeen, A., Wepster, B. M.: Helv. Chim. Acta *43,* 1889 (1960)
370. Haselbach, E.: Helv. Chim. Acta *53,* 1526 (1970)
371. Hoefnagel, M. A., Van Veen, A., Wepster, B. M.: Rec. Trav. Chim. *88,* 562 (1969)
372. Haselbach, E., Henriksson, A., Schmelzer, A., Berthou, H.: Helv. Chim. Acta *56,* 705 (1973)
373. Fabian, J., Sühnel, J.: J. prakt. Chem. *315,* 307 (1973)
374. Kogan, V. A., Kharabaev, N. N., Osipov, O. A., Yurchenko, E. N.: Zh. Obshch. Khim. *45,* 362 (1975)
375. Yagupol'ski, L. M., Gandel'sman, L. S.: ibid. *35,* 1252 (1965); *37,* 2101 (1967)
376. Korolev, B. A., Titova, S. P.: Zh. Org. Khim. *7,* 1188 (1971)
377. Bolton, P. D., Ellis, J., Fleming, K. A., Lantzke, I. R.: Austr. J. Chem. *26,* 1005 (1973)
378. Bershtein, I. Ya., Ginsburg, O. F.: Zh. Org. Khim. *3,* 2032 (1967)
379. Liler, M.: In: Adv. phys. organic chem. (ed. V. Gold, D. Bethell), Vol. 11, p. 267. London: Acad. Press 1975
380. Chu, K. Y., Griffiths, J.: Tetrahedron Lett. *1976,* 405
380a. Yagupol'skii, L. M., Gandel'sman, L. Z.: Ukr. Khim. Zh. *45,* 145 (1979)
381. Sühnel, J., Hartmann, H., Fabian, J.: Z. Chem. *18,* 183 (1978)
382. Bigelow, R. W.: J. Phys. Chem. *80,* 2694 (1976)
383. Bigelow, R. W.: J. Phys. Chem. *79,* 2411 (1975)
384. Gerson, F., Schulze, J., Heilbronner, E.: Helv. Chim. Acta *41,* 1463 (1958)
385. Kazitshyna, N. B., Kupletskaya, N. B., Ptitenya, V. A., Bochkereva, M. N., Reutev, O. A.: Zh. Org. Khim. *2,* 565, 571 (1966)
386. Baumann, H., Hensel, H. R.: Topics Curr. Chem. *7,* 643 (1966/1967)
387. Král, M.: Chem. Listy *69,* 435 (1975)
388. Trombetti, A.: Canad. J. Phys. *46,* 1005 (1968)
389. Ahrens, W., Berndt, A.: Angew. Chem. *85,* 657 (1973), Int. Edit. *12,* 655 (1973)
390. Marsh, F. D., Hermes, M. E.: J. Amer. Chem. Soc. *87,* 1819 (1965)
391. Huang, P. C., Kosower, E. M.: J. Amer. Chem. Soc. *89,* 3910 (1967)
392. Schantl, J., Karpellus, P.: Monatsh. Chem. *109,* 1081 (1978)

393. Huang, S. J., Paneccasio, V., DiBattista, F., Picker, D., Wilson, G.: J. Org. Chem. *40,* 124 (1975)
394. Gerson, F., Heilbronner, E.: Helv. Chim. Acta *42,* 1877 (1959)
395. Skulski, S., Waclawek, W., Müller, B.: Bull. Acad. Pol. Sci. *19,* 329 (1971)
396. Skulski, L., Urbanski, T.: Rocz. Chem. *34,* 141 (1960)
397. Birnbaum, P. P., Linford, J. H., Style, D. W. G.: Trans. Faraday Soc. *49,* 735 (1953)
398. Gore, P. H., Wheeler, O. H.: J. Org. Chem. *26,* 3295 (1961)
399. Kleps, J., Prot, T.: Rocz. Chem. *48,* 1525 (1974)
400. Sawicki, E.: J. Org. Chem. *22,* 915 (1957)
401. Izmail'skii, V. A., Koshevnika, A., Fedorov, E. I.: Dokl. Akad. Nauk SSSR *183,* 341 (1968)
402. Bloor, J. E., Copley, D. B.: Chem. Ind. *1960,* 526
403. Grammaticakis, P.: Compt. Rend. *245* B, 1145 (1957)
404. Bridgeman, I., Peters, A. T.: J. Soc. Dyers Colour. *1970,* 519
405. Griffith, J., Roozpeikar, B.: J. Chem. Soc., Perkin Trans. 1, *1976,* 42
406. Ueno, K., Akiyoshi, S.: J. Amer. Chem. Soc. *76,* 3667 (1954)
407. Chekalin, M. A., Zhukova, H. A.: Zh. Prikl. Spektrosk. *10,* 1008 (1969)
408. Kniese, W.: Thesis, Univers. Marburg 1963
409. Ospenson, J. N.: Acta Chem. Scand. *5,* 491 (1951)
410. Perekalin, V. V., Sokolina, Z. S.: Zh. Obshch. Khim. *24,* 683 (1954)
411. Fierz-David, H. E., Blangey, L., Merian, E.: Helv. Chim. Acta *34,* 846 (1951)
412. Perėkalin, V. V., Savost'yanova, M. V.: Zh. Obshch. Khim. *21,* 1329 (1951)
413. Hofer, E., Uffmann, H.: Tetrahedron Lett. *1971,* 3241
414. Badger, G. M., Buttery, R. G., Lewis, G. E.: J. Chem. Soc. *1954,* 1888
415. Pentimalli, L.: Gazz. Chim. Ital. *90,* 1203 (1960)
416. Heckemann, H.-J.: Thesis, Techn. Univers. Dresden 1973
417. Grayshan, P. H., Peters, A. T.: J. Heterocycl. Chem. *10,* 699 (1973)
418. Grayshan, P. H., Kadhim, A. M., Peters, A. T.: J. Heterocycl. Chem. *11,* 33 (1974)
419. Zahradník, R., Nepraš, M., Arient, J., Koutecký, J.: Collect. Czechoslov. Chem. Commun. *31,* 1180 (1966)
420. Nepraš, M., Arient, J.: ibid. *35,* 2211 (1970)
421. Arient, J., Marhan, J., Havličková, L.: J. Soc. Dyers Colour. *85,* 246 (1969)
422. Grayshan, P. H., Mak, K. H. R., Peters, A. T.: J. Appl. Chem. Biotechnol. *24,* 121 (1974)
423. Kadhim, A. M., Peters, A. T.: J. Soc. Dyers Colour. *90,* 189 (1974)
424. Asquith, R. S., Bridgeman, I., Peters, A. T.: ibid. *81,* 439 (1965)
425. Godfrey, M., Murrell, J. N.: Proc. Roy. Soc. A *278,* 71 (1964)
426. Millefiori, S., Favini, G., Millefiori, A., Grasso, D.: Spectrochim. Acta *33A,* 21 (1977)
427. Milliaresi, E. E., Iz'mailskii, B. A.: Zh. Obshch. Khim. *35,* 766 (1965), and references therein.
428. Bell, M. G. W., Day, M., Peters, A. T.: J. Soc. Dyers Colour *82,* 410 (1966)
429. Day, M., Peters, A. T.: ibid. *85,* 8 (1969)
430. Grison, E.: Acta Cryst. *2,* 410 (1949)
431. Peters, A. T.: J. Appl. Chem. Biotechnol. *23,* 751 (1973), *26,* 131 (1976), and references therein
432. Merian, E.: Angew. Chem. *72,* 766 (1960)
433. Asquith, R. S., Blair, H. S., Hamill, A. C.: J. Soc. Dyers Colour. *90,* 185 (1974)
434. Asquith, R. S., Campbell, B.: ibid. *79,* 678 (1963)
435. Labana, S. S., Labana, L. L.: Chem. Revs. *67,* 1 (1967)
436. Zaharia, C. N., Tarabasanu-Mihaila, C.: Rev. Roum. Phys. *20,* 769 (1975)
437. Engelhardt, K.-P.: Thesis, Univers. Göttingen 1976
438. Zaharia, C. N., Tarabasanu-Mihaila, C.: Rev. Roum. Phys. *21,* 359 (1976)
439. Schroeder, W. A., Wilcox, P. E., Trueblood, K. N., Dekker, A. O.: Anal. Chem. *23,* 1740 (1951)
440. Scott, A. I.: Interpretation of the ultraviolet spectra of natural products. Oxford: Pergamon Press 1964
441. Thomson, R. H.: Naturally occurring quinones. New York. Acad. Press 1971

442. Fabian, J., Nepraš, M.: to be published
443. Peover, M. E.: J. Chem. Soc. *1962,* 4540
444. Swenson, J. R., Hoffmann, R.: Helv. Chim. Acta *53,* 2331 (1970)
445. Cowan, D. O., Gleiter, R., Hashmall, J. A., Heilbronner, E., Hornung, V.: Angew. Chem. *83,* 405 (1971). Int. Edit. *10,* 401 (1971)
446. Hoffmann, R.: Acc. Chem. Res. *4,* 1 (1971)
447. Lauer, G., Schäfer, W., Schweig, A.: Chem. Phys. Letters *33,* 312 (1975)
448. Dougherty, D., McGlynn, S. P.: J. Amer. Chem. Soc. *99,* 3234 (1977)
449. Hug, W., Kuhn, J., Seibold, K. J., Labhart, H., Wagniere, G.: Helv. Chim. Acta *54,* 1451 (1971)
450. Trommsdorff, H. P.: J. Chem. Phys. *56,* 5358 (1972)
451. Merienne-Lafore, M. F., Trommsdorff, H. P.: J. Chem. Phys. *64,* 3791 (1976)
452. Haselbach, E., Schmelzer, A.: Helv. Chim. Acta *55,* 1745 (1972)
453. Zander, M.: Ber. Bunsenges. Phys. Chem. *71,* 424 (1967)
454. Olbrich, G., Polansky, O. E., Zander, M.: Ber. Bunsenges. Phys. Chem. *81,* 692 (1977)
455. Nepraš, M., Novák, A.: Collect. Czechoslov. Chem. Commun. *42,* 2343 (1977)
456. Shcheglova, N. A., Shigorin, D. N., Dokunikhin, N. S.: Zh. Fiz. Khim. *42,* 2724 (1968)
457. Schorer, M., Dörr, F.: Ber. Bunsenges. Phys. Chem. *68,* 478 (1964)
458. Dörr, F.: In: Opt. Anreg. organ. Systeme (ed. W. Foerst), p. 732. Weinheim: Verlag Chemie 1966
459. Kuboyama, A., Matsuzaki, S., Takagi, H., Arano, H.: Bull. Chem. Soc. Japan *47,* 1604 (1974)
460. Hoffmann, H., Rasch, G.: Z. Physikal. Chem. [Leipzig] *257,* 689 (1976)
461. Novák, A., Titz, M., Nepraš, M.: Collect. Czechoslov. Chem. Commun. *39,* 1532 (1974)
462. Leibovici, C.: An. Fis. *46,* 105 (1970)
463. Stevenson, P. E.: J. Phys. Chem. *76,* 2424 (1972)
464. Remko, M., Polčin, J.: Monatsh. Chem. *108,* 1313 (1977)
465. Galasso, V.: Gazz. Chim. Ital. *106,* 571 (1976)
466. Bunce, N. J., Ridley, J. E., Zerner, M. C.: Theoret. Chim. Acta *45,* 283 (1977)
467. Fabian, J., Mehlhorn, A.: Z. Chem. *18,* 338 (1978)
468. Wood, M. H.: Theoret. Chim. Acta *36,* 345 (1975)
469. Becker, L.; Thesis, Techn. Univers. München 1971
470. Flaig, W., Salfeld, J.-Ch., Baume, E.: Liebigs Ann. Chem. *618,* 117 (1958)
471. Stevenson, P. E.: J. Chem. Phys. *43,* 27 (1965); J. Mol. Spectrosc. *23,* 191 (1967)
472. Stevenson, P. E.: J. Mol. Spectrosc. *17,* 58 (1965)
473. Balint, A., Simon, Z.: Rev. Roum. Chim. *13,* 1427, 1635 (1968)
474. Flaig, W., Ploetz, Th., Küllmer, A.: Z. Naturforsch. *10B,* 668 (1955)
475. Teubner, H.-J., Staiger, G.: Chem. Ber. *88,* 802 (1955)
476. Volod'ko, L. V., Komyak, A. I., Min'ko, A. A., Tatarinov, B. A., Matusevich, P. A.: Zh. Prikl. Spektrosk. *24,* 1009 (1976)
477. Dähne, S., Seebacher, M.: Z. Chem. *12,* 141 (1972)
478. Leupold, D., Dähne, S.: Theoret. Chim. Acta *3,* 1 (1965)
479. Wallenfels, K., Draber, W.: Tetrahedron *20,* 1889 (1964)
480. Buckley, D., Henbest, H. B., Slade, P.: J. Chem. Soc. *1957,* 4891
481. Corbett, J. F.: J. Chem. Soc. B *1969,* 818
482. Corbett, J. F., Gamson, E. P.: J. Chem. Soc., Perkin 2, *1972,* 1531
483. Nogami, T., et al.: Bull. Chem. Soc. Japan *48,* 3709 (1975)
483a. Issa, I. M., El Samahy, A. A., Issa, R. M., El Kashef, H. S.: Rev. Roum. Chim. *23,* 617 (1978)
484. Smith, W. F.: Tetrahedron *20,* 671 (1964)
485. Lurie, A. P., Brown, G. H., Thirtle, J. R., Weissberger, A.: J. Amer. Chem. Soc. *83,* 5015 (1961)
486. Hünig, S., Richters, P.: Liebigs Ann. Chem. *612,* 282 (1958)
487. Musso, H., Matthies, H.-G.: Chem. Ber. *90,* 1814 (1957)
488. Butenandt, A.: Angew. Chem. *69,* 16 (1957)
489. Schäfer, W., Geyer, I.: Tetrahedron *28,* 5261 (1972)

490. Tai, J. C., Craig, W. S., Gaber, B. P.: J. Amer. Chem. Soc. *98*, 7925 (1976)
491. Zerweck, W., Ritter, H., Schubert, M.: Angew. Chem. *60*, 141 (1948)
492. Solodar, W. E., Monahan, A. R.: Canad. J. Chem. *54*, 2909 (1976)
493. Chen, I., Lardon, M., Weinberger, L.: J. Chem. Phys. *47*, 2873 (1967)
494. Tinland, B.: Theoret. Chim. Acta *12*, 85 (1968)
495. Pugin, A.: Chimia, Suppl. *1968*, 54
496. Urbanski, T., Szyc-Lewanska, K.: Bull. Acad. Pol. Sci. *4*, 775 (1956); *5*, 203 (1957)
497. Adam, F. G., Simpson, W. T.: J. Mol. Spectrosc. *3*, 363 (1959)
498. Wähnert, M., Dähne, S.: J. prakt. Chem. *318*, 321 (1976)
499. Bohn, C. R., Campbell, T. W.: J. Org. Chem. *22*, 458 (1957)
500. Volod'kin, A. A., Ershov, V. V., Kudinova, L. I.: Izv. Akad. Nauk SSSR, Ser. Khim. *1978*, 512
501. West, R., Zecher, D. C.: J. Amer. Chem. Soc. *92*, 155 (1970)
502. Koster, S. K., West, R.: J. Org. Chem. *40*, 2300 (1975)
503. Kuboyama, A.: Bull. Chem. Soc. Japan *33*, 917 (1960)
504. Wild, U.: Chimia *22*, 473 (1968)
505. Kortüm, G.: Ber. Bunsenges. Phys. Chem. *78*, 391 (1974)
506. Hartmann, H., Lorenz, E.: Z. Naturforsch. *7a*, 360 (1952)
507. Morton, R. A., Earlam, W. T.: J. Chem. Soc. B *1941*, 159.
508. Singh, I., et al.: Tetrahedron *24*, 6053 (1968)
509. Casádo, J., Peleteiro, J., Rios, M. A.: In: Application of MO theory in organic chemistry (ed. I. G. Csizmadia), p. 392. Amsterdam: Elsevier 1977
509a. Chu, K.-Y., Griffiths, J.: J. Chem. Research *1978*, 180
509b. Chu, K.-Y., Griffiths, J.: J. Chem. Soc., Perkin Trans. 1, *1978*, 1083
509c. Chu, K.-Y., Griffiths, J.: J. Chem. Soc., Perkin Trans. 1, *1979*, 696
510. Hill, R. R., Mitchell, G. H.: J. Chem. Soc. B *1969*, 61
511. Drott, H. R., Dearman, H. H.: J. Chem. Phys. *47*, 1896 (1967)
512. Scott, J. D., Watson, W. H.: J. Chem. Phys. *49*, 4246 (1968)
513. Platonova, N. V., Popov, K. R., Smirnov, L. V.: Opt. Spectrosc. *26*, 357 (1969)
514. El Ezaby, M. S., Salem, T. M., Zewait, A. H., Issa, R.: J. Chem. Soc. *1970* B, 1293
515. Giovanelli, K. H., Dehler, J., Hohlneicher, G.: Ber. Bunsenges. Phys. Chem. *75*, 864 (1971)
516. Edwards, T. G.: Theoret. Chim. Acta *30*, 267 (1973)
517. Peters, R. H., Summer, H. H.: J. Chem. Soc. *1953*, 2101
518. Yoshida, Z., Takabayashi, F.: Tetrahedron *24*, 933 (1968)
519. Fain, V. Ya.: Tables of electronic spectra of anthraquinone and its derivatives. Leningrad: Khimiya 1970.
520. Inoue, H., Hoshi, T., Yoshino, J., Tanizaki, Y.: Bull. Chem. Soc. Japan *45*, 1018 (1972)
521. Inoue, H., Hoshi, T., Yoshino, J., Tanazaki, Y.: ibid. *46*, 380 (1973)
522. Pilipenko, A. T., Savranskii, L. I.: Opt. Spektr. *28*, 808 (1970)
523. Zaitsev, B. E., Allenov, V. M.: Zh. Prikl. Spektrosk. *20*, 109 (1974)
524. Davies, A. K., McKellar, J. F., Phillips, G. O.: Proc. Roy. Soc. A *323*, 69 (1971)
525. Popov, K. R., Platonova, N. V., Smirnov, L. V.: Opt. Spektr. *32*, 1097 (1973)
526. Egerton, G. S., Roach, A. G.: J. Soc. Dyers Colour. *74*, 401 (1958)
527. Sinclair, R. S., McAlpine, E.: ibid. *91*, 399 (1975)
528. Labhart, H.: Chimia *15*, 20 (1961)
529. Labhart, H.: Helv. Chim. Acta *40*, 1410 (1957)
530. Moran, J. J., Stonehill, N. I.: J. Chem. Soc. *1957*, 765
531. Metwally, S. A. M., Bredereck, K.: Bull. Chem. Soc. Japan *48*, 2939 (1975)
532. Nitzl, K., Dörr, F.: Melliands Textilber. *45*, 893 (1964)
533. Weinstein, J., Merritt, Ch.: J. Amer. Chem. Soc. *81*, 3759 (1959)
534. Bailay, M.: Acta Cryst. *8*, 182 (1955)
534a. Matsuoka, M., Kishimoto, M., Kitao, T.: J. Soc. Dyers Colour. *94*, 435 (1978)
535. Titz, M., Nepraš, M.: Collect. Czechoslov. Chem. Commun. *37*, 2674 (1972)

536. Kuboyama, A., Kobayashi, F., Morokuma, S.: Bull. Chem. Soc. Japan *48*, 2145 (1975)
537. Kuboyama, A., Arano, H.: ibid. *49*, 1401 (1976)
537a. Raasch, M. S.: J. Org. Chem. *44*, 632 (1979)
537b. Ayyangar, R. N., Purao, S. R., Tilak, B. D.: Indian J. Chem. Sect. 3, *16B*, 673 (1978)
538. Nurmukhametov, R. N., Bondareva, L. V.: Zh. Fiz. Khim. *37*, 1143 (1963)
539. Efimov, A. A., Nurmukhametov, R. N.: Opt. Spektr. *28*, 58 (1970)
540. Madhavan Nair, P., Manjrekar, T. G., Unni, M. K.: Chimia *20*, 300 (1966)
541. Durie, R. A., Shannon, J. S.: Austral. J. Chem. *11*, 168 (1958)
542. Figueras, J., Scullard, P. W., Mack, A. R.: J. Org. Chem. *36*, 3497 (1971)
543. Issa, I. M., El-Shafei, A. K., Etaiw, S. H., El-Kashef, H. S.: J. prakt. Chem. *320*, 557 (1978)
544. Davis, M. M., Hetzer, H. B.: Analyt. Chem. *38*, 451 (1966)
545. Pilipenko, A. T., Savranskii, L. I., Skorokhov, E. G.: Zh. Prikl. Spektrosk. *13*, 751 (1970)
546. Handa, T.: Bull. Chem. Soc. Japan *28*, 483 (1955)
547. Nepraš, M., et al.: Collect. Czechoslov. Chem. Commun. *38*, 1003 (1973)
548. Nepraš, M., et al.: ibid. *38*, 2397 (1973)
549. Schwarzenbach, G.: Z. Elektrochem. *47*, 40 (1941)
550. Kratochvil, V., Nepraš, M.: Collect. Czechoslov. Chem. Commun. *37*, 1533 (1972)
551. Peters, R. H., Summer, H. H.: J. Soc. Dyers Colour. *72*, 77 (1956)
552. Cooper, C. J., Stonehill, H. T.: J. Chem. Soc. *1958*, 3353
553. Braude, E. A.: J. Chem. Soc. *1945*, 490
554. Hammond, P. R.: J. Chem. Soc. *1964*, 471
555. Eugster, C. H., Bosshard, P.: Helv. Chim. Acta *46*, 815 (1963)
556. Ukai, S., Hirose, K.: Chem. Pharm. Bull. *16*, 195 (1968)
557. Corbett, J. F.: J. Chem. Soc. B *1970*, 1502
558. Vittum, P. W., Brown, G. H.: J. Amer. Chem. Soc. *68*, 2235 (1946)
559. Vittum, P. W., Brown, G. H.: ibid. *69*, 152 (1946)
560. Houston, D. F., Kester, E. B., DeEds, F.: ibid. *71*, 3816, 3819 (1949)
561. Stužka, V., Golovina, A. P., Alimarin, I. P.: Collect. Czechoslov. Chem. Commun. *34*, 221 (1969)
562. Wanzlick, H.-W., Pankow, B., Daneke, J.: Liebigs Ann. Chem. *727*, 106 (1969)
563. Hughes, G. M. K., Saunders, B. C.: J. Chem. Soc. *1956*, 3814
564. Gorelik, M. V., Gvon, Kh. I., Vinogradova, N. P., Korolev, B. A.: Zh. Org. Khim. *12*, 2593 (1976)
565. Macdonald, A. L., Trotter, J.: J. Chem. Soc., Perkin 2, *1973*, 476
566. Hagen, K., Hedberg, K.: J. Chem. Phys. *59*, 158 (1973)
567. Gaultier, J., Hauw, Ch.: Acta Cryst. *18*, 179 (1965)
568. Londsdale, K., Milledge, H. J., Sayed, K. E.: ibid. *20*, 1 (1966)
569. Ross, D. L., Blanc, J.: In: Photochromism (ed. G. H. Brown), Chap. 5. New York: J. Wiley 1971
569a. Paramonov, V. D., Mostoslavskii, M. A., Shevchuk, I. I.: Zh. Fiz. Khim. *52*, 2676 (1978)
570. Wille, E., Lüttke, W.: Angew. Chem. *83*, 853 (1971), Int. Edit. *10*, 803 (1971)
571. Sieghold, H., Lüttke, W.: Angew. Chem. *87*, 63 (1975), Int. Edit. *14*, 52 (1975)
572. Van Alphen, J.: Rec. Trav. Chim. *59*, 289 (1940)
573. Junek, H., Fischer-Colbrie, H., Sterk, H.: Chem. Ber. *110*, 2276 (1977)
574. Wille, E., Lüttke, W.: Chem. Ber. *106*, 3240 (1973)
575. Lüttke, W., Klessinger, M.: Chem. Ber. *97*, 2342 (1964)
576. Klessinger, M.: Tetrahedron *22*, 3355 (1966)
577. Luhmann, U.: Thesis, Univers. Göttingen 1974; Luhmann, U., Wentz, F. G., Knieriem, B., Lüttke, W.: Chem. Ber. *111*, 3233 (1978)
578. Sühnel, J., Gustav, K., Paetzold, R., Fabian, J.: Z. Physikal. Chem. [Leipzig] *259*, 17 (1978); Sühnel, J., Gustav, K.: J. prakt. Chem. *320*, 917 (1978)
578a. Tidow, K.: Thesis, University of Göttingen 1977
579. Känzig, H.: Thesis, ETH Zürich 1973
579a. Pfeiffer, G., Bauer, H.: Chem. Ztg. *102*, 323 (1978)

580. Pummerer, R., Meininger, F., Schrott, G., Wagner, H.: Liebigs Ann. Chem. *590,* 195 (1954)
581. Sieghold, H.: Thesis, Univers. Göttingen 1973
582. Wizinger, R.: Chimia *19,* 339 (1965)
583. Sadler, P. W.: Spectrochim. Acta *16,* 1094 (1960)
584. Holt, S. J., Sadler, P. W.: Proc. Roy. Soc. B *148,* 495 (1958)
585. Tarabasanu-Mihaila, C., Deascu, C.: Stud. Cercet. Chim. *21,* 417 (1973)
586. Sadler, P. W., Warren, R. L.: J. Amer. Chem. Soc. *78,* 1251 (1956)
586a. Süsse, P., Krampe, C.: Naturwiss. *66,* 110 (1979)
586b. Banerji, K. D., Mazumdar, A. K. D.: Proc. Nat. Acad. Sci. India, Sect *A47,* 67 (1978)
587. Lüttke, W., Hunsdiecker, D.: Chem. Ber. *99,* 2146 (1966)
588. Meier, H.: Thesis, Univers. Göttingen 1971
589. Kaul, B. L.: Helv. Chim. Acta *57,* 2664 (1974)
590. Nederlof, A.: Rec. Trav. Chim. *82,* 75 (1963)
591. Götze, J., Kampfer, H.: Chem. Ber. *99,* 1851 (1966)
592. Stewart, F. H. C.: J. Chem. Soc. *1963,* 701
593. Paterson, W., Proctor, G. R.: J. Chem. Soc. *1962,* 3468, cf. Cooke, R. G., Russell, I. M.: Tetrahedron Lett. *1968,* 4587
594. Thomson, R. H.: Angew. Chem. *86,* 355 (1974), Int. Edit. *13,* 305 (1974)
595. Chioccara, F., Prota, G.: Tetrahedron Lett. *1975,* 811
596. Chioccara, F., Prota, G., Thomson, R. H.: Tetrahedron *32,* 1407 (1976)
597. Dimroth, P., Reicheneder, F.: Angew. Chem. *81,* 742 (1969), Int. Edit. *8,* 751 (1969)
598. Lamchen, M., Mittag, T.: Proc. Chem. Soc. *1963,* 302
599. Schelz, D.: Helv. Chim. Acta *60,* 2082 (1977)
600. Punkt, J.: Thesis, Univers. Leipzig 1978
601. Weissenfels, M., Punkt, J.: Tetrahedron *34,* 311 (1978)
601a. Wentz, F. G.: Thesis, Univers. Göttingen 1976
602. Jaffe, E. E., Matrick, H.: J. Org. Chem. *33,* 4004 (1968)
603. Gündel, W., Pummerer, R.: Liebigs Ann. Chem. *529,* 11 (1938)
604. Felder, E., Maffei, S., Pietra, S., Pitre, D.: Helv. Chim. Acta *43,* 888 (1960)
605. Seitz, G., Mann, K., Matusch, R.: Arch. Pharm. *308,* 792 (1975)
606. Eggerding, D., West, R.: J. Org. Chem. *41,* 3904 (1976)
607. von Eller-Pandraud, H.: Acta Cryst. *13,* 936 (1960), *14,* 901 (1961)
608. Wille, E.: Thesis, Univers. Göttingen 1974
609. Hartmann, H., Scheithauer, St.: J. prakt. Chem. *311,* 827 (1969)
610. Hartmann, H.: unpublished
611. Elvidge, J. A., Golden, J. H.: J. Chem. Soc. *1956,* 4144
612. Kornilov, M. Yu., Makovetskii, V. P., Shtilerman, V. N.: Ukr. Khim. Zh. *40,* 629 (1974)
613. Kornilov, M. Yu., Makovetskii, V. P.: ibid. *40,* 879 (1974)
614. Ballantine, J. A., Fenwick, R. G.: J. Chem. Soc. C *1970,* 2264
615. Schlenk, W., Blum, O.: Liebigs Ann. Chem. *433,* 95 (1923)
616. Measured by the authors
617. Fang, C. S., Bergmann, W.: J. Org. Chem. *16,* 1231 (1951)
618. Klingsberg, E.: Chem. Revs. *54,* 59 (1954)
619. Paetzold, R., Hauke, G., Erler, M.: Univers. Jena, unpublished results
620. Raffa, L.: Gazz. Chim. Ital. *72,* 549 (1942)
621. Goldschmidt, St., Wessbecher, H.: Chem. Ber. *61,* 372 (1928)
622. Bauer, H., Pfeiffer, G.: Chem.-Ztg. *100,* 373 (1976)
623. Haucke, G.: Thesis B, Univers. Jena 1977
624. Barba, F., Soler, A., Varea, J.: Tetrahedron Lett. *1976,* 557
625. Mostoslavskii, M. A., Yugai, G. A.: Ukr. Khim. Zh. *42,* 1219 (1976)
626. Yugai, G. A., Mostoslavskii, M. A., Paramonov, V. D.: Teor. Eksp. Khim. *12,* 700 (1976)
627. Scheibe, G., Dörfling, H., Assmann, J.: Liebigs Ann. Chem. *544,* 240 (1940)
628. Klessinger, M., Lütke, W.: Chem. Ber. *99,* 2136 (1966)
629. Hermann, H., Lüttke, W.: Chem. Ber. *101,* 1715 (1968)
630. Fitjer, L., Lüttke, W.: Chem. Ber. *105,* 919 (1972)

631. Monahan, A. R., Kuder, J. E.: J. Org. Chem. *37,* 4182 (1972)
632. Kunz, C.: Thesis, Univers. Göttingen 1971
633. Brode, W. R., Wyman, G. M.: J. Amer. Chem. Soc. *73,* 4267 (1951)
634. Kagan, G. I., Kosobyzkii, V. A., Bel'akov, V. K., Tarakanov, C. G.: Khim. Geterots. Soed. *1972,* 794
635. Fabian, J.: Tetrahedron *29,* 2449 (1973)
636. Cromwell, N. H., Watson, W. R.: J. Org. Chem. *14,* 411 (1949)
637. von Eller, H.: Bull. Soc. Chim. France *1955,* 1444
638. Pandraud, H.: Acta Cryst. *15,* 1131 (1962)
639. Süsse, P., Kunz, G., Lüttke, W.: Naturwissenschaften *60,* 49 (1973)
640. Hermann, J. A., Ammon, H. L., Gibson, R. E.: Tetrahedron Lett. *1969,* 2559
640a. Ammon, H. L., Hermann, H.: J. Org. Chem. *43,* 4581 (1978)
641. Luhmann, U., Wentz, F. G., Lüttke, W.: Chem. Ber. *110,* 1421 (1977)
642. Wyman, G. H., Zenhäusern, A. F.: J. Org. Chem. *30,* 2348 (1965)
643. Friedländer, P., Woroshzow, N.: Liebigs Ann. Chem. *388,* 1 (1922)
644. Tróyanov, I. A., Oksengendler, G. M., Kostomarova, E. F.: Ukr. Khim. Zh. *28,* 367 (1962)
645. Brooker, L. G. S.: In: Theorie of photographic process (ed. C. E. K. Mees), p. 371, London: McMillan 1954
646. Koutecký, J., Paldus, J.: Collect. Czechoslov. Chem. Commun. *28,* 1483 (1963); and papers cited therein .
647. Tseng, K. L., Michl, J.: J. Amer. Chem. Soc. *98,* 6138 (1976)
648. Gomes de Mesquita, A. H., Mac Gillavry, C. H., Eriks, K.: Acta Cryst. *18,* 437 (1965)
649. Hoffmann, R., Bissell, R., Farnum, D. G.: J. Phys. Chem. *73,* 1789 (1969)
650. Grinter, R., Mason, S. F.: Proc. Chem. Soc. *1961,* 386, Trans. Faraday Soc. *60,* 264 (1964)
651. Zahradník, R., Kröhn, A., Pancíř, J., Šnobl, J.: Collect. Czechoslov. Chem. Commun. *34,* 2553 (1965)
652. Němcová, I., Malát, M., Zahradník, R.: ibid. *34,* 2880 (1969)
653. Horák, V., Párkányi, C., Pecka, J., Zahradník, R.: ibid. *32,* 2272 (1967)
654. Hart, H., Sulzberg, T., Schwendeman, R. H., Young, R. H.: Tetrahedron Lett. *1967,* 1337
655. Häfelinger, G., Streitwieser, A.: Chem. Ber. *101,* 657 (1968)
656. Perkampus, H.-H.: In: Adv. phys. organ. chem. (ed. V. Gold), Vol. 4, p. 195. New York: Academic Press 1966
657. Olah, G. A., Pittman, Ch. U., Waack, R., Doran, M.: J. Amer. Chem. Soc. *88,* 1488 (1966)
658. Sondermann, J., Kuhn, H.: Chem. Ber. *99,* 2491 (1966)
659. Häfelinger, G., Streitwieser, A.: Chem. Ber. *101,* 2785 (1968)
660. Häfelinger, G., Streitwieser, A., Wright, J.: Ber. Bunsenges. Phys. Chem. *73,* 456 (1973)
661. Zahradník, R., Pancíř, J.: private communication
662. Wasilewski, J.: Acta Phys. Polon. *A 45,* 143 (1974)
662a. Grif, V. Kh., Nikitchenko, V. M., Lavrushin, V. F.: Zh. Org. Khim. *14,* 1293 (1978)
663. Lewis, G. M., Bigeleisen, J.: J. Amer. Chem. Soc. *65,* 2102, 2107 (1943)
664. Albrecht, A. C., Simpson, W. T.: ibid. *77,* 4454 (1955)
665. Adam, F. C.: J. Molec. Spectrosc. *4,* 359 (1960)
666. Ng, E. K., Adam, F. C.: Canad. J. Chem. *42,* 810 (1964)
667. Davidsson, A., Norden, B.: Chem. Scripta *11,* 68 (1977)
668. Wizinger, R., Arndt, A.: cit. in Grinter, R., Heilbronner, E.: Helv. Chim. Acta *45,* 2496 (1962)
669. Gandhi, S. S., Hallas, G., Thomasson, J.: J. Soc. Dyers Colour. *93,* 451 (1977)
670. Dubinin, N. V., Blinov, L. M., Lutsenko, E. L., Rozenshtein, L. D.: Org. Poluprovodn. (ed. M. V. Kurik) p. 37, Akad. Nauk Ukr. SSR, Inst. Fiz. 1976
671. Holmes, E. O.: J. Phys. Chem. *73,* 273 (1969)
672. Barker, C. C.: In: Steric effects in conjugated systems (ed. G. W. Gray), Chap 4. London: Butterworth 1958
673. Barker, C. C., Hallas, G., Stamp, A.: J. Chem. Soc. *1960,* 3790
674. Barker, C. C., Hallas, G.: J. Chem. Soc. *1961,* 1529
675. Barker, C. C., Stamp, A.: J. Chem. Soc. *1961,* 3445

676. Aaron, C., Barker, C. C.: J. Chem. Soc. *1963,* 2655
677. Hallas, G.: J. Soc. Dyers Colour. *83,* 368 (1967)
678. Hallas, G., Hirst, G. A.: ibid. *85,* 310 (1969)
679. Barker, C. C., Hallas, G.: J. Chem. Soc. *B 1969,* 1068
680. Hallas, G., Waring, D. R.: J. Chem. Soc. *B 1970,* 979
681. Hallas, G., Grocock, D. E., Hepworth, J. D.: J. Soc. Dyers Colour. *86,* 200 (1970)
682. Castelino, R. W., Hallas, G.: J. Chem. Soc. *B 1971,* 1471.
683. Ferguson, A. S., Hallas, G.: J. Soc. Dyers Colour. *87,* 187 (1971)
684. Hallas, G.: J. Chem. Soc. *B 1971,* 319
685. Hallas, G., Paskins, K. N., Waring, D. R.: J. Chem. Soc., Perkin Trans. 2, *1972,* 2281
686. Castelino, R. W., Hallas, G., Taylor, D. C.: J. Soc. Dyers Colour. *88,* 25 (1972)
687. Ferguson, A. S., Hallas, G.: ibid. *89,* 22 (1973)
688. Hallas, G., Potts, R. M.: J. Chem. Soc., Perkin Trans. 2, *1974,* 59
689. Hallas, G., et al.: ibid. *1977,* 450
690. Dewar, M. J. S.: In: Steric effects in conjugated systems (ed. G. W. Gray), Chap. 5. London: Butterworth 1958
691. Schmidt, H., Wizinger, R.: Liebigs Ann. Chem. *623,* 204 (1959)
692. Tuemmler, W. B., Wildi, B. S.: J. Amer. Chem. Soc. *80,* 3772 (1958)
693. Lees, W. A., Burawoy, A.: Tetrahedron *19,* 419 (1963)
694. Gronowska, J., Heldt, J.: Rocz. Chem. *37,* 145 (1963)
695. Brieskorn, C. H., Kallmayer, H.-J.: Arch. Pharm. *304,* 799 (1971), *305,* 141 (1972); Liebigs Ann. Chem. *741,* 124 (1970)
696. Fabian, J., Hartmann, H., Fabian, K.: Tetrahedron *29,* 2609 (1973)
697. Lavrushin, V. F., Zukerman, S. V., Syrovatka, I. G.: Zh. Obshch. Khim. *31,* 1275 (1961)
698. Mikhailenko, F. A., Shevchuk, U. J.: Synthesis *1973,* 621
699. Hartmann, H.: J. prakt. Chem. [2], *36,* 50 (1967)
700. Hartmann, H.: Habilitationsschrift, Techn. Univers. Dresden 1970
701. Malhotra, S. S., Whiting, M. C.: J. Chem. Soc. *1960,* 3812
702. Novak, Th. J., et al.: J. Org. Chem. *41,* 870 (1976)
703. Jutz, Ch.: Tetrahedron Lett. *1959,* 1
704. Margerum, J. D., et al.: J. Phys. Chem. *66,* 2434 (1962)
705. Friedrich, H. J., Scheibe, G.: Z. Elektrochem. *65,* 767 (1961)
706. Friedrich, H. J., Hohlneicher, G.: Angew. Chem. *74,* 466 (1962)
707. Leubner, J., Dehler, J., Hohlneicher, G.: Ber. Bunsenges. Phys. Chem. *71,* 560 (1967)
708. Jaffé, H. H.: J. Chem. Phys. *22,* 1430 (1954)
709. Bugai, P. M.: Zh. Obshch. Khim. *26,* 1729 (1956)
710. Peters, A. T.: J. Soc. Dyers Colour. *86,* 77 (1970)
711. Dawidson, J. M., French, C. M.: J. Chem. Soc. *1958,* 114
712. Neunhoeffer, O., Heitmann, P.: Chem. Ber. *92,* 245 (1959)
713. Pestemer, M., Brück, D.: In: Methoden der organischen Chemie (Houben-Weyl, ed. E. Müller), Bd. 3/2, p. 648/649
714. Brooker, L. G. S., Sprague, R. H.: J. Amer. Chem. Soc. *63,* 3214 (1941)
715. Figueras, J.: J. Amer. Chem. Soc. *93,* 3255 (1971)
716. Havemann, R., Nutsch, E., Pietsch, H.: Z. Physikal. Chemie [Leipzig] *219,* 171 (1961)
717. Kern, J., Dörr, F.: Z. Naturforsch. *16a,* 363 (1961)
717a. Platonova, N. V., Popov, K. R.: Zh. Prikl. Spektroskop. *30,* 464 (1979)
718. Barker, A., Barker, C. C.: J. Chem. Soc. *1954,* 1307
719. Brown, D. A., Dewar, M. J. S.: J. Chem. Soc. *1954,* 2134
720. Tani, T., Kikuchi, S.: Rep. Inst. Ind. Sc. Univ. Tokyo *18,* 87 (1968)
720a. Ito, H., I'Haya, Y. J.: Int. J. Quantum Chem. *2,* 5 (1968)
720b. Leska, J., Zahradník, R.: Chem. Zwesti *23,* 895 (1969)
720c. Hammond, P. R.: J. Photochem. *10,* 467 (1979)
721. Drexhage, K. H.: In: Topics appl. phys., 1, p. 144. Berlin, Heidelberg, New York: Springer Verlag 1973
722. Buttgereit, G., Scheibe, G.: Ber. Bunsenges. Phys. Chem. *69,* 301 (1965)

723. Kemula, W., Axt, A.: Bull. Acad. Pol. Sci. *15,* 43 (1967)
724. Nizamov, N., Zakhidov, U., Atakhodzhaev, A. K.: Izv. Akad. Nauk SSSR, Ser. Fiz. *42,* 579 (1978)
725. Feichtmayr, F., Schlag, J.: Ber. Bunsenges. Phys. Chem. *68,* 95 (1964)
725a. Buncel, E., Menon, B.: J. Chem. Soc., Chem. Comm. *1978,* 758; J. Org. Chem. *44,* 317 (1979)
726. Rohatgi, K. K., Mukhopadhyay, A. K.: J. Phys. Chem. *76,* 3970 (1972)
727. Chambers, R. W., Kajiwara, T., Kearns, D. R.: J. Phys. Chem. *78,* 380 (1974)
728. Muto, J.: J. Phys. Chem. *80,* 1342 (1976)
729. Obermueller, G., Bojarski, C.: Acta Phys. Polon. *A 52,* 431 (1977)
730. Selwyn, J. E., Steinfeld, J. I.: J. Phys. Chem. *76,* 762 (1972)
731. Zanker, V.: Z. Physikal. Chem. *199,* 225 (1952)
732. Bergmann, K., O'Konski, C. T.: J. Phys. Chem. *67,* 2169 (1963)
733. Stork, W. H. J., Lippits, G. J. M., Mandel, M.: ibid. *76,* 1772 (1972)
734. Gianneschi, L. P., Cant, A., Kurucsev, T.: J. Chem. Soc., Faraday Trans. 2, *1977,* 664
734a. Brignoli, C. A., DeVoe, H.: J. Phys. Chem. *82,* 2507 (1978)
734b. Urrecha, I., Castano, F.: J. Chem. Soc., Faraday Trans. *74,* 1094 (1978)
734c. Kozlov, I. N., Kotyava, V. T., Popechits, V. I., Samtsov, M. P., Sarzhevskii, A. M.: Zh. Prikl. Spektrosk. *30,* 271 (1979)
735. Hünig, S., Schweeberg, R., Schwarz, H.: Liebigs Ann. Chem. *587,* 132 (1954)
736. Hünig, S., Schwarz, H.: ibid. *599,* 131 (1956)
736a. Kolling, O. W.: Anal. Chem. *50,* 212 (1978)
737. Bodforss, S., Ahrland, S., Cigén, R.: Z. Physikal. Chem. [Leipzig] *203,* 73 (1954)
738. Ioffe, I. S., Otten, V. F.: Zh. Org. Khim. *1,* 343 (1965)
738a. Rosenthal, I., Peretz, P., Muzkat, K. A.: J. Phys. Chem. *83,* 350 (1979)
739. Knauer, K. H., Gleiter, R.: Angew. Chem. *89,* 116 (1977), Int. Edit. *16,* 113 (1977)
740. Ferguson, J., Mau, A. W. H.: Chem. Phys. Letters *17,* 543 (1972)
741. Lewis, G. N., Bigeleisen, J.: J. Amer. Chem. Soc. *65,* 1144 (1943)
742. Sommer, U., Kramer, H. E. A.: Photochem. Photobiol. *13,* 387 (1971)
742a. Stolyarov, K. P., Vinogradova, N. V.: Fiz. Khim. Metody Anal. *2,* 50 (1977)
743. Rayez, J.-C., Chalvet, O. J.: J. Chim. Phys., Physicochim. Biol. *69,* 1545 (1972)
744. Jakobi, H., Kuhn, H.: Z. Elektrochem. *66,* 46 (1962)
745. Schwarzenbach, G., Mohler, H., Sorge, J.: Helv. Chim. Acta *21,* 1636 (1938)
746. Mohler, H., Forster, H., Schwarzenbach, G.: ibid. *20,* 654 (1937)
747. Farrell, P. G., Grinter, R., Mason, S. F.: In: Optische Anregung organ. Systeme (ed. W. Foerst), p. 88. Weinheim: Verlag Chemie 1966
748. Hafner, K., Goliasch, K.: Angew. Chem. *74,* 118 (1962)
749. Sondermann, J.: Thesis Univers. Marburg 1964
750. Hart, H., Sulzberg, T., Rafos, R. R.: J. Amer. Chem. Soc. *85,* 1800 (1963)
751. Kothe, G., et al.: Ber. Bunsenges. Phys. Chem. *74,* 977 (1970)
752. Wizinger-Aust, R.: Quart. Rep. Sulfur Chem. *5,* 191 (1970)
753. Il'chenko, A. Ya., Koval'chik, R. E., Yagupol'skii, L. M.: Zh. Org. Khim. *9,* 1744 (1973)
754. Van der Lugt, W. Th., Buck, H. M., Oosteroff, L. J.: Ind. Chim. Belge *32* (Spec. No.), 161 (1967)
755. Mason, C. D., Nord, F. F.: J. Org. Chem. *16,* 722 (1951)
756. Ghaisas, V. V., Kane, B. J., Nord, F. F.: J. Org. Chem. *23,* 560 (1958)
757. Zubarovskii, V. M., Khodot, G. P.: Khim. Geterots. Soed. *1973,* 1037
758. Behringer, H., Ramert, R.: Liebigs Ann. Chem. *1975,* 1272
759. Granick, S., Michaelis, L.: J. Amer. Chem. Soc. *69,* 2983 (1947)
760. Hamer, F. M.: In: Chemistry of heterocyclic compounds (ed. A. Weissberger), Vol. 18, p. 1. New York, London: Interscience 1964
760a. Kiprianov, A. I., Colour and structure of cyanine dyes (Russ). Kiev: Naukova Dumka 1979
761. West, W.: Photogr. Sci. Eng. *18,* 35 (1974) (review paper)
762. Scheibe, G.: In: Opt. Anreg. organ. Systeme (ed. W. Foerst), p. 109. Weinheim: Verlag Chemie 1966
763. Scheibe, G.: Ind. Chim. Belge *32* (Spec. No.), 6 (1967)

764. Sturmer, D. M.: In: Chemistry of heterocyclic compounds (ed. A. Weissberger and E. C. Taylor), Vol. 30, p. 441. New York: John Wiley 1977
765. Ayyangar, N. R., Tilak, B. D.: In: Chemistry of synthetic dyes (ed. K. Venkataraman), Vol. 4, p. 103. New York, London: Academic Press 1971
766. Raue, R.: Textilveredlung *7,* 799 (1972)
767. Fischer, E.: Topics Curr. Chem. *7,* 605 (1967)
768. Epperlein, J., Hofmann, B., Stopperka, K.: J. Signalaufzeichnungsmaterialien *4,* 155 (1970)
769. Wizinger, R.: Angew. Chem. *66,* 80 (1954), Z. Naturforsch. *9B,* 729 (1954) Chimia, Suppl. *1968,* 82
770. Baumann, H., Hensel, H. R.: Topics Curr. Chem. *7,* 643 (1967)
771. Hooper, W. D.: Rev. Pure Appl. Chem. *19,* 221 (1969)
772. Fabian, J., Hartmann, H.: J. Signalaufzeichnungsmaterialien *2,* 457 (1974)
773. Scheibe, G., Daltrozzo, E., Wörz, O., Heiss, J.: Z. Physikal. Chem. NF *64,* 97 (1969)
774. Lemke, R.: Ber. Bunsenges. Phys. Chem. *78,* 663 (1974)
775. Dähne, S., Radeglia, R.: Tetrahedron *27,* 3673 (1971)
776. Labhart, H.: J. Chem. Phys. *27,* 957 (1957)
777. Tyutyulkov, N. N.: personal communication
778. Dähne, S., Gürtler, O.: J. prakt. Chem. *315,* 786 (1973)
779. Scheibe, G., Heiss, J., Feldmann, K.; Ber. Bunsenges. Phys. Chem. *70,* 52 (1966)
780. Nikolajewski, H. E., Dähne, S., Leupold, D., Hirsch, B.: Tetrahedron *24,* 6685 (1968)
781. Leupold. D.: Z. Physikal. Chem. [Leipzig] *223,* 405 (1963)
782. Kendall, J. D.: IX. Congr. Internat. Photogr. sci. appl. (ed. L. P. Clere), p. 227. Paris 1936
783. Fabian, J., Hartmann, H.: Tetrahedron *29,* 2597 (1973)
784. Elkomoss, S. G.: J. Mol. Spectrosc. *7,* 14 (1961)
785. Brooker, L. G. S., et al.: Chem. Revs. *41,* 325 (1947)
786. Falk, H., Hofer, O.: Monatsh. Chem. *106,* 115 (1975)
787. Guy, R. W., Jones, R. A.: Austral. J. Chem. *18,* 363 (1965)
788. Falk, H., Hofer, O.: Monatsh. Chem. *106,* 97 (1975)
789. Falk, H., Grubmayr, K., Schlederer, Th.: ibid. *109,* 1191 (1978)
789a. Lehner, H., Braslavsky, S. E., Schaffner, K.: Liebigs Ann. Chem. *1978,* 1990
789b. Pasternak, R., Wagnière, G.: J. Amer. Chem. Soc. *101,* 1662 (1979)
790. Sheldrick, W. S.: J. Chem. Soc., Perkin Trans. 2, *1976,* 1457
791. Fuhrhop, J. H., et al.: Liebigs Ann. Chem. *1975,* 1131
792. Sugimoto, T., Oishi, M., Suzuki, H.: J. Phys. Soc. Japan *43,* 619 (1977)
793. Tolmachev, A. I., Derevyanko, N. A., Kudina, M. A.: Khim. Geterots. Soed. *1975,* 617
794. Levkoev, I. I., Sveshnikov, N. N., Barvyn, N. S., Krasnova, T. V.: Zh. Obshch. Khim. *30,* 291 (1960)
795. Sych, E. D., Shulezko, A. A., Kiprianov, A. I.: Ukr. Khim. Zh. *28,* 213 (1962)
796. Yagupol'skii, L. M., et al.: Zh. Obshch. Khim. *37,* 191 (1967)
797. Krainer, Z. Ya., Guds, P. F., Yagupol'skii, L. M.: Khim. Geterots. Soed. *1973,* 235
798. Lifshits, E. B., et al.: Dokl. Akad. Nauk SSSR *236,* 1375 (1977)
799. Grigor'eva, N. E., Gintse, I. K., Severina, A. P.: Zh. Obshch. Khim. *26,* 3096 (1956)
800. Hishiki, Y., Ogo, K., Nonaka, G.: Nippon Shashin Gakkai Kaishi *33,* 1 (1971)
801. Tolmachev, A. I., Kudinova, M. A.: Dopov. Akad. Nauk Ukr. RSR, Ser. B, Geol. Khim. Biol. Nauki *1977,* 50
802. Shibata, K.: Bull. Chem. Soc. Japan *25,* 378 (1952)
803. Yoshioka, H., Nakatsu, K.: Chem. Phys. Letters *11,* 255 (1971)
804. Honda, C., Hada, H.: Photogr. Sci. Eng. *20,* 15 (1976)
805. Nys, J., VanDormael, A.: Bull. Soc. Chim. Belges *65,* 809 (1956)
806. Scheibe, G., Heiß, J., Feldmann, K.: Angew. Chem. *77,* 545 (1965), Int. Edit. *4,* 525 (1965)
807. Baumgärtner, F., Günther, E., Scheibe, G.: Z. Elektrochem., Ber. Bunsenges. Phys. Chem. *60,* 570 (1965)
808. Scheibe, G., Friedrich, H. J., Hohlneicher, G.: Angew. Chem. *73,* 383 (1961)
809. Brooker, L. G. S.: In: Recent progress in chemistry of natural and synthetic colouring matters and related field (ed. T. S. Gore et al.) p. 573. New York: Academic Press 1962

810. Slominskii, Yu. L., Shulezko, L. M.: Ukr. Khim. Zh. *40,* 625 (1974); Slominskii, Yu. L., Skul'bidenko, A. L., Tolmachev, A. I.: Ukr. Khim. Zh. *40,* 1166 (1974)
811. Tolmachev, A. I., Slominskii, Yu. L., Kiprianov, A. I.: Dokl. Akad. Nauk SSSR *177,* 869 (1967); Babichev, F. S., Romanov, N. N., Slominskii, Yu. Lu., Tolmachev, A. I.: Ukr. Khim. Zh. *41,* 1165 (1975)
812. Slominskii, Yu. L., Skul'bidenko, A. L., Tolmachev, A. I.: Zh. Org. Khim. *11,* 392 (1975)
813. Reynolds, G. A., Drexhage, K. H.: J. Org. Chem. *42,* 885 (1977)
814. Williams, J. K., Wiley, D. W., McKusick, B. C.: J. Amer. Chem. Soc. *84,* 2216 (1962)
815. Strell, M., Braunbruck, W. B., Fühler, W. F., Huber, O.: Liebigs Ann. Chem. *587,* 177 (1954)
816. Bossa, M., Ciullo, G., Sgamellotti, A.: Tetrahedron *25,* 1191 (1969)
817. Pollitt, R. J., Saunders, B. C.: J. Chem. Soc. *1964,* 1132
817a. Gamon, N., Reichardt, Ch.: Tetrahedron Letters *1979,* 225
818. Strauss, M. J.: Chem. Revs. *70,* 667 (1970)
819. Balli, H.: Textilveredlg. *4,* 37 (1969)
820. Mee, J. D.: J. Amer. Chem. Soc. *96,* 4712 (1974)
821. Smith, D. L., Luss, H. R.: Acta Cryst. B *31,* 402 (1975)
822. Mee, J. D., Sturmer, D. M.: J. Org. Chem. *42,* 1041 (1977)
823. Falk, H., Hofer, O., Leodolter, A.: Monatsh. Chem. *106,* 983, 991 (1975)
824. Abramenko, P. I., Kosobutskii, V. A.: Khim. Geterots. Soed. *1978,* 626
825. Tanizaki, Y., Ono, H.: Bull. Chem. Soc. Japan *33,* 1207 (1960)
826. Řehák, V., Novák, A., Titz, M.: Chem. Phys. Letters *52,* 39 (1977)
827. Tani, T., Kikuchi, S.: Photogr. Sci. Eng. *11,* 129 (1967)
828. Tani, T., Kikuchi, S., Honda, K.: ibid. *12,* 80 (1968)
829. Tani, T.: ibid. *13,* 231 (1969)
830. Tani, T.: Bull. Soc. Phot. Sc. Techn. Japan *1971,* 15.
831. Il'chenko, A. Ya.: Teor. Eksp. Khim. *7,* 421 (1971)
832. Elwood, J. K.: J. Org. Chem. *38,* 2430 (1973)
833. Minegishi, T., Kondo, E., Yamanouchi, T., Kinjo, K.: Z. Physikal. Chem. N.F. *91,* 13 (1974)
834. Dietz, F., Müller, G., Bach, G., von Großmann, J.: J. Signalaufzeichnungsmaterialien *3,* 35 (1975)
835. Mishra, N., Patnaik, L. N., Rout, M. K.: Indian J. Chem. *15A,* 951 (1977)
836. Fischer, H., Rewicki, D.: In: Progr. organ. chem. (ed. J. Cook, W. Carruthers), Vol. 7, p. 116. London: Butterworths 1968
837. Brooker, L. G. S., Keyes, G. H., Williams, W. W.: J. Amer. Chem. Soc. *64,* 199 (1942)
838. Brooker, L. G. S., et al.: J. Amer. Chem. Soc. *67,* 1875 (1945)
839. Beilenson, B., Fisher, N. I., Hamer, F. M.: Proc. Roy. Soc. *163A,* 138 (1937)
840. Sych, E. D., Belaya, Zh. N., Dyadyusha, G. G.: Ukr. Khim. Zh. *30,* 1065 (1964)
841. Il'chenko, A. Ya.: Ukr. Khim. Zh. *42,* 162 (1976)
842. Brooker, L. G. S., et al.: J. Amer. Chem. Soc. *73,* 5332 (1951)
843. Grahn, W.: Tetrahedron *32,* 1931 (1976)
844. Herz, A. H.: Photogr. Sci. Engng. *18,* 207 (1974)
845. Gasco, A., Barni, E., DiModica, G.: Tetrahedron Letters *1968,* 5131
846. Föhlisch, B., Bürgle, P., Krokenberger, D.: Chem. Ber. *101,* 4004 (1968)
847. Gompper, R., Wagner, H.-U.: Tetrahedron Lett. *1968,* 165
848. Tanchev, S.: Dokl. Bolg. Akad. Nauk *24,* 1337 (1971)
849. Harper, K. A.: J. Appl. Chem. Biotechnol. *23,* 261 (1973)
850. Kurtin, W. B., Song, P. E.: Tetrahedron *24,* 2255 (1968)
851. Bendz, G., Martensson, O., Nilsson, E.: Arkiv Kemi *27,* 65 (1967)
852. Mabry, T. J., Dreiding, A. S.: In: Rec. adv. in phytochemistry (ed. T. J. Mabry), p. 145. Amsterdam: North Holland 1968
853. Knott, E. B., Williams, L. A.: J. Chem. Soc. *1951,* 1586
854. Bailey, J., Williams, L. A.: In: Chemistry of synthetic dyes (ed. K. Venkataraman), Vol. 4, p. 341. New York, London: Academic Press 1971

855. Brown, G. H., Graham, B., Vittum, P. W., Weissberger, A.: J. Amer. Chem. Soc. *73,* 919 (1951)
856. Knott, E. B., Pauwels, P. J. S.: J. Org. Chem. *33,* 2120 (1968)
857. Brown, G. H., et al.: J. Amer. Chem. Soc. *79,* 2919 (1957)
858. Smith, W. F.: J. Phys. Chem. *68,* 1501 (1964)
859. Arriau, J., Deschamps, J., Parmentier, F.: Tetrahedron *28,* 5755 (1972)
860. Johnson, A. W., McCaldin, P. J.: J. Chem. Soc. *1957,* 3470
861. West, W., Pearce, S., Grum, F.: J. Phys. Chem. *71,* 1316 (1967)
862. Steiger, R., Kitzing, R., Hagen, R., Stoeckli-Evans, H.: J. Photogr. Sci. *22,* 151 (1974)
863. Allmann, R., et al.: Chem. Ber. *109,* 3005 (1976)
864. Henrichs, P. M., Gross, S.: J. Amer. Chem. Soc. *98,* 7169 (1976)
865. Permogorov, V. I., Serdyukova, L. A., Lifshits, E. B., Levkoev, I. I.: Opt. Spektrosk. *31,* 890 (1971)
866. Reichardt, Ch.: Liebigs Ann. Chem. *715,* 74 (1968)
867. Grahn, W., Mrosek, W., Reichardt, Ch.: Chem. Ber. *110,* 1674 (1977)
867a. Slominskii, Yu. L., Radchenko, I. D., Tolmachev, A. I.: Zh. Org. Khim. *15,* 400 (1979)
868. Il'chenko, A. Ya., Lifshits, E. B., Yagupol'skii, L. M.: Ukr. Khim. Zh. *44,* 942 (1978)
869. Yagupol'skii, L. M., Kul'chitskii, M. M., Il'chenko, A. Ya.: Zh. Org. Khim. *10,* 1321 (1974)
870. Slominskii, Yu. L., et al.: Ukr. Khim. Zh. *40,* 1163 (1974)
871. Shulezhko, A. A., Rozhdestvenskaya, I. T., Kiprianov, A. I.: Zh. Org. Khim. *6,* 2118 (1970)
872. Hafner, K., Vöpel, K. H., Ploss, G., König, C.: Liebigs Ann. Chem. *661,* 52 (1963)
873. Brasen, W. R., Holmquist, K. E., Benson, R. E.: J. Amer. Chem. Soc. *83,* 3125 (1961)
874. Sonnet, P. E.: J. Org. Chem. *37,* 925 (1972)
875. von Doering, W. E., Knox, L. H.: J. Amer. Chem. Soc. *72,* 2305 (1954)
876. Doub, L., Vandenbelt, J. M.: J. Amer. Chem. Soc. *71,* 2414 (1949)
877. Hartmann, H.: J. Signalaufzeichnungsmaterialien *7,* 101, 181 (1979)
878. Stepanov, F. N., Lukashina, L. I.: Zh. Obshch. Khim. *30,* 2850 (1960)
879. Kiprianov, A. I., Shulezko, A. A.: Zh. Org. Khim. *2,* 1852 (1966)
880. Freimanis, Ya. F., Vanag, G. Ya.: Zh. Obshch. Khim. *31,* 1945 (1961)
881. Grimm, B., Dähne, S., Bach, G.: J. prakt. Chem. *317,* 161 (1975)
882. Hafner, K., Bangert, K. F., Orfanos, V.: Angew. Chem. *79,* 414 (1967), Int. Edit. *6,* 451 (1967)
883. Ficken, G. E., Kendall, J. D.: Chimia *15,* 110 (1961)
884. Hünig, S., Wolff, E.: Chimia *22,* 33 (1963)
884a. Abramenko, P. I., Gusarov, A. N., Kosobutskii, V. A.: Zh. Prikl. Spektrosk. *31,* 91 (1979)
885. Kiprianov, A. I., Mushkalo, I. L.: Ukr. Khim. Zh. *35,* 1175 (1969)
886. Kiprianov, A. I., Buryak, V. Ya.: Zh. Org. Khim. *9,* 1257 (1973)
887. Kiprianov, A. I., Lesenko, G. A.: Zh. Org. Khim. *9,* 2587 (1973)
888. Mushkalo, I. L., Turova, L. S.; Ukr. Khim. Zh. *43,* 742, 953 (1977)
889. Herz, A. H.: Photogr. Sc. Eng. *18,* 323 (1974)
890. Mikhailenko, F. A., Il'chenko, I. I.: Ukr. Khim. Zh. *40,* 1331 (1974)
891. Kiprianov, A. I., Dyadyusha, G. G.: Ukr. Khim. Zh. *35,* 609 (1969)
892. Mushkalo, I. L.: Ukr. Khim. Zh. *41,* 930 (1975)
892a. Borisevich, Yu. E.: Dokl. Akad. Nauk. SSSR *241,* 1359 (1978)
893. Blinov, L. M., et al.: Dokl. Akad. Nauk SSSR *220,* 860 (1975)
894. Mikhailenko, F. A., Fridman, S. G., Dyadyusha, G. G., Savina, T. I.: Zh. Org. Khim. *10,* 2446 (1974)
894a. Rogovik, L. I., Chernyuk, I. N., Pridan, V. E.: Zh. Org. Khim. *15,* 637 (1979)
895. Mikhailenko, F. A., Dyadyusha, G. G., Boguslavskaya, A. N.: Khim. Geterots. Soed. *1975,* 370
896. Mikhailenko, F. A., Mushkalo, I. L., Boguslavskaya, A. N.: Ukr. Khim. Zh. *40,* 253 (1974)
897. Reichardt, Ch.: Tetrahedron Lett. *1967,* 4327
898. Reichardt, Ch., Mormann, W.: Chem. Ber. *105,* 1815 (1972)
899. Mormann, W.: Thesis, Univers. Marburg 1971
900. Sprenger, H.-E., Ziegenbein, W.: Angew. Chem. *80,* 541 (1968) Int. Edit. *7,* 530 (1968)

901. Hamer, F. M., Rathborne, R. J., Winston, B. S.: J. Chem. Soc. *1947, 1434; 1948,* 1872
902. Van Dormael, A., Nys, J.: Ind. Chim. Belge (Edit. Spec.) *2,* 512 (1959)
903. Horner, M., Hünig, S.: Angew. Chem. *89,* 424 (1977), Int. Edit. *16,* 410 (1977)
904. West, W., Geddes, A. L.: J. Phys. Chem. *68,* 837 (1964)
905. Lifshits, E. B.: Dokl. Akad. Nauk SSSR *179,* 596 (1968)
906. Minegishi, T., Kondo, E., Yamanouchi, T., Kinjo, K.: Z. Physikal. Chem. N.F. *91,* 1 (1974)
907. Wähnert, M., Dähne, S., Radeglia, R., Al'perovich, M. A., Levkoev, I. I.: Adv. Mol. Relax. Interact. Proc. *11,* 263 (1977)
908. Mehlhorn, A.: J. Signalaufzeichnungsmaterialien *6,* 211 (1978)
909. Daltrozzo, E., Scheibe, G., Gschwind, K., Haimerl, F.: Photogr. Sci. Eng. *18,* 441 (1974)
910. Hada, H., Honda, Ch., Tanemura, H.: ibid. *21,* 83 (1977); Honda, Ch., Hada, H.: ibid. *21,* 97 (1977)
911. Dietz, F.: J. Signalaufzeichnungsmaterialien *1,* 157, 237, 381 (1973)
912. West, W., Pearce, S.: J. Phys. Chem. *69,* 1894 (1965)
913. Kiprianov, A. I.: Usp. Khim. *29,* 1336 (1960)
914. Brooker, L. G. S., Keyes, G. H., Heseltine, D. W.: J. Amer. Chem. Soc. *73,* 5350 (1951)
915. Izmail'skii, V. A., Mostoslavskii, M. A.: Dokl. Akad. Nauk SSSR *139,* 601 (1961)
916. McRae, E. G.: Spectrochim. Acta *12,* 192 (1958)
917. Brooker, L. G. S., Craig, A. C., Heseltine, D. W., Jenkins, P. W., Lincoln, L. L.: J. Amer. Chem. Soc. *87,* 2443 (1965)
918. Kosower, E. M.: ibid. *80,* 3253. 3261 (1958)
919. Reichardt, Ch., Dimroth, K.: Topics Curr. Chem. *11,* 1 (1968)
920. Dähne, S., Schob, F., Nolte, K.-D., Radeglia, R.: Ukr. Khim. Zh. *41,* 1170 (1975)
921. Reichardt, Ch.: Lösungsmittel-Effekte in der organ. Chem. Weinheim: Verlag Chemie 1969
922. Radeglia, R.: J. prakt. Chem. *315,* 1121 (1973); and literature cited therein
923. Schneider, S.: Ber. Bunsenges. Phys. Chem. *80,* 218 (1976)
924. Dähne, S., Nolte, K.-D.: Acta Chim. Acad. Sc. Hung. *97,* 147 (1978)
925. Subarovskii, V. M.: Khim. Geterots. Soed. *1972,* 1579
926. Voevodskaya, M. V., Kuz'min, V. A., Khudyakov, I. V.: Izv. Akad. Nauk SSSR, Ser. Khim. *1976,* 1991
927. Leubner, I. H.: J. Org. Chem. *38,* 1098 (1973)
928. Lauerer, D., Coenen, M., Pestemer, M., Scheibe, G.: Z. Physikal. Chem. N.F. *10,* 236 (1957)
929. Reichardt, Ch., Grahn, W.: Chem. Ber. *103,* 1072 (1970), Grahn, W.: Thesis, Univers. Marburg 1971
930. Föhlisch, B., Krockenberger, D.: Chem. Ber. *101,* 3990 (1968)
931. Kuder, J. E., Wychick, D.: Chem. Phys. Lett. *24,* 69 (1974)
932. Fabian, J., Hartmann, H.: Z. Chem. *12,* 349 (1972)
933. Scheibe, G., Seiffert, W., Wengenmayr, H., Jutz, C.: Ber. Bunsenges. Phys. Chem. *67,* 560 (1963)
934. Scheibe, G., Buttgereit, G., Daltrozzo, E.: Angew. Chem. *75,* 1023 (1963)
935. Romanov, N. N., Babichev, F. S., Slominskii, A. I., Tolmachev, A. I.: Dopol. Akad. Nauk Ukr. RSR, Ser. B., Geol. Khim. Biol. Nauki *1977,* 42
935a. Avdeeva, V. I., Al'perovich, M. A., Levkoev, I. I.: Dokl. Akad. Nauk SSSR *241,* 1341 (1978)
936. Tolmachev, A. I., Kornilov, M. Yu., Karaban, E. F.: Teor. Eksp. Khim. *12,* 817 (1976)
937. Feldmann, L. H., Herz, A. H., Regan, T. H.: J. Phys. Chem. 72, 2008 (1968)
938. Beretta, P., Jaboli, A.: Photogr. Sci. Engng. *18,* 197 (1974)
939. Dynes, P. J., Chapman, G. S., Kebede, E., Schneider, F. W.: J. Amer. Chem. Soc. *94,* 6356 (1972)
940. Hada, H., Honda, C.: Photogr. Sci. Engng. *19,* 363 (1975)
941. Al'perovich, M. A., Raikhina, R. D., Avdeeva, V. I., Kolodkin, F. L., Levkoev, I. I.: Dokl. Akad. Nauk SSSR *239,* 1350 (1978)

942. Brouillard, R., Dubois, J.-E.: J. Amer. Chem. Soc. *99*, 1359 (1977)
943. Daltrozzo, E., Hohlneicher, G., Scheibe, G.: Ber. Bunsenges. Phys. Chem. *69*, 190 (1965)
944. Daltrozzo, S., Scheibe, G., Smits, J.: Chimia *19*, 325 (1965)
945. Scheibe, G., Daltrozzo, E.: Diquinolylmethane and its analogs. In: Adv. heterocyclic chem. (ed. A. R. Katritzky), Vol. 7, p. 153. New York, London: Academic Press 1966
946. Cooper, W.: Photogr. Sci. Engng. *17*, 3 (1973)
947. Wizinger, R., Wenning, H.: Helv. Chim. Acta *23*, 247 (1940)
948. Gugglielmetti, R., Metzger, J.: Bull. Soc. Chim. France *1967*, 2824
949. Gugglielmetti, R., Mossé, M., Metras, J.-Ch., Metzger, J.: J. Chim. Phys. Physicochim. Biol. *65*, 454 (1968)
950. Bertelson, R. C.: Photochromic procresses involving heterocyclic cleavage. In: Photochromism (ed. G. E. Brown), p. 45. New York: Interscience 1971
951. Le Beuze, A., Botrel, A., Samat, A., Appriou, P., Gugglielmetti, R.: J. Chim. Phys., Physicochim. Biol. *75*, 255, 267 (1978)
952. Tinland, B., Decoret, C.: Gazz. Chim. Ital. *101*, 792 (1971); J. Molec. Struct. *17*, 414 (1973)
953. Tinland, B., Gugglielmetti, R., Chalvet, O.: Tetrahedron *29*, 665 (1973)
954. Gugglielmetti, R.: J. Photogr. Sci. *22*, 77 (1974)
955. Pommier, H., Samat, A., Metzger, J., Gugglielmetti, R.: J. Chim. Phys., Physicochim. Biol. *72*, 589 (1975)
956. Tyutyulkov, N., Stojanov, S., Taseva, M., Schuster, P.: J. Signalaufzeichnungsmaterialien *3*, 435 (1975)
956a. Simkin, B. Ya., Minkin, V. I., Nivorozhkin, L. E.: Khim. Geterots. Soed. *9*, 1180 (1978)
956b. Danilov, V. V., Dyadyusha, G. G., Rykov, A. A.: Dokl. Akad. Nauk SSSR *245*, 639 (1979)
957. Gürtler, O., Dähne, S.: Z. Physik. Chem. [Leipzig] *255*, 501 (1974)
958. van Dormael, A.: Chimia *15*, 67 (1961)
959. Makin, S. M., Monich, N. V., Shavrygina, O. A., Berezhnaya, M. I., Kheiferts, S. A.: Tetrahedron *25*, 4939 (1969)
960. Mezheritskaya, L. V., Matskovskaya, E. S., Dorofeenko, G. N.: Khim. Geterots. Soed. *1977*, 595
961. Müller, E.: Thesis, TH München 1968
962. Leifer, A., Boedner, M., Dougherty, P., Fusco, A., Koral, M., Lu Valle, J. E.: Appl. Spectr. *20*, 150 (1966)
963. Wizinger, R., Dürr, O.: Helv. Chim. Acta *46*, 2167 (1963)
964. Keyes, G. H., Brooker, L. G. S.: J. Amer. Chem. Soc. *59*, 74 (1937)
965. Fischer, G., Poppe, E. J.: Veröffentl. Wiss. Photolab. Wolfen *10*, 123 (1965)
966. Neunhoeffer, O., Weigel, W.: Liebigs Ann. Chem. *647*, 101 (1961)
967. Levkoev, I. I., Durmashkina, V. V.: Zh. Obshch. Khim. *15*, 215 (1945)
968. Tani, T.: J. Photogr. Sci. *19*, 161 (1971)
969. Kampfer, H.: Mitt. Forsch. Lab. Agfa-Gevaert (München) *4*, 386 (1964)
970. Dieckmann, U.: Diplomarbeit, Techn. Univers. Dresden 1970
971. Förster, E.: Thesis, Techn. Univers. Dresden 1971
972. Glauert, R. H., Mann, F. G., Wilkinson, A. J.: J. Chem. Soc. *1955*, 30
973. Gold, H.: Angew. Chem. *72*, 956 (1960)
974. Ziegler, H.: Ind. Chem. Belge *22*, 533 (1957)
975. Reynolds, G. A., Van Allen, J. A.: J. Heterocycl. Chem. *14*, 539 (1977)
976. Schwarzenbach, G., Weber, R.: Helv. Chim. Acta *25*, 1628 (1942)
977. Jacques, P., Faure, J.: J. Chim. Phys., Physicochim. Biol. *70*, 653 (1973)
978. Jauer, E. A.: Thesis, Techn. Univers. Dresden 1966
979. Quast, H.: Thesis, Univers. Marburg 1963
980. Dimroth, K., Greif, N., Klapproth, A.: Liebigs Ann. Chem. *1975*, 373
981. Märkl, G., Lieb, F.: Tetrahedron Lett. *1967*, 3489
982. Gamon, N., Reichardt, Ch.: Angew. Chem. *89*, 418 (1977), Int. Edit. *16*, 404 (1977)
983. Brooker, L. G. S., Sprague, R. H., Cressman, H. W. J.: J. Amer. Chem. Soc. *67*, 1889 (1945)

984. Fabian, K.: Thesis, Tech. Univers. Dresden 1972.
985. Phillips, A. P.: J. Org. Chem. *13,* 622 (1948); *14,* 302 (1949)
986. Jutz, C., Voitenleitner, F.: Chem. Ber. *97,* 29, 1337 (1964)
987. Reichardt, Ch., Halbritter, K.: Chem. Ber. *104,* 822 (1971); Halbritter, K.: Thesis, Univers. Marburg 1971
987a. Kul'chitskii, M. M.: Ukr. Khim. Zh. *45,* 872 (1979)
988. Vompe, A. F., Monich, N. V., Spasokukotskii, N. S.: Zh. Org. Khim. *10,* 1980 (1974)
989. Mormann, W.: Thesis, Univers. Marburg 1971
990. Shulezko, A. A.: Zh. Org. Khim. *6,* 2122 (1970); *4,* 1089 (1968)
991. Stepanov, F. N., Lukashina, L. I.: Zh. Obshch. Khim. *29,* 2792 (1959)
992. Babichev, F. S., Babicheva, A. F.: Khim. Geterots. Soed. *1967,* 917
993. Kiprianov, A. I., Khilya, W. P.: Zh. Org. Khim. *2,* 1474 (1966)
994. Sprenger, H.-E., Ziegenbein, W.: Angew. Chem. *79,* 581 (1967), Int. Edit. *6,* 553 (1967)
995. Sprenger, H.-E., Ziegenbein, W.: Angew. Chem. *78,* 937 (1966), Int. Edit. *5,* 894 (1966)
996. Neuse, E. W., Green, B. R.: J. Amer. Chem. Soc. *97,* 3987 (1975)
997. Radeglia, R., Dähne, S.: J. Molec. Structure *5,* 399 (1970)
998. Lauher, J. W., Ibers, J. A.: J. Amer. Chem. Soc. *95,* 5148 (1973), and references quoted therein
999. Almlöf, J.: Int. J. Quantum Chem. *8,* 915 (1974)
1000. Furhop, J.-H.: Angew. Chem. *86,* 363 (1974), Int. Edit. *13,* 321 (1974)
1001. Spangler, D., Maggiora, G. M., Shipman, L. L., Christoffersen, R. E.: J. Amer. Chem. Soc. *99,* 7470 (1977)
1002. Chantrell, S. J., McAuliffe, C. A., Munn, R. W., Pratt, A. C.: Coord. Chem. Rev. *16,* 259 (1975)
1003. Oth, J. F. M., Baumann, H., Gilles, J.-M., Schröder, G.: J. Amer. Chem. Soc. *94,* 3498 (1972)
1004. Aronoff, S.: Chem. Revs. *47,* 175 (1950)
1005. Falk, J. E.: Porphyrines and metalloporphyrines. Amsterdam: Elsevier 1964
1006. Lever, A. B. P.: The phthalocyanines. In: Adv. inorgan. chem., Vol. 7, p. 27. New York, London: Academic Press 1965
1007. Gurinovich, G. P., Sevchenko, A. N., Solov'ev, K. N.: Spectroscopy of chlorophyll and related compounds (Russ). Minsk: Nauka i Technika 1968
1008. Smith, K. M.: Porphyrins and metalloporphyrins. Amsterdam: Elsevier 1975
1009. Robertson, J. M.: J. Chem. Soc. *1935,* 615, *1936,* 1195
1010. Hoskins, B. F., Mason, S. A., White, J. C. B.: Chem. Commun. *1969,* 554
1011. Schlechtman, B. H., Spicer, W. E.: J. Mol. Spectrosc. *33,* 28 (1970)
1012. Edwards, L., Gouterman, M.: J. Mol. Spectrosc. *33,* 292 (1970)
1013. Clark, P. F., Elvidge, J. A., Linstead, R. P.: J. Chem. Soc. *1954,* 2490
1014. Gouterman, M.: J. Mol. Spectrosc. *6,* 138 (1961)
1015. Gouterman, M., Wagniere, G., Snyder, L. C.: J. Mol. Spectrosc. *11,* 108 (1963)
1016. Weiss, C., Kobayashi, H., Gouterman, M.: J. Mol. Spectrosc. *16,* 415 (1965)
1017. McHugh, A. J., Gouterman, M., Weiss, C.: Theoret. Chim. Acta *24,* 346 (1972)
1018. Weiss, C.: J. Mol. Spectrosc. *44,* 37 (1972)
1019. Knop, J. V., Knop, A.: Z. Naturforsch. A *25,* 1720 (1970); Knop, J. V., Thesis, Univers. Düsseldorf 1969
1020. Knop, J. V., Stichtenoth, H.: Z. Naturforsch. A *27,* 639 (1972)
1021. Sundbom, M.: Acta Chem. Scand. *22,* 1317 (1968)
1022. Henrikson, A., Sundbom, M.: Theoret. Chim. Acta *27,* 213 (1972)
1023. Maggiora, G. M., Weimann, L. J.: Chem. Phys. Letters *22,* 297 (1973); Maggiora, G. M., Weimann, L. J.: Int. J. Quantum Chem., Quantum Biol. Symp. *1,* 179 (1974)
1024. Chantrell, S. J., McAuliffe, C. A., Munn, R. W., Prett, A. C., Weaver, R. F.: Bioinorg. Chem. *7,* 283 (1977)
1025. Kuz'mitskii, V. A., Solov'ev, K. N.: Zh. Prikl. Spektrosk. *27,* 724 (1977)
1026. Försterling, H. D., Kuhn, H.: Int. J. Quantum Chem. *2,* 413 (1968)
1027. Tomono, K., Nishimoto, K.: Bull. Chem. Soc. Japan *49,* 1179 (1976)

1028. Yip, K. L., Duke, C. B., Salaneck, W. R., Plummer, E. W., Loubriel, G.: Chem. Phys. Letters *49,* 530 (1977)
1029. Petke, J. D., Maggiora, G. M., Shipman, L. L., Christoffersen, R. E.: J. Mol. Spectrosc. *71,* 64 (1978)
1030. Knop, J. V., Knop, A.: Z. Naturforsch. A *25,* 1726 (1970)
1030a. Jung, Ch.: Chem. Phys. Lett. *61,* 630 (1979)
1030b. Austin, E., Gouterman, M.: Bioinorg. Chem. *9,* 281 (1978); and former literature quoted therein
1031. Spanjaard, C., Berthier, G.: J. Chim. Phys., Physicochim. Biol. *58,* 169 (1961)
1032. Millie, P., Veillard, A.: ibid. *62,* 20 (1965)
1033. Zerner, M., Gouterman, M.: Theoret. Chim. Acta *4,* 44 (1966)
1034. Roos, B., Sundbom, M.: J. Mol. Spectrosc. *36,* 8 (1970)
1035. Nagae, H., Yomosa, S.: J. Phys. Soc. Japan *42,* 988 (1977)
1035a. Chang, C. K.: J. Heterocycl. Chem. *14,* 1285 (1977)
1035b. Little, R. G.: J. Heterocycl. Chem. *15,* 203 (1978)
1036. Rimington, C., Mason, S. F., Kennard, D.: Spectrochim. Acta *12,* 65 (1958)
1037. Eisner, U., Lichtarowicz, A., Linstead, R. P.: J. Chem. Soc. *1957,* 733
1038. Dorough, G. D., Miller, J. R., Huennekeus, F. M.: J. Amer. Chem. Soc. *73,* 4315 (1951)
1039. Barrett, P. A., Linstead, R. P., Rundall, F. G., Tuey, G. A. P.: J. Chem. Soc. *1940,* 1079
1040. Eisner, U.: J. Chem. Soc. *1957,* 3461
1041. Stern, A., Pruckner, F.: Z. Physik. Chem. *178A,* 420 (1937)
1042. Elvidge, J. A., Golden, J. H., Linstead, R. P.: J. Chem. Soc. *1957,* 2466
1043. Anderson, J. S., Bradbrook, E. F., Cook, A. H., Linstead, R. P.: J. Chem. Soc. *1938,* 1151
1044. Whalley, M.: J. Chem. Soc. *1961,* 866
1045. Edwards, L., Dolphin, D. H., Gouterman, M.: J. Mol. Spectrosc. *35,* 90 (1970)
1046. Loyd, D., Sneezum, J. S.: Tetrahedron *3,* 334 (1958)
1047. Fabian, J., Mehlhorn, A.: to be published
1048. Kosower, E. M., Ramsey, B. G.: J. Amer. Chem. Soc. *81,* 856 (1959)
1049. Schweig, A.: Z. Naturforsch. A *22,* 724 (1967)
1050. Fabian, J.: J. prakt. Chem., in press
1051. Kröhnke, F.: Ber. Dtsch, Chem. Ges. *83,* 253 (1950)
1052. Lloyd, D., Singer, M. I. C.: J. Chem. Soc. C *1971,* 2939
1053. Kreizberga, Ya. N., Kampar, V. E., Neiland, O. Ya.: Zh. Org. Khim. *11,* 1508 (1975)
1054. Kreizberga, Ya. N., Neiland, O. Ya., Kampar, V. E.: Zh. Org. Khim. *11,* 1941 (1975)
1055. Freeman, B. H., Lloyd, D., Singer, M. I. C.: Tetrahedron *28,* 343 (1972)
1056. Yoshida, Z., Iwata, K., Yoneda, S.: Tetrahedron Letters *1971,* 1519
1057. Evleth, E. M., Berson, J. A., Manatt, S. L.: J. Amer. Chem. Soc. *87,* 2908 (1965)
1058. Stafford, W. H.: J. Chem. Soc. *1952,* 580
1059. Surpateanu, G., Catteau, J. P., Karafiloglou, P., Lablache-Combier, A.: Tetrahedron *32,* 2647 (1976)
1060. Neiland, O. Ya.: Latv. PSR. Zinat. Akad. Vestis, Kim. Ser. *1971,* 73
1061. Kampar, V. E., Kokars, V. E., Neiland, O. Ya.: Khim. Dicarbonyl'nykh Soed. Tezisy Dokl. Vses. Konf. 4th 1975, 64 – 5, C. A. *87,* 54505 h (1977)
1061a. Kokars, V. E., Kampar, V. E., Neiland, O. Ya.: Latv. PSR Zinat Akad. Vestis, Khim. Ser. *1978,* 742
1062. Junek, H., Hermetter, A., Fischer-Colbrie, H.: Angew. Chem. *86,* 380 (1974), Int. Edit. *13,* 343 (1974)
1063. Kampar, V. E., Liepa, I., Pukitis, G., Neiland, O. Ya.: Latv. PSR. Zinat. Akad. Vestis, Kim. Ser. *1978,* 357
1064. Kampar, V. E., Neiland, O. Ya.: Latv. PSR. Zinat. Akad. Vestis, Kim. Ser. *1975,* 727
1065. Balode, D., et al.: Latv. PSR. Zinat. Akad. Vestis, Fiz. Teh. Zinat. Ser. *1977,* 3.
1066. Kuhn, R., Bär, F.: Liebigs Ann. Chem. *516,* 155 (1935)
1067. Amiel, J., Ploquin, J., Sparfel, L., LeBaut, G., Floc'h, R.: Bull. Soc. Chim. France *1974,* 2154

1068. Dimroth, K., Reichardt, Ch., Siepmann, Th., Bohlmann, F.: Liebigs Ann. Chem. *661,* 1 (1963)
1068a. Reichardt, Ch.: Angew. Chem. *91,* 119 (1979); Int. Edit. *18,* 98 (1979)
1069. Dimroth, K., Reichardt, Ch.: Liebigs Ann. Chem. *727,* 93 (1969)
1070. Reichardt, Ch., Müller, R.: ibid. *1976,* 1937
1071. Tak, B. K., Saxena, J. P.: J. Indian. Chem. Soc. *47,* 791 (1970)
1072. Saxena, J. P., Stafford, W. H., Stafford, W. L.: J. Chem. Soc. *1959,* 1579
1073. Saxena, J. P., Tak, B. K., Agrawal, T. N.: J. Indian. Chem. Soc. *49,* 255 (1972)
1074. Gibson, H. W., Bailey, F. C.: Tetrahedron *30,* 2043 (1974)
1075. Kiprianov, A. I., Lasukina, L. A.: Zh. Org. Khim. *4,* 1480 (1968)
1076. Titova, S. P., Ufimtsev, V. N.: Zh. Org. Khim. *7,* 1050 (1971)
1077. De Schryver, F. C., Boens, N., Put, J.: In: Adv. photochemistry (ed. J. N. Pitts et al.), Vol. 10, p. 359. New York: Interscience 1977
1078. Schäfer, F. D.: In: Topics appl. phys., Dye Lasers (ed. F. D. Schäfer), Vol. 1, p. 144. Berlin: Springer 1973

Subject Index

Reactivity and Structure

Concepts in Organic Chemistry

Editors: K. Hafner, J.-M. Lehn, C. W. Rees, P. v. Ragué Schleyer, B. M. Trost, R. Zahradnik

This series will not only deal with problems of the reactivity and structure of organic compounds but also consider synthetical-preparative aspects.
Suggestions as to topics will always be welcome.

Volume 1: J. Tsuji
Organic Synthesis
by Means of Transition Metal Complexes
A Systematic Approach
1975. 4 tables. IX, 199 pages
ISBN 3-540-07227-6

Volume 2: K. Fukui
Theory of Orientation and Stereoselection
1975. 72 figures, 2 tables. VII, 134 pages
ISBN 3-540-07426-0

Volume 3: H. Kwart, K. King
d-Orbitals in the Chemistry of Silicon, Phosphorus and Sulfur
1977. 4 figures, 10 tables. VIII, 220 pages
ISBN 3-540-07953-X

Volume 4: W. P. Weber, G. W. Gokel
Phase Transfer Catalysis in Organic Synthesis
1977. 100 tables. XV, 280 pages
ISBN 3-540-08377-4

Volume 5: N. D. Epiotis
Theory of Organic Reactions
1978. 69 figures, 47 tables. XIV, 290 pages
ISBN 3-540-08551-3

Volume 6: M. L. Bender, M. Komiyama
Cyclodextrin Chemistry
1978. 14 figures, 37 tables. X, 96 pages
ISBN 3-540-08577-7

Volume 7: D. I. Davies, M. J. Parrott
Free Radicals in Organic Synthesis
1978. 1 figure. XII, 169 pages
ISBN 3-540-08723-0

Volume 8: C. Birr
Aspects of the Merrifield Peptide Synthesis
1978. 62 figures, 6 tables. VIII, 102 pages
ISBN 3-540-08872-5

Volume 9: J. R. Blackborow, D. Young
Metal Vapour Synthesis in Organometallic Chemistry
1979. 36 figures, 32 tables. XIII, 202 pages
ISBN 3-540-09330-3

Volume 10: J. Tsuji
Organic Synthesis with Palladium Compounds
1980. 9 tables. XII, 207 pages
ISBN 3-540-09767-8

Volume 11:
New Syntheses with Carbon Monoxide
Editor: J. Falbe
1980. 118 figures, 127 tables.
Approx. 490 pages
ISBN 3-540-09674-4

Lecture Notes in Chemistry

Edited by G. Berthier, M. J. S. Dewar, H. Fischer, K. Fukui, H. Hartmann, H. H. Jaffé, J. Jortner, W. Kutzelnigg, K. Ruedenberg, E. Scrocco, W. Zeil

Volume 1: G. H. Wagnière
Introduction to Elementary Molecular Orbital Theory and to Semiempirical Methods
1976. 33 figures. V, 109 pages
ISBN 3-540-07865-7

Volume 2: E. Clementi
Determination of Liquid Water Structure
Coordination Numbers for Ions and Solvation for Biological Molecules
1976. 32 figures, 18 tables. VI, 107 pages
ISBN 3-540-07870-3

Volume 3: S. R. Niketić, K. Rasmussen
The Consistent Force Field
A Documentation
1977. 24 figures, 26 tables. IX, 212 pages
ISBN 3-540-08344-8

Volume 4: A. Graovac, I. Gutman N. Trinajstić
Topological Approach to the Chemistry of Conjugated Molecules
1977. 34 figures, 2 tables. IX, 123 pages
ISBN 3-540-08431-2

Volume 5: R. Carbó, J. M. Riera
A General SCF Theory
1978. 25 figures, 19 tables. XII, 210 pages
ISBN 3-540-08535-1

Volume 6: I. Hargittai
Sulphone Molecular Structures
Conformation and Geometry from Electron Diffraction and Microwave Spectroscopy; Structural Variations
1978. 40 figures. VII, 175 pages
ISBN 3-540-08654-4

Volume 7
Ion Cyclotron Resonance Spectrometry
Editors: H. Hartmann, K.-P. Wanezek
1978. 66 figures, 32 tables. V, 326 pages
ISBN 3-540-08760-5

Volume 8: E. E. Nikitin, L. Zülicke
Selected Topics of the Theory of Chemical Elementary Processes
1978. 41 figures, 6 tables. IX, 175 pages
ISBN 3-540-08768-0

Volume 9: A. Julg
Crystals as Giant Molecules
1978. 8 figures, 21 tables. VII, 135 pages
ISBN 3-540-08946-2

Volume 10: J. Ulstrup
Charge Transfer Processes in Condensed Media
1979. 32 figures. VII, 419 pages
ISBN 3-540-09520-9

Volume 11: F. A. Gianturco
The Transfer of Molecular Energies by Collision: Recent Quantum Treatments
1979. 17 figures, 6 tables. VIII, 327 pages
ISBN 3-540-09701-5

Volume 12:
The Permutation Group in Physics and Chemistry
Editor: J. Hinze
1979. 20 figures, 13 tables. VI, 230 pages
ISBN 3-540-09707-4

Volume 13: G. Del Re, G. Berthier, J. Serre
Electronic States of Molecules and Atom Clusters
Foundations and Prospects of Semi-empirical Methods
1980.
ISBN 3-540-09738-4

Volume 14: E. W. Thulstrup
Aspects of the Linear and Magnetic Circular Dichroism of Planar Organic Molecules
1980. 27 figures, 5 tables.
VI, 100 pages
ISBN 3-540-09754-6

Volume 15:
Steric Fit in Quantitative Structure-Activity Relations
By A. T. Balaban, A. Chiriac, I. Motoc, Z. Simon
1980. Approx. 20 figures, approx. 41 tables.
VII, 178 pages
ISBN 3-540-09755-4

Springer-Verlag
Berlin Heidelberg New York

FSC
www.fsc.org
MIX
Papier aus verantwortungsvollen Quellen
Paper from responsible sources
FSC® C105338

If you have any concerns about our products,
you can contact us on
ProductSafety@springernature.com

In case Publisher is established outside the EU,
the EU authorized representative is:
Springer Nature Customer Service Center GmbH
Europaplatz 3, 69115 Heidelberg, Germany

Printed by Libri Plureos GmbH
in Hamburg, Germany